Multidisciplinary Analysis of Digital Transformation and Global Market Dynamics

Ahdi Hassan
Global Institute for Research Education and Scholarship, The Netherlands

Shakir Ullah
Southern University of Science and Technology, China

Sergey E. Barykin
Peter the Great St. Petersburg Polytechnic University, Russia

Elena de la Poza
Universitat Politècnica de València, Spain

IGI Global
Publishing Tomorrow's Research Today

Published in the United States of America by
IGI Global
701 E. Chocolate Avenue
Hershey PA, USA 17033
Tel: 717-533-8845
Fax: 717-533-8661
E-mail: cust@igi-global.com
Web site: https://www.igi-global.com

Copyright © 2025 by IGI Global. All rights reserved. No part of this publication may be reproduced, stored or distributed in any form or by any means, electronic or mechanical, including photocopying, without written permission from the publisher.
Product or company names used in this set are for identification purposes only. Inclusion of the names of the products or companies does not indicate a claim of ownership by IGI Global of the trademark or registered trademark.

Library of Congress Cataloging-in-Publication Data

CIP PENDING

ISBN13: 979-8-3693-3423-2
Isbn13Softcover: 979-8-3693-5007-2
EISBN13: 979-8-3693-3424-9

Vice President of Editorial: Melissa Wagner
Managing Editor of Acquisitions: Mikaela Felty
Managing Editor of Book Development: Jocelynn Hessler
Production Manager: Mike Brehm
Cover Design: Phillip Shickler

British Cataloguing in Publication Data
A Cataloguing in Publication record for this book is available from the British Library.

All work contributed to this book is new, previously-unpublished material.
The views expressed in this book are those of the authors, but not necessarily of the publisher.

Table of Contents

Preface xxii

Chapter 1
"Product as a Service" Business Model in the Context of the Development of
a Digital Circular Economy: The Transition to a Circular Economy 1
    Maria A. Vetrova, Saint Petersburg State University, Russia
    Dinara V. Ivanova, Saint Petersburg State University, Russia

Chapter 2
Application of the Business Ecosystem Approach to the Transformation
of Business Models in Industry: The Example of the Light Industry
Transformation 25
    Sviatlana Hrytsevich, Belarusian National Technical University, Belarus
    Muhammad Asad Tahir, University of Leicester, UK

Chapter 3
Desk Research Methods for Place Reputation Monitoring and Evaluation in
the Digital Economy: Place Reputation Monitoring of the Region 43
    Valeriia V. Kulibanova, Institute for Regional Economic Studies,
        Russian Academy of Science, Russia
    Tatiana R. Teor, Saint Petersburg Electrotechnical University "LETI",
        Russia
    Irina A. Ilyina, Saint Petersburg Electrotechnical University "LETI",
        Russia

Chapter 4
Digital Technologies in Ensuring Sustainable Development of Oil and Gas
Territories in the Russian Federation 65
    Danil P. Egorov, St. Petersburg State University, Russia
    Anatoliy I. Chistobaev, St. Petersburg State University, Russia
    Kirill P. Egorov, St. Petersburg State University, Russia

Chapter 5
Digital Transformation Features of Mining Industry Enterprises: Mining 4.0
Technologies Development Trends 87
    Alisa Torosyan, ITMO University, Russia
    Olga Tcukanova, ITMO University, Russia
    Natalia Alekseeva, Peter the Great St. Petersburg Polytechnic
        University, Russia
    Abdul Hameed Qureshi, Pakistan Telecommunication Authority,
        Islamabad, Pakistan
    Shafiya Qadeer Memon, Department of Software Engineering, UET
        Jamshoro, Pakistan

Chapter 6
Digitalization of Business Sector and Commercial Banks Interaction in the
Russian Federation: Results of the Study 113
    Elena I. Generalnitskaia, Dar-Cosmetic Ltd., Russia
    Natalya A. Loginova, Russian Customs Academy the St. Petersburg
        Branch Named After Vladimir Bobkov, Russia

Chapter 7
Factor Analysis of the Economic Security of Regions: The Level of
Economic Security of Regions 135
    Yulia V. Granitsa, National Research Lobachevsky State University of
        Nizhny Novgorod, Russia

Chapter 8
Human-Centered Technologies in the Development of Digital Energy:
Integrated Power Management Systems 157
    Anatolii Gladkikh, Ulyanovsk State Technical University, Russia
    Rafael Sayfutdinov, Ulyanovsk Civil Aviation Institute, Russia
    Artem Nichunaev, Ulyanovsk State Technical University, Russia
    Tatyana Ulasyuk, Ulyanovsk Civil Aviation Institute, Russia

Chapter 9
Impact of Reward Crowdfunding on Entrepreneurial Nature of Crowdfunded
Ventures: Model of Client-Entrepreneur Relationship 183
    Ivan D. Kotliarov, Peter the Great St. Petersburg Polytechnic University,
        Russia

Chapter 10
Improving the Efficiency of Using Equity Captial in Oil and Gas Companies of Russia: The Study of the Financial Performance Indicators 201
    Irina Filimonova, Institute of Economics and Industrial Engineering SB RAS, Novosibirsk State University, Russia
    Anna Komarova, Institute of Economics and Industrial Engineering SB RAS, Novosibirsk State University, Russia
    Irina Provornaya, Novosibirsk State University, Russia
    Timofey Mikheev, Novosibirsk State University, Russia

Chapter 11
Involvement of Participation in Network Communication as a Factor of Its Sustainability: The Public Policy and Conditions of Its Formation 221
    Anna M. Kuzmina, St. Petersburg State University, Russia
    Alexey E. Kuzmin, The Russian Presidential Academy of National Economy and Public Administration, Russia

Chapter 12
Management of the Petroleum Projects on the Russian Arctic Shelf: Project Management Aspects 241
    Anna Komarova, Trofimuk Institute of Petroleum Geology and Geophysics SB RAS, Novosibirsk State University, Russia
    Irina Filimonova, Trofimuk Institute of Petroleum Geology and Geophysics SB RAS, Novosibirsk State University, Russia
    Vasily Nemov, Trofimuk Institute of Petroleum Geology and Geophysics SB RAS, Novosibirsk State University, Russia
    Mohamed M. Adel A. M., Novosibirsk State University, Russia

Chapter 13
Prospects of Human-Centric Technology's Influence on Economic and Non-Economic Aspects of Public Life: The Influence of Modern Technologies on Economic and Non-Economic Life 261

    Oksana Pirogova, Peter the Great St. Petersburg Polytechnic University, Russia

    Vladimir Plotnikov, St. Petersburg State University of Economics, Russia

    Sergey Barykin, Peter the Great St. Petersburg Polytechnic University, Russia

    Anna Karmanova, Peter the Great St. Petersburg Polytechnic University, Russia

    Irina Kapustina, Peter the Great St. Petersburg Polytechnic University, Russia

    Natalya Golubetskaya, St. Petersburg University of Management Technologies and Economics, Russia

    Alexander Igorevich Puchkov, St. Petersburg University of Management Technologies and Economics, Russia

    Anna Sedyakina, Peter the Great St. Petersburg Polytechnic University, Russia

    Mikhail Loubochkin, Saint Petersburg State University, Russia

    Ahdi Hassan, Global Institute for Research Education and Scholarship, The Netherlands

Chapter 14
Risk Assessment of Intellectual Captial of the University by Ranking Methods: Digital Decisions and Expert Methods of Assessment 279

    Anush G. Airapetova, St. Petersburg State University of Economics, Russia

    Vladimir V. Korelin, St. Petersburg State University of Economics, Russia

    Galiya R. Khakimova, St. Petersburg State University of Economics, Russia

Chapter 15
Integrated Management System Implementation Prospects at the Aerospace Complex Enterprises: Enterprise's Digital-Integrated Management System 301

    Ekaterina M. Messineva, Moscow Aviation Institute (National Research University), Russia

    Alexander G. Fetisov, Moscow Aviation Institute (National Research University), Russia

Chapter 16
Top Trends of the Transport-and-Logistic Activity Development in the
Conditions of Digitalization: Identification of the Main Trends in Logistics.... 321
    *Elena Yu. Vasilyeva, Moscow State University of Civil Engineering,
       Russia*

Compilation of References 341

About the Contributors 401

Index 403

Detailed Table of Contents

Preface xxii

Chapter 1
"Product as a Service" Business Model in the Context of the Development of
a Digital Circular Economy: The Transition to a Circular Economy 1
    Maria A. Vetrova, Saint Petersburg State University, Russia
    Dinara V. Ivanova, Saint Petersburg State University, Russia

The COVID-19 pandemic has brought additional threats to the achievement of sustainable development goals from the health, economic and social crisis. In a dominant linear economy, achieving zero emissions and waste targets and enhancing sustainability in general is difficult. Therefore, the transition to a circular economy model, which has been actively developing in theory and practice in the last decade due to significant environmental, economic and social benefits, is of acute relevance. The development of the circular economy is supported by digital technologies that optimize costs and help implement circular business models into practice. The article focuses on the product as a service business model, which is the first step towards the transition to a circular economy. Product-as-a-service finds application in practice among leading companies and brings significant positive effects for producers, consumers and helps to achieve the goal of sustainable development.

Chapter 2
Application of the Business Ecosystem Approach to the Transformation of Business Models in Industry: The Example of the Light Industry Transformation 25
> *Sviatlana Hrytsevich, Belarusian National Technical University, Belarus*
> *Muhammad Asad Tahir, University of Leicester, UK*

Traditional industries that have dominated the market for decades are now facing new competitors that are changing the boundaries of industries and business models. These competitors are taking leadership positions due to the growth of digital technologies. One of the possible ways of successful functioning in the digital world can be the use of the concept of business ecosystems (BES), namely the introduction of a business ecosystem approach (BEA) to the formation of a new strategy for managing traditional business models. The article aims to reveal the relevance of the development of the BES concept and the use of BEA as a transitional element to the modern transformation of traditional industries. In this article, the author presents key conclusions on the refinement and use of criteria characterizing the functioning of business ecosystems for the organization of transformational processes, considering digitalization and intercompany integration. The author examines the example of the transformation of the light industry (textile and clothing) in the Republic of Belarus.

Chapter 3
Desk Research Methods for Place Reputation Monitoring and Evaluation in the Digital Economy: Place Reputation Monitoring of the Region 43
> *Valeriia V. Kulibanova, Institute for Regional Economic Studies, Russian Academy of Science, Russia*
> *Tatiana R. Teor, Saint Petersburg Electrotechnical University "LETI", Russia*
> *Irina A. Ilyina, Saint Petersburg Electrotechnical University "LETI", Russia*

A positive place reputation is the important competitive advantage, which allows it to attract resources and to facilitate mutually beneficial partnership with stakeholder groups. The main task is not only to identify the mechanism of the place reputation formation, but to analyze the factors influencing this process and their real-time dynamics. Now the information obtained by monitoring social networks, electronic media, and other looks more reliable and unbiased. Desk research has recently come to the fore, as it allows a fairly rapid analysis of large amounts of information. Another trend is a shift of focus from quantitative indicators' study to the identification of qualitative ones. The article reveals methods of conducting desk research using various digital platforms. On the example of St. Petersburg, the authors analyzed possibilities of monitoring the place reputational capital, allowing to assess the reputation characteristic expressed quantitatively, and a qualitative characteristic in the form of an attitude reflection towards the region expressed in publications' tone.

Chapter 4
Digital Technologies in Ensuring Sustainable Development of Oil and Gas
Territories in the Russian Federation 65
    Danil P. Egorov, St. Petersburg State University, Russia
    Anatoliy I. Chistobaev, St. Petersburg State University, Russia
    Kirill P. Egorov, St. Petersburg State University, Russia

The introduction of digital technologies in oil and gas industry helps to reduce the anthropogenic burden on the environment and ensures sustainable development of territories in Russia. The article assesses the conditions for the introduction of digital technologies in the oil and gas business, identifies barriers to the spread of innovative methods for making managerial decisions, including everyday issues of people's lives and operating oil and gas industry facilities. Based on the statistical analysis, authors of the article identified the main causes of emergency situations in the oil and gas industry and developed a list of recommendations for its prevention or elimination of consequences. The study is shown that when using digital technologies the costs of the management process are reduced, bureaucratic delays in the preparation of the necessary documents for the population are eliminated; the opportunities for members of society to make managerial decisions are realized. The institutional imperfections and other factors were identified as the main barriers for innovation.

Chapter 5
Digital Transformation Features of Mining Industry Enterprises: Mining 4.0 Technologies Development Trends 87
Alisa Torosyan, ITMO University, Russia
Olga Tcukanova, ITMO University, Russia
Natalia Alekseeva, Peter the Great St. Petersburg Polytechnic University, Russia
Abdul Hameed Qureshi, Pakistan Telecommunication Authority, Islamabad, Pakistan
Shafiya Qadeer Memon, Department of Software Engineering, UET Jamshoro, Pakistan

The concept of the 4th industrial revolution is becoming a strategic driver of sustainability, success, and competitiveness in the modern mining sector. An important factor is that the industry in question has its own features in operational and production processes, which are changing due to the development and implementation of digital technologies. The purpose of this article is to determine the features of the digital transformation of enterprises in the mining industry and to study the role of digital technologies implemented in companies. To conduct the analysis, the method of content analysis of scientific studies and articles related to digital transformation in the mining sector was used. The results of the analysis are the formulation of the main directions for the development of digital technologies for mining enterprises, the definition of their role in the mining industry under consideration, and clarification of the features of digital transformation. The results of the study can be used in subsequent work aimed at further emplementing modern digital technologies.

Chapter 6
Digitalization of Business Sector and Commercial Banks Interaction in the
Russian Federation: Results of the Study 113
    Elena I. Generalnitskaia, Dar-Cosmetic Ltd., Russia
    Natalya A. Loginova, Russian Customs Academy the St. Petersburg
        Branch Named After Vladimir Bobkov, Russia

Based on the analysis of the Bank of Russia statistics, as well as data collected by the National Research University Higher School of Economics, the authors of the article identified 12 trends describing the business and banking sector interaction in the context of digital environment development. Based on questioning business sector and commercial banks representatives, the identified trends were ranked according to the degree of importance for each of the groups of respondents, which helped to identify their development priorities. Moreover, it helped to indicate one of the most important problems of digitalization of commercial banks and business sector interaction. Thus, it has been pointed out that credit institutions, characterized by a high degree of digitalization at present time, can become a tool for the diffusion of innovations for the business sector, in the case of corresponding business sector adaptive capabilities. The authors proposed a set of measures to increase the digitalization of economic interaction between credit institutions and the business sector.

Chapter 7
Factor Analysis of the Economic Security of Regions: The Level of
Economic Security of Regions... 135
    Yulia V. Granitsa, National Research Lobachevsky State University of
        Nizhny Novgorod, Russia

The level of economic security is assessed on the basis of resource, labor, financial, innovation, investment, budgetary and regional security, which together represent the socio-economic security or socio-economic potential of the regions. In the study, the regions of Russia are ranked by the level of economic security on the basis of the selected group of indicators for assessing the socio-economic potential of the regions. The tests carried out by Mann-Whitney, Fligner-Keelin, and split testing allowed us to conclude that the selected regional groups in terms of economic security are heterogeneous, that is, the distribution of indicators in groups and their mean values differ significantly. An adequate logit model has been formed, which makes it possible to predict the class of economic security of a region. The receiver operating characteristic has been constructed, and the quality of the model has been assessed by calculating the area under the curve. The key indicators of economic security are determined by the method of principal components.

Chapter 8
Human-Centered Technologies in the Development of Digital Energy:
Integrated Power Management Systems 157
    Anatolii Gladkikh, Ulyanovsk State Technical University, Russia
    Rafael Sayfutdinov, Ulyanovsk Civil Aviation Institute, Russia
    Artem Nichunaev, Ulyanovsk State Technical University, Russia
    Tatyana Ulasyuk, Ulyanovsk Civil Aviation Institute, Russia

The fundamental factor in ensuring the development of digital energy at the present stage is the quality of life of the population, where human-oriented technologies are fundamental in the information society. Integrated Power Management Systems (IEMS) are one of the most effective areas of digital technologies and processes in the energy sector. The most important task of IEMS is timely identification, reduction and elimination of hazards and risks in the system of energy generating enterprises. In the conditions of digital energy as an integral part of the digital economy, data on the state of managed subsystems become necessary for the IEMS resources management, where the role of human-oriented technologies is significantly increasing. These directions essentially determine the main content of the concept of intelligent energy systems, which are based on information and control technologies using a distributed data processing system. The paper proposes a method for protecting data from errors, which is based on the method of cluster partitioning of the space of code combinations.

Chapter 9
Impact of Reward Crowdfunding on Entrepreneurial Nature of Crowdfunded
Ventures: Model of Client-Entrepreneur Relationship 183
    Ivan D. Kotliarov, Peter the Great St. Petersburg Polytechnic University, Russia

The goal of the paper consists in identification of changes in entrepreneur-customer relations generated by the emergence of crowdfunding. The research is based on the concept of on-demand economy. It introduces taxonomy of rewards used in crowdfunding based on the type of value provided to backers and amount of this value. The analysis of reward crowdfunding demonstrates that backers actually finance ventures, not products. Crowdfunded venture is established in response to backers' needs and backers become customers of this venture. It means that such venture can be considered as venture on demand. Rise of this new type of ventures transforms the model of relations between customers and entrepreneurs. Customers become active stakeholders of business development and create demand for new ventures (not for new products). Thanks to this new model of relations customers can better satisfy their needs while entrepreneurs can reduce the level of risks. The paper contains a conceptual framework that describes this evolution of customers within the system of customer-entrepreneur relations.

Chapter 10
Improving the Efficiency of Using Equity Captial in Oil and Gas Companies
of Russia: The Study of the Financial Performance Indicators 201
> *Irina Filimonova, Institute of Economics and Industrial Engineering SB*
> *RAS, Novosibirsk State University, Russia*
> *Anna Komarova, Institute of Economics and Industrial Engineering SB*
> *RAS, Novosibirsk State University, Russia*
> *Irina Provornaya, Novosibirsk State University, Russia*
> *Timofey Mikheev, Novosibirsk State University, Russia*

The paper is devoted to the study of the indicators of the financial performance of the Russian oil and gas companies in the era of the digital transformation of markets. Nowadays oil and gas companies operate against the background of the financial and technological sanctions, the global trends for decarbonization, volatile oil prices, as well as challenges imposed by the changes in the structure and quality of the resource base. The paper aims to propose the decomposition of the coefficient of the sustainability of economic growth for the petrolium companies. The calculations were carried out for the major Russian petroleum companies in 2012-2020. It was revealed that there are two main drivers for the changes in the coefficient mainly connected with the capital and profit ratios. The results of the study justify the complex evaluation of the factors of the companies' development, as well as the formation of suitable managerial decisions. It is possible to increase the coefficient of sustainability of economic growth within the framework of management activities in different ways.

Chapter 11
Involvement of Participation in Network Communication as a Factor of Its
Sustainability: The Public Policy and Conditions of Its Formation 221
    Anna M. Kuzmina, St. Petersburg State University, Russia
    Alexey E. Kuzmin, The Russian Presidential Academy of National
        Economy and Public Administration, Russia

The paper deals with the phenomenon of public policy and conditions of its formation through the contractual choice. Purpose of the research is to characterize the process of involving the target audience in the communication of non-profit organizations in social networks from the point of view of digitalization tasks. Research objectives are to determine a set of methods and tools in the media communication practice of working with the audience in social networks to increase the degree of involvement of the target audience in communication; to characterize parameters of audience reflection in various digital systems for its impact on engagement. Authors prove the hypothesis that network communication is necessary for providing the conventional basis for public policy formation and determines the sustainability of network interactions through the prism of contract choice. In addition, the authors determine the need to support the contract choice within the framework of communication of public sphere actors by the mechanism of their involvement in the digital communication process.

Chapter 12

Management of the Petroleum Projects on the Russian Arctic Shelf: Project Management Aspects 241

> Anna Komarova, *Trofimuk Institute of Petroleum Geology and Geophysics SB RAS, Novosibirsk State University, Russia*
> Irina Filimonova, *Trofimuk Institute of Petroleum Geology and Geophysics SB RAS, Novosibirsk State University, Russia*
> Vasily Nemov, *Trofimuk Institute of Petroleum Geology and Geophysics SB RAS, Novosibirsk State University, Russia*
> Mohamed M. Adel A. M., *Novosibirsk State University, Russia*

The study is devoted to a relevant problem of management and evaluation of the petroleum projetcs in Russian Arctic shelf. The significance of the region is supported by a number of factors, such as a vast resource base, important geopolitical location, and others. The Arctic region presents unique logistics, transportation, production and market challenges. The aim of the research is the development of the multi-criteria decision-making approach for the evaluation of the Arctic petroleum projects. Both qualitative and quantitative indicators (economic, technological, logistical) of the efficiency of the project implementation were considered and the difference in their evaluation was studied. Results of the evaluation of the forteen petroleum projects showed the influence of the qualitative criteria on the overall problem which in turn pointed out their importance when evaluating projects in the Arctic region as financial evaluation does not have the sufficient efficiency. For the given set of objects, the preliminary sequence of development was determined under current conditions.

Chapter 13
Prospects of Human-Centric Technology's Influence on Economic and Non-Economic Aspects of Public Life: The Influence of Modern Technologies on Economic and Non-Economic Life 261

>   Oksana Pirogova, Peter the Great St. Petersburg Polytechnic University, Russia
>   Vladimir Plotnikov, St. Petersburg State University of Economics, Russia
>   Sergey Barykin, Peter the Great St. Petersburg Polytechnic University, Russia
>   Anna Karmanova, Peter the Great St. Petersburg Polytechnic University, Russia
>   Irina Kapustina, Peter the Great St. Petersburg Polytechnic University, Russia
>   Natalya Golubetskaya, St. Petersburg University of Management Technologies and Economics, Russia
>   Alexander Igorevich Puchkov, St. Petersburg University of Management Technologies and Economics, Russia
>   Anna Sedyakina, Peter the Great St. Petersburg Polytechnic University, Russia
>   Mikhail Loubochkin, Saint Petersburg State University, Russia
>   Ahdi Hassan, Global Institute for Research Education and Scholarship, The Netherlands

Human-centric technologies are one of the emerging trends in technological progress. The problems of its formation have not been studied well enough at the moment, which opens up a field for scientific research. In particular, the question of the impact of human-centric technologies on the economic and non-economic life of society remains open. Empirical methods of scientific research are used, in particular, to give disclosure the phenomena of scientific and technological progress at the present stage in the context of the study of human-centric technologies, and include observation, description, and measurement. As a result, it was possible to form and apply a basic approach to assessing the impact of human-centric technologies on economic and social life. The characteristics of their basic directions of development are given. The authors substantiate and finally formulate the conclusion that the formation of human-centric technologies is based on subjective and economically supported objective factors.

Chapter 14
Risk Assessment of Intellectual Captial of the University by Ranking
Methods: Digital Decisions and Expert Methods of Assessment...................... 279
> *Anush G. Airapetova, St. Petersburg State University of Economics, Russia*
> *Vladimir V. Korelin, St. Petersburg State University of Economics, Russia*
> *Galiya R. Khakimova, St. Petersburg State University of Economics, Russia*

Risk management is one of the main areas of the theory of managing complex social and economic systems, the need for the development of which is due to the complexity of the structure of enterprises as risk management objects and the high level of uncertainty in the external socio-economic and internal environment of enterprises. Thanks to the rapid development of software and hardware, today it is possible to use all the potential capabilities of digital technologies in risk assessment. Also, computer technology allows you to quickly process this information, keep it in a secure form. There are different management decision-making processes in conditions of uncertainty (mainly due to incomplete or inaccurate risk information). Some approaches do not allow the use of a well-developed device for simulating dangerous situations, which affects the conclusions of the algorithmic and software-information software and reduces the validity of decisions on risk management of industrial enterprises. The article describes the risk assessment procedure for allocating risks to individual groups.

Chapter 15
Integrated Management System Implementation Prospects at the Aerospace Complex Enterprises: Enterprise's Digital-Integrated Management System..... 301
    Ekaterina M. Messineva, Moscow Aviation Institute (National Research
        University), Russia
    Alexander G. Fetisov, Moscow Aviation Institute (National Research
        University), Russia

A comprehensive study of the current situation related to certification on ISO functional management systems (ISO 9001, ISO 14001 and ISO 45001) was carried out to assess the prospects for the integrated management systems creation and implementation in aerospace industry. This study analyzes complex data on the certificates number issued for aerospace enterprises in the world leading aerospace countries during the period 2009-2019. Paper demonstrates the current trends in the implementation and certification of quality, environmental, occupational health, and safety management systems by enterprises in this countries. During the investigation, both global trends and the domestic Russian trends were identified. Then Pearson correlation coefficients were calculated between the time series of the number of ISO 9001 and ISO 14001 certificates. Based on the obtained results, a basic scheme was proposed for the development of a digital integrated management system for enterprises in the aerospace industry, taking into account the specifics of the Russia.

Chapter 16
Top Trends of the Transport-and-Logistic Activity Development in the
Conditions of Digitalization: Identification of the Main Trends in Logistics.... 321
   *Elena Yu. Vasilyeva, Moscow State University of Civil Engineering,
     Russia*

The factors, which influenced the activity of the transport-and-logistic companies most significantly, are analysed in the article. The top trends, peculiar to transport-and-logistic activity in modern conditions, are pointed out. According to the results of the analysis that was carried out, nowadays digitalization is the main subject in logistics. In particular, paperless registration of cargo transportation, robotization during freight processing, the use of unmanned vehicles for transportation, and the introduction of obligatory marking for separate types of commodities are revealed. The author finds out the logical communication between the development of electronic commerce and the adaptation of logistics for it, and concludes, that only the complex logistics will be able to conform to modern requirements. At the same, some factors which slow down the digitalization of the transport-and-logistic sphere in Russia are revealed. The results of the analysis and the conclusions, drawn by the author, can be useful for the further development of the transport-and-logistic complex.

Compilation of References 341

About the Contributors 401

Index 403

Preface

This scholarly monograph, «Multidisciplinary Analysis of Digital Transformation and Global Market Dynamics» stands as an epitome of intellectual endeavors to dissect the myriad dimensions of digital transformation and its ramifications on markets in the global context. The monograph comprises the crystallized insights and empirical findings presented during the International Scientific Conference Global Challenges of Digital Transformation of Markets (GDTM) international conference series, spanning annually during the period from 2021 to 2023 years (on the platform of Peter the Great St.Petersburg Polytechnic University, Russia). It addresses the contemporary state of affairs and aims to shed light on future trajectories and emergent paradigms.

The monography «Multidisciplinary Analysis of Digital Transformation and Global Market Dynamics» values scientific integrity and fosters cooperation among esteemed academics, researchers, and specialists globally. Amid its borders, appraisers will divulge miscellaneous issues essential for digital mechanisms in our daily lives, such as the evolution of digitalization and its utilization in diverse domains, such as the finance sector and the configuration of multiple integrated systems in digital energy. Moreover, considering elements closely connected with sustainable development and using innovative tools take the lead role in captivating more and more scientists to investigate such processes. Each handout offers a unique outlook on how the digital era and revolution in the diligence efforts in the elaboration of business models has primarily altered regular and confirmed market models, accompanied by both uncommon and occasion and complicated issues.

Moving on to more ubiquitous digital transformation applications, the monograph ventures into public life and energy sectors. It scrutinizes the human elements in technological adaptations and offers a pragmatic perspective on sustainable development, especially in energy production and distribution.

Of particular note is incorporating newer methodologies for evaluating social and economic phenomena – ranging from the reputation of places in digital spaces to innovative financial models that empower entrepreneurial ventures. Moreover, the text offers an in-depth analysis of business models reshaping traditional industries, highlighting the role of business ecosystems in this transformative journey.

The monograph «Multidisciplinary Analysis of Digital Transformation and Global Market Dynamics» caters to a diverse range of audiences, offering valuable insights and perspectives across various subject areas and job functions. Some principal audiences for this publication include:

- Academic Researchers and Scholars: The book is particularly relevant to researchers and scholars across disciplines such as Management, human-centric technologies, Logistics, Marketing, and Economics. They will find original research contributions that delve into the complexities of digital transformation's impact on markets and society.
- Practitioners and Professionals: Professionals working in the fields of technology, urban planning, sustainability, marketing, and supply chain management can gain practical insights from the book's exploration of real-world applications and strategies in the context of digital transformation.
- Policy Makers and Government Officials: Those involved in policy-making related to urban development, sustainable practices, and economic growth will benefit from the book's analysis of how digital transformation influences policy decisions and shapes future initiatives.
- Business Executives and Managers: Business leaders and managers can find valuable perspectives on how digital transformation impacts business models, market strategies, and customer engagement, as well as how to navigate the challenges and opportunities presented by these changes.
- Students and Educators: Students studying disciplines like technology, business administration, urban planning, and environmental sciences can gain a comprehensive understanding of the multidimensional aspects of digital transformation and its effects on markets and society. Educators can also use the book as a reference for teaching material.
- Environmentalists and Sustainability Advocates: The book's coverage of sustainable development and its intersections with digital transformation is relevant to individuals and organizations focused on promoting environmentally conscious practices.
- Technology Innovators and Entrepreneurs: Innovators and entrepreneurs exploring opportunities in the digital realm will find inspiration and insights into emerging trends, innovative business models, and potential areas for disruption.
- Market Analysts and Researchers: Market analysis and research professionals can benefit from the book's exploration of how digital transformation reshapes market dynamics, consumer behavior, and industry trends.

- Consultants and Advisers: Consultants and advisers in technology, business strategy, and sustainable development can leverage the book's knowledge to provide informed guidance to their clients.

This reference monography is designed to cater to a broad audience, including researchers, technology professionals, students, and industry experts who are deeply engaged in understanding and navigating the digital transformation across various global sectors. This monography serves as a comprehensive guide, offering valuable insights into the multidimensional impacts of digital transformation on industries, economies, and societies at large. As editors, we believe that the wealth of knowledge encapsulated in this volume will significantly contribute to ongoing discussions and advancements in digital technologies, emphasizing their pivotal role in shaping the future of global interactions.

ORGANIZATION OF THE MONOGRAPHY

- **Chapter 1**: "Product as a Service Business Model in the Context of the Development of a Digital Circular Economy: The Transition to a Circular Economy" by **Maria Vetrova** and **Dinara Ivanova** explores how the circular economy is evolving due to digital technologies and the product-as-a-service business model. This chapter highlights how digital solutions support sustainability goals by reducing emissions and waste.
- **Chapter 2**: "Application of the Business Ecosystem Approach to the Transformation of Business Models in Industry: The Example of the Light Industry Transformation" by **Sviatlana Hrytsevich** delves into the transformation of traditional industries like textiles and clothing in Belarus through business ecosystem strategies that leverage digital advancements for competitive growth.
- **Chapter 3**: "Desk Research Methods for Place Reputation Monitoring and Evaluation in the Digital Economy: Place Reputation Monitoring of the Region" by **Valeriia Kulibanova**, **Tatiana Teor**, and **Irina Ilyina** presents methodologies for desk research in monitoring a region's reputation using digital platforms, with an emphasis on the qualitative and quantitative aspects of place reputational capital.
- **Chapter 4**: "Digital Technologies in Ensuring Sustainable Development of Oil and Gas Territories in the Russian Federation" by **Danil Egorov**, **Anatoliy Chistobaev**, and **Kirill Egorov** examines the role of digital technologies in improving sustainability and reducing environmental impacts in

Russia's oil and gas sector, offering recommendations for overcoming barriers to innovation.

- **Chapter 5**: "Digital Transformation Features of Mining Industry Enterprises: Mining 4.0 Technologies Development Trends" by **Alisa Torosyan, Olga Tcukanova**, and **Natalia Alekseeva** investigates the impact of digital technologies on the mining sector, focusing on the development of Mining 4.0 and the specific challenges and trends in digital transformation within this industry.
- **Chapter 6**: "Digitalization of Business Sector and Commercial Banks Interaction in the Russian Federation: Results of the Study" by **Elena Generalnitskaia** and **Natalya Loginova** discusses the interaction between the business sector and commercial banks in Russia, identifying key trends and development priorities in digitalization that can foster innovation and economic growth.
- **Chapter 7**: "Factor Analysis of the Economic Security of Regions: The Level of Economic Security of Regions" by **Yulia Granitsa** assesses the socio-economic potential of Russian regions by analyzing key indicators related to economic security and offering predictive models based on these metrics.
- **Chapter 8**: "Human-Centered Technologies in the Development of Digital Energy: Integrated Power Management Systems" by **Anatolii Gladkikh, Rafael Sayfutdinov, Artem Nichunaev**, and **Tatyana Ulasyuk** emphasizes the importance of human-centric approaches in digital energy systems, highlighting the role of integrated power management systems in ensuring safety and efficiency in energy generation and distribution.
- **Chapter 9**: "Impact of Reward Crowdfunding on Entrepreneurial Nature of Crowdfunded Ventures: Model of Client-Entrepreneur Relationship" by **Ivan Kotliarov** explores how reward crowdfunding transforms the relationship between entrepreneurs and customers, creating a new venture model that responds directly to customer needs, with backers becoming key stakeholders in business development.
- **Chapter 10**: "Improving the Efficiency of Using Equity Capital in Oil and Gas Companies of Russia: The Study of the Financial Performance Indicators" by **Irina Filimonova, Anna Komarova, Irina Provornaya**, and **Timofey Mikheev** analyzes financial performance indicators in Russia's oil and gas sector, proposing strategies to enhance economic growth and financial sustainability within this key industry.
- **Chapter 11**: "Involvement of Participation in Network Communication as a Factor of Its Sustainability: The Public Policy and Conditions of Its Formation" by **Anna Kuzmina** and **Alexey Kuzmin** addresses how network

communication and digital tools can strengthen public policy formation by increasing engagement and participation through digital platforms.
- **Chapter 12**: "Management of the Petroleum Projects on the Russian Arctic Shelf: Project Management Aspects" by **Anna Komarova**, **Irina Filimonova**, **Vasily Nemov**, and **Mohamed M. Adel A. M.** discusses the unique challenges of managing petroleum projects on the Russian Arctic shelf and proposes a multi-criteria decision-making approach to evaluate project efficiency and performance in this complex environment.
- **Chapter 13**: "Prospects of Human-Centric Technology's Influence on Economic and Non-Economic Aspects of Public Life: The Influence of Modern Technologies on Economic and Non-Economic Life" by **Oksana Pirogova**, **Vladimir Plotnikov**, **Sergey Barykin**, **Anna Karmanova**, **Irina Kapustina**, **Natalya Golubetskaya**, **Alexander Puchkov**, **Anna Sedyakina**, and **Mikhail Loubochkin** assesses the impact of human-centric technologies on both economic and social aspects of life, offering insights into how these technologies will shape future developments.
- **Chapter 14**: "Risk Assessment of Intellectual Capital of the University by Ranking Methods: Digital Decisions and Expert Methods of Assessment" by **Anush Airapetova**, **Vladimir Korelin**, and **Galiya Khakimova** outlines digital tools and expert methods for assessing the intellectual capital of universities, focusing on risk management in academic environments.
- **Chapter 15**: "Integrated Management System Implementation Prospects at the Aerospace Complex Enterprises: Enterprise's Digital-Integrated Management System" by **Ekaterina Messineva** and **Alexander Fetisov** explores the implementation of integrated management systems in the aerospace sector, with an emphasis on quality, environmental, and safety standards and their correlation with digital technologies.
- **Chapter 16**: "Top Trends of the Transport-and-Logistic Activity Development in the Conditions of Digitalization: Identification of the Main Trends in Logistics" by **Elena Vasilyeva** highlights the impact of digitalization on the logistics sector, discussing trends such as paperless transactions, automation in freight handling, and the use of unmanned vehicles.

In Conclusion, this edited reference monography provides a diverse range of perspectives on digital transformation, encapsulating its challenges, opportunities, and future directions. Through the contributions of esteemed researchers, academics, and industry professionals, this monography offers a comprehensive analysis of the digital transformation landscape. It provides readers with the theoretical foundations and practical insights needed to understand the complexities of digitalization across sectors and its impact on the global economy.

The interdisciplinary nature of the monography is reflected in its coverage of topics ranging from sustainability and finance to human-centric technologies and public policy, making it a valuable resource for a wide array of stakeholders. Whether you are a researcher, professional, or student, we hope this monography will stimulate critical thinking, inspire innovation, and encourage further exploration into the transformative power of digital technologies.

As editors, it is our sincere hope that the insights shared within these pages will foster meaningful dialogue, promote collaboration, and contribute to the continued evolution of digital transformation on a global scale. We extend our deepest gratitude to all contributors for their invaluable expertise and dedication to advancing the understanding of digital transformation in today's interconnected world.

In sum, the monography «Multidisciplinary Analysis of Digital Transformation and Global Market Dynamics» serves as a comprehensive guide for academics, researchers, policymakers, and practitioners interested in the ever-evolving landscape of digital transformation. It amalgamates a vast range of topics and methodologies into a single cohesive narrative, seeking to provide a holistic understanding of the challenges and opportunities wrought by the digital age.

Sergey E. Barykin

Doctor of Economic Sciences, Professor, Graduate School of Service and Trade, Peter the Great St. Petersburg Polytechnic University, Russia

Shakir Ullah

Professor of Faculty of International Studies, Henan Normal University, Xinxiang, China

Chapter 1
"Product as a Service" Business Model in the Context of the Development of a Digital Circular Economy:
The Transition to a Circular Economy

Maria A. Vetrova
Saint Petersburg State University, Russia

Dinara V. Ivanova
Saint Petersburg State University, Russia

ABSTRACT

The COVID-19 pandemic has brought additional threats to the achievement of sustainable development goals from the health, economic and social crisis. In a dominant linear economy, achieving zero emissions and waste targets and enhancing sustainability in general is difficult. Therefore, the transition to a circular economy model, which has been actively developing in theory and practice in the last decade due to significant environmental, economic and social benefits, is of acute relevance. The development of the circular economy is supported by digital technologies that optimize costs and help implement circular business models into practice. The article focuses on the product as a service business model, which is the first step towards the transition to a circular economy. Product-as-a-service finds application in practice among leading companies and brings significant positive effects for

DOI: 10.4018/979-8-3693-3423-2.ch001

Copyright © 2025, IGI Global. Copying or distributing in print or electronic forms without written permission of IGI Global is prohibited.

producers, consumers and helps to achieve the goal of sustainable development.

NOVELTY STATEMENT

This chapter introduces a pioneering analysis of the "Product as a Service" (PaaS) business model, emphasizing its critical role in advancing the digital circular economy. By integrating digital innovation with circular economy principles, the research explores how PaaS enables sustainable business practices, extending product lifecycles, enhancing resource efficiency, and minimizing waste. This approach offers fresh perspectives on the transition towards a circular economy, positioning PaaS as a key driver in achieving long-term environmental and economic sustainability.

1. INTRODUCTION

The shift to circular economy has become a major concern to counter environmental degradation. With dwindling resources and challenges, linear economy model is no longer sufficient. Therefore, with a broader framework, "Product as a Service" business model has emerged in the economic arena deriving more attention from scholars and practitioners alike. It is an innovative business model of an upgraded strategy under the current conditions of need for circular economy conversion via opportunities of digital transformation. The subject of this chapter is the application of the PaaS business model to navigate through the concept of digital circular economy and through its integration create more sustainable economy and look forward to environmental-friendly economic future.

1.1 The Circular Economy: A Paradigm Shift

The circular economy is an innovative concept that departs from the prevailing linear model of economic growth and entails "take-make-dispose" approach move towards the model that revitalize resources, reuse and restore natural stocks (OECD, 2020). This change of approach is due to increasing awareness that the existing linear model, based on the extraction of non-renewable resources and the production of waste, is unsustainable in the long-term (Jensen, 2015). The circular economy aims at closing the loop of product generation cycle in the context of designing product and systems that will have less waste output, with increasing life cycles of materials, and ingeniously turning resources into closed loops. At its core, the circular economy is built on three key principles: eliminating the discarded waste and pollution, utilizing goods and materials to their optimal utility, and restoring ecosystems (Vezzoli & Manzini 2008). These principles challenge the normal perception about economic development which is relieved of being responsible for causing disposal of resources and pollution in the environment after the process of product consumption. On the contrary, the circular economy supports the belief that economic activity of consumption can be perpetual and enriched by preserving the value of products, materials and resources in the economy for as long as possible (Tischner et al., 2000).

The circular economy's model hinges upon the design phase as one of its most crucial pillars. All products and services are developed considering their whole life cycle, thus using strategies as modularity, durability, reparability or recyclability (Jensen, 2015). This approach not only leads to the minima-libation of the ecological footprint of a product but it also opens up new business approaches, like the Product as a Service model which is discussed in the subsequent sections. Moreover, the shift to CE needs to be underpinned by a change of the social paradigm concerning resources. This change requires a number of processes such as revision of value chains, integration across sector, as well as influencing consumers to change their patterns of consumption (Defra, 2013). The eligible actors or stakeholders that are involved in enabling circular economy are governments, businesses, and civil society in regard to formulating enabling policies, establishing circular infrastructure, and cognitively pushing for an adoption of circular economy. Debates on circular economy models are in the process of perpetual discussion as the prevailing global problems of resource depletion, environmental pollution and climate change call for this kind of approach. Thus, circular economy principles will be critical for constructing a long-lasting, sustainable, and prosperous world that can withstand such challenges in the future (Iqbal et al., 2021).

1.2 Understanding the "Product as a Service" (PaaS) Business Model

One of the business models functioning within the contemporary global markets could be described as "Product as a Service", or in terms of the abbreviation, the PaaS business model – the model that transforms the conventional possession of product ownership to the possession of service ownership. As to PaaS, the seller allows the customers to use a system under the condition that the customer consumption of the product is transferred to the seller for some period (Pupentsova et al., 2020). This shift to service-based consumption is thus indeed very much in tune with the circular economy model to the extent that such a strategy optimizes use of resources, reduces the levels of waste and optimizes life cycle of products (Khun, 2021). Unlike the other similar models like SaaS, IaaS, and DaaS, the model of PaaS is more customer centric and the purpose is to co-create value with the consumers over the long period.

It lowers the desire for creating products that are disposable, irreparable, and un-upgradable which is something the circular economy wants to eliminate because it is resource-wasteful (Firoiu et al., 2019). It has been used in technology and fashion industries among others, which proves that it has its efficiency and extensiveness in respective fields (Belmonte-Ureña et al., 2021).

Moreover, PaaS within the automotive industry helps to decrease CO_2 emissions and the industry's negative impact on the environment by encouraging people to use service-oriented transport instead of unsustainable vehicles. This model does not only avoid the constant presence of cars on the roads but it also practices the rationality of using vehicles that has the highest utilization rate (Sinkevičius, 2020). Technology firms such as Microsoft and Adobe have moved from selling licenses to basic software to customer subscription (Silkina, 2019). Some of these are familiar examples, such as Microsoft Office where clients no longer buy a license but subscribe to the software. It also provides consumers the facility to constantly update their products which in turn enhances their functionality. This gives consumers a complete control of the life cycle of the product (Vetrova & Ivanova 2021). Thus, the software is kept updated, or generally, companies can guarantee that there will be few reasons for customers to upgrade and thus contribute to the generation of more e-waste.

PaaS model is not restricted to large industries it also has its applicability in consumer-products industry. For instance, Rent the Runway and Patagonia are known to provide clothing as rental based services since customers do not need to own the items, despite getting to wear quality clothes (Kharlamova et al., 2020). It can be safely said that this model is sustainable since it allows for the wearing of garments for longer periods and consequently, less garments are discarded. Among the factors that define the demands for PaaS, one of the most significant is the shift in consumer preferences, most evident among which is today's youth who would rather rent and experience than own the product (Hernandez, 2019). The PaaS model also enables such companies to meet changing consumer needs while at the same time help realize the objectives of the circular economy i.e., lower utilization of resources and the creation of less waste. Furthermore, PaaS provides businesses with prospects of improving customer ties and offering consistent value in the form of services (Özkan & Yücel, 2020).

Nevertheless, the PaaS model also has its drawbacks. There is an essential need for organizations that put up the time and effort to ensure the right features, channels and networks to accommodate for service-related offers, and coordinate the various stages involved in product return, repair, and redistribution (Nurulin et al., 2020). Furthermore, businesses have to adjust their ways of making money i.e., moving from transactional income to recurring one, which might entail profound changes in a company's operations. However, the PaaS model has quite significant potential for improvement of the circular economic conditions. PaaS model encourages to popularize durative, reusable, and upgradable products and services, reduce raw material consumption, and prolonging the life cycle of products. With the advancement of the digital technologies and its parallel shift of consumers moving from owning assets to obtaining access to them, the PaaS model will be one of the enablers of the CE transition (Brezet & van Hemel, 1997).

1.3 The Role of Digitalization in Enabling PaaS

Several authors demonstrated that digitalization is strategically important for the execution and upscaling of the "Product as a Service" (PaaS) business model within the framework of the circular economy. This upgradation of strategies have become possible because of advancement in digital technologies, including Internet of Things (IoT), big data analytics, Artificial Intelligence, and Blockchain into how businesses manage their products right from design, production, usage, maintenance and disposal or recycling (Ibn-Mohammed et al., 2021). These technologies do not only help to improve the operational efficiency but also form the basis for new forms of value propositions that are based on sustainability and resource effectiveness. Probably the best benefit of digitalization to PaaS offers is the use of devices to monitor and manage the products. IoT products can constantly gather data regarding the usage, state, and performance of products which in turn enable service providers to provide timely, preventive, and prescriptive services (Wulf et al., 2021). For instance, bearings attached to industrial equipment are collected as data on the signs of wear and tear thereby notifying service providers before a failure happens, thus elongating the life of the product and minimizing the disposal cycle by the customer (Nurulin et al., 2020). This does not only complement the focus on the value offered by the PaaS model, but also supports the circular economy which aims for optimization of resource utilization.

Big data analytics and AI also enhances the features of the PaaS model as they help players to process the enormous amount of data derived from IoT devices or other online channels. By embracing the use of analytics, firms manage to understand customers, services, and products, and the market as a whole. This enables the managers to design improved services, organize personalized services, and even better products (Firoiu et al., 2019). In specific, it is able to find out a product that is likely to need repair in future according to the past data, using habits, or environmental factors and therefore makes provision that can help companies to offer timely services, reduce loss-time, and increase customers satisfaction. It also helps to maintain the PaaS model because it opens the way to block chain technology, which ensures confidentiality, transparency and exchangeability of the transactions and history of the PaaS products. As products flow in a PaaS ecosystem and experience several ownership changes and service operations, block chain provides all concerned parties with tamper-proof information concerning the original and subsequent origins of the product, the services to which it has been subjected, or its condition for a given operation (Hernandez, 2019). Such level of transparency is highly relevant for ensuring institutional credibility among the customers, providers of services and other members of the value chain especially for industries where issues of quality, safety and compliance assume paramount importance.

In the same regard, digital platforms are the last enabler of the PaaS model. They enable easy consumption of the products and services in the business. Similarly, the social context of service selling is accentuated through digital platforms such as Uber and Airbnb where real-time betwixt users and service-providers are inter-related (Liu et al., 2021). The digital performs can be related to renting, sharing, leasing, tools and mechanisms to manage subscriptions, payments for products and services in the context of PaaS. In addition, these platforms can easily benefit from the user data to provide customer tailored recommendations, reward-based programs as well as other over and above services to enrich the value of the customer experience, hence promoting organizational growth (Matsumoto et al., 2017). The significance of digitalization for supporting PaaS does not end with business technological change; it also involves a cultural and structural change of the firm.

Firoiu et al., (2019) posited that for organizations to effectively leverage on the benefits of digitalization, they have to invest in technology, transform business practices, create capabilities, and champion creativity and cooperation. Furthermore, companies have to be prepared for the new risks that come with implementing digital strategies like data protection, privacy issues, and frequent technological updates.

1.4 Challenges and Opportunities in Implementing PaaS

The operationalization of "Product as a Service" business model encounters important challenges as well as promising possibilities. It means that despite the fact that the PaaS model can be considered as a shift towards the principles of the circular economy and provides a direction for the sustainable and efficient usage of resources, the transition to the PaaS model is not without its challenges. On the other hand, numerous possible advantages for the firm, buyers, and the ecosystem provide a strong incentive that encourages organizations to adopt this new model.

(A) Challenges in Implementing PaaS

One of the main issues that can be observed when it comes to the implementation of the PaaS model is the fact that it requires definite rearrangements in the approaches that companies use. Legacy of tangible goods is modified in the companies. They have to adapt to the paradigm shift from selling products to providing services and to do that by means of redefining their value proposition, revenue streams, and customers (Ardolino et al., 2018). This shift entails resources' commitment to the development of new competencies, e. g. digital environment, data analysis or customer relationship management, which may require significant and sometimes difficult innovations (Gesvindr & Buhnova, 2016). There are also other challenges which are of a strategic nature for example; the challenge of culture change in organizations. In many ways, the PaaS model requires planning for the long-term, for product longevity, product support, and customer interaction, and it does not support merely acquiring the quick sales. This encourages the promotion of organizational culture and climate of learning, teamwork and creativity which can be a challenge especially within the context that rewards focus on short-term, mainly financial KPIs (Garcia-Gomez et al., 2012; Li et a., 2024). Also, the PaaS model requires management shifts in leadership and organizational structure because this model emphasizes the delivery of services and customers. The final threat to the growth of PaaS is consumer acceptance or the lack of it, as many consumers are not willing or ready to adopt this change. On the one hand, the newer generations are open to the idea of service-based consumption, while others will remain against the concept of not owning things (Anouar et al., 2017). This resistance can be especially notorious in the markets, where access to services is somehow linked with ownership or where there are reasons not to trust the availability and reliability of services. As much as there are benefits of using PaaS, consumers remain skeptical and have not fully embraced the services because the companies have not ensured educating their consumers enough about the benefits. They have not bought their trust yet and that the services provided are trust-worthy and reliable as they are inherently in the best interest of consumers (Ikundi et al., 2017). The last but not the least, the legal and political factor is another significant aspect that can enhance or restrain PaaS. The current laws regulating industries in diverse areas are formulated for the conventional product-based industries and do not necessarily provide sufficient solutions to the peculiarities which PaaS businesses deliver, including the concerns of legal responsibility, taxation, and consumer rights (Freet et al., 2015). Businesses and decision-makers also have a significant responsibility to support governments and provide an appropriate legal and regulatory environment for the introduction and adoption of PaaS.

(B) Opportunities in Implementing PaaS

The potential of the PaaS model is vast, as potential clients can see all the benefits of the business without wading through unnecessary layers. PaaS model presents multiple opportunities to new and growing stream of business as it has shifted towards building a closer relationship with customers, unleashing an opportunity to stand out in today's relatively saturated markets (Anouar et al., 2017). As the customer in today's society owns the product, firms are able to continually interact with customers and offer upgrades, maintenance, additional services, which are all capable of improving the loyalty of the customer. Speaking of the environmental benefits of the PaaS model, it can be noted that the offered model complies with the circular economy objectives in terms of resource utilization, waste management, and the reduction of environmental footprint. Creating products that are long-lasting with the approach that the simpler the product the easier it is to maintain, leads to reduction in the chances of those products to find their way into the dump. Furthermore, inculcating the capacity to recycle this with refurbishing, remanufacturing or recycling at the end of its useful service also supports sustainability (Garcia-Gomez et al., 2012). Manufacturers can create opportunities that foster formation of new business ecosystems that link manufacturers, service providers, technology companies among others to provide value for a customer need based on evolutionary measures and approach (Ardolino et al., 2018). Such partnerships can lead to further innovation by using technologies, thereby increasing the effectiveness of work, and provide value for all stakeholders in the value chain.

1.5 The Future of PaaS in a Digital Circular Economy

With the global economy in a quest to mitigate problems of resource deficit, environmental pollution, and climate change effects, PaaS business model is positioned to define the future of sustainable economy for digital circular economy. The PaaS model, based as it is on the sharing perspective rather than the ownership one, can be considered to fit in the logic of the circular economy, as it provides the framework for minimizing waste, managing the resources and encouraging innovation at the product and service levels. Several more trends and development will continue to shape the trend and diffusion of PaaS in a digital circular economy in the future.

(1) Technological Advancements and Integration

The future prospects of practicing PaaS is anchored to the steady progression and further adoption of digital technologies. Advances in Internet of Things (IoT), artificial intelligence (AI), block chain and big data analytics will improve on the

PaaS models whereby companies will be able to provide better and high service level to their clients. As depicted in figure 2 below, concept such as artificial intelligence will facilitate predictive analytics which will enable more accurate anticipation of maintenance requirements leading to less time out of service and further elongation of product's life-cycle (Ikundi et al., 2017). Likewise, through the use of blockchain technology there will be improved transparency to track and trace ability of products throughout their life cycle. This will build trust among consumers and other stakeholders (Sharma & Santharam, 2013). With time, the adoption of these technologies will cause great progress for the induction of PaaS model in businesses. Since the barriers to implementing the technology are reducing, more companies especially SMEs will be in a position to implement the model. This democratization of technology will most definitely create a shift towards PaaS market in the competitive environment and lead to the creation of new services that fit the divergent needs of customers (Liu et al., 2023).

(2) Evolving Consumer Preferences

Consumers' behavior and their preferences are also going to be influential in determining the course of the PaaS marketplace. It is noteworthy that a recent tendency of 'experience economy,' perhaps led by the younger generations of consumers, is gradually shifting from ownership to access. This change in the consumer attitude is in tandem with PaaS model where the offer is more of services than tangible products (Che et al., 2015). With an increasing consciousness about the adverse effects of environmental pollution, consumers will increasingly look for non-conventional ways of ownership; thereby, making the PaaS market quite firmly grounded. Moreover, there is a rising increase in the use of the subscription model in other fields, as people migrate from owning a product to subscribing to a product either from media, software and various other products such as home appliances and personal vehicles (Sharma & Santharam, 2013). Organizations that can articulate the value proposition of PaaS in terms of costs, ease of use and the green economy will be in a good position to tap into this rapidly growing market segment.

(3) Policy and Regulatory Support

Future Patterns of PaaS will also depend upon the policy and regulatory sector that is active in the IT industry. States as well as global institutions are starting to pay attention to the utilization of circular economy models in attaining sustainable development goals proposed by the United Nations in the SDG (Lüthi & Prässler, 2011). Consequently, variations of circular economy will probably be increasing with policy support to business models such as PaaS. This support could act as a form of carrots, as reimbursing companies that adopt PaaS in a similar way. These companies take initiatives like compensating individuals for renovating energy saving houses or encouraging legislations that enforce adoption of circular principles in product design and production (Knill et al., 2012). Nevertheless, the rules and regulations need to be changed correspondingly to meet the new requirements and challenges connected with PaaS, including questions of responsibility, consumers' protection, and data protection. Simple and constant provisions will be vital for gaining consumers' confidence and avoiding that PaaS providers work in an unethical and unsustainable way (Lampkin et al., 1999).

(4) Expanding Business Ecosystems

As the PaaS model evolves in the future, we shall be likely to witness the emergence of more elaborate and inter-connected business value webs. Businesses will rely on solution providers as it becomes a fashionable way to call for the involvement of suppliers, technology vendors and service providers to meet the ever-changing customer needs. These ecosystems will enable the sharing of funds, information, and the best practices within the sector and the industries making the field more dynamic and productive (Frosch & Gallopoulos, 1989). In this regard, platform-based business models will increase, as organizations will be able to deliver PaaS at a larger scale and with lower rigidity. Pervasive digital technologies supporting the exchange and coordination with customers and the provision of real-time insights about the services will be vital to the future of PaaS in a digital circular economy (Taylor et al., 2019).

2. LITERATURE REVIEW

2.1 Overview of Circular Economy Principles

Circular Economy has become an important concept in order to respond to the global crises in the areas of resource scarcity, environmental pollution and climate change. In contrast to the current conventional linear economy model which is typified by "take-make-dispose", the circular economy supports an economy where resources are used correctly in the shortest time possible and after which they should be circulated in the economy. This model is underpinned by three core principles: elimination of waste and pollution, use of products and materials, and reinvestment of natural capital or stock (Tischner et al., 2000). The first of them, eliminate the need for waste and pollution requires the reconsideration of the process of producing a product. Some of the classical linear value chains and value webs fail to consider the backstage effects of waste production and pollution. In a circular economy, waste is considered a problem to be annulled by enhancing the design and improving processes. Products themselves are developed and have features of utilizing resources during and after their use aiming to minimize waste, prolong the products' use, have easy means of repair and refurbishment or recycling (Vetrova & Ivanova 2021). Besides, this approach is aimed at the reduction of the environmental impact of resources and products and the elaboration of new economic activities based on the principles of sustainable usage of resources.

The second principle of circular economy, which is to keep products and materials in use concern with the usage of products and the effective utilization of materials (Liu et al., 2023). In circular economy, the input materials and products are kept for as long as possible within the economy without disposal by selling for further use, repairing, remanufacturing, or recycling. This is a stark contrast to the linear model in which after use, products are mostly disposed, hence, tremendous wastage is observed. The circular economy minimizes the virgin resource use and decomposition costs since it promotes the reuse of products and materials (Silkina, 2019). Not only that, this principle also provides impetus to adoption of new business models like PaaS where businesses own the physical products and offer them to be used by customers and maintained by the businesses themselves to avoid creation of waste products.

The third focusses on regenerating natural systems, stresses the replenishment of the environment and not exploiting it (Belmonte-Ureña et al., 2021). In contrast to the linear model's reliance on linear economy which results in the depletion of natural ecosystems, the goal of a circular economy is to establish environmental-friendly loops within the framework of which flow of resources is beneficial to living things and support ecosystem's life forms (Sinkevičius, 2020).

The impetus for circular economy is as a result of realization of the fact that the linear economy cannot be sustained in the future. Dependence on the finite resources and the detrimental effect being caused to the environment by waste and pollution is a threat both to the long-term economic growth as well as short-term prosperity. It cannot proceed along with sustainability of the environment. Firoiu et al., (2019) argued that circular economy insists to make a paradigm shift at business level, supply chain, as well as consumer level. Many companies still remain trapped in the 'shareholder-value' model that encourages the maximization of short-term profits at the cost of long-term value, which has important implications for redesigning consumables and other materials, producing and using them responsibly and disposing of them. This includes prescriptive measures such as putting in place policies that support effective utilization of resources, policies that encourage sustainable business practices and policies that discourage wastage of resources (Khun, 2021).

2.2 The Emergence of Product as a Service (PaaS) Models

Within the circular economy context, the PaaS model is one of the organizations' strategic options to replace the traditional product ownership trend. Consumers are only able to use a product, paying for the product's service while the product's ownership and responsibilities regarding its wearing out or damage lies to the manufacturer (Pupentsova et al., 2020). In contrast to the traditional business models of production and consumption, the PaaS model prolongs the utility lifespan of the objects and minimizes wastage (Sinkevičius, 2020). The PaaS model have evolved from the more generic concept of servitization— under which manufacturing firms began to transition from companies that delivered tangible goods to service providers. Liu et al. (2023) contended that servitization is a business model transition wherein firms integrate products and services to deliver more value to the clients. This approach also adds customer value along with improving resource utilization as it lowers the demand for new products and extends the use of existing ones.

Subscription-based models has become a particular preference among various industries and that is because PaaS models have developed the niche for such choices in industries where there are high costs at the initial stage, where the maintenance is complex, or where the technologies are changing rapidly. As in the automotive industry, liberal models such as car-sharing and subscription services appear to be liberal because they enable consumers to use automobiles without the fiscal and organizational costs accruing to owning a car (Giessmann & Stanoevska-Slabeva, 2012). This business model not only delivers upgrades and new functions for customers on a regular basis but also avoids physical delivery and packaging belonging to the principles of circular economy (Classen et al., 2019). It has also become evident that SaaS is the dominant model of consumption for software solutions and turned the industry into the cloud, that utilizes the model of software as a service provider. This is a significant step forward towards the creation of sustainable business models.

2.3 Digitalization as a Catalyst for PaaS

The role of digitalization should not be underestimated as the enabler of the PaaS model, as discussed in the prior discussions. Wortmann et al., (2024) has pointed out that IoT, big data, and recently developed technologies have been employed in enhancing the manner that products are created, supervised, and regulated at different stages of production. These technologies supplied the base and equipment necessary in the PaaS model that could allow the company to offer more efficient, prompt and diverse services (Jensen, 2015). In this regard, IoT can be noted to have a significant contribution to the PaaS, especially where tracking of usage and performance of a particular product in real-time is desirable. IoT devices also enable the state of the product to be known thus having service providers offer maintenance and other services that tend to make the product to have prolonged life cycle (Villaroel, 2018). It also facilitates the reduction of the total cost of ownership for the consumers and is closer to the goals of circular economy.

Big data analytics and AI enrich the PaaS model by facilitating the analysis of customer and product experience and other tendencies. These technologies allow the companies to enhance their service delivery system, tailor solutions for consumers, and adapt products based on consumer performance data (Tohanean & Toma, 2024). For instance, AI can forecast when a product has a tendency to develop some complications that may necessitate either repairs or overhaul, and accordingly, organizations can take measures to undo the complications thus preventing complete failure of the product. Blockchain technology is also useful in the PaaS model since it enables safe and transparent record of past dealings and record of the products as well. All the participants in a PaaS ecosystem are interested in comprehensive and objectively true information about the product's origin, operations, and service history in the customer environment, as well as the history of service interventions applied to it. This necessary data is provided by blockchain (Iqbal et al., 2021). Such level of transparency is crucial in establishing the confidence of the customers, suppliers and other stake-holders in the chain.

2.4 Challenges and Barriers to PaaS Adoption

However, there are some challenges and barriers to the adoption of the PaaS model as follows. One of the biggest problems is that the core issue necessitates a major change of course in business planning and function. In some sectors, companies who predominantly depended on selling a product now need to shift to a service-based model. Service-based model requires different skills such as digital support, data support, and customer support (Saygıner, 2023). It is more often than not a time-consuming and costly affair, especially if the movement is undertaken under circumstances that may not be financially in a position to afford the new capabilities. Another hindrance to the adoption of PaaS is cultural resistance within organizations. The PaaS model is long-term oriented, which means product longevity, maintenance, and customer relations are preferred over the product selling (Liu et al., 2023). For companies that tend to set short-term financial goals, this can definitely prove to be a big challenge. Also, the PaaS model may disrupt the organization's functional reporting due to the shift from product-centricity to service delivery and heightened customer orientation.

Another important consideration that can directly affect PaaS is consumer acceptance. While the youth may embrace service-based consumption, other consumers may not entertain the idea of using a product without owning it (Alghamdi et al., 2021).

This resistance can be especially evident in the markets that are associated with status or that involve ownership and where people do not trust the availability and reliability of the services. Businesses also need to engage consumers with a clear message on the advantages of PaaS and gain their confidence in service provision. Regulatory constraints are also significant in influencing PaaS personnel adoption. Currently, many regions already have established rules for regulating product-based business models; and updated rules targeting the peculiarities of PaaS including key issues like liability, taxation, and consumer protection which are still neglected (Osembe & Padayachee, 2016). It is therefore the duty of governments and policymakers to coordinate efforts towards extending a favorable legal and regulatory environment that supports the adoption of PaaS and eliminating obstacles to its deployment.

2.5 Opportunities and Future Directions

The opportunities the PaaS model offers are massive. From the perspective of businesses, it brings opportunities for generating new and additional revenue streams, expanding customer ties and setting themselves apart in more saturated markets (Compagnoni & Stadler, 2021). In businesses who adopt PaaS, service-providers can have constant communication with the buyers offering them new versions, repairs and complementary increased customer retention. In the context of circular economy (which is efficient in the use of resources) and in relation to the reduction of adverse effects on the environment, the PaaS model is more consistent with the ideals of eco-friendly approach to deliver services for business. Durable products which are easy to maintain do not find their way to the landfills and where possible products can be refurbished, remanufactured, or recycled at the end of their useful life cycle. This will add to sustainability (Tohanean & Toma, 2024). Furthermore, it will not only lead to the promotion of CSR activities but also to the saturation of the demands of consumers and investors for the development of the company's more sustainable activity.

The PaaS model will also remain prominent in the future as the economy expands and shares more on its online presence, employing better technologies and leading to customer's transition towards being environmentally conscious. Therefore, due to IoT, AI, blockchain and other technologies embedded in the system, PaaS models will advance and improve the provision of complex goods and services (Phaphoom et al., 2015). As more companies seek sustainable business models, more governments and international organizations are extending their support to firms that promote the circular economy. It is increasing policy support for PaaS and other circular business models.

In general, it can be noted that the PaaS model may be considered as one of the possible trends to implement the initiatives of the circular economy. Despite the hurdles that have been observed as the main reasons that hinders the widespread adoption of green supply chain management, the benefits that accrue to the business, consumer and the environment are numerous. In the light of the present-day advancement in the digital technology and conducive policies of the PaaS model, it is likely to emerge as a standard model of the global economy in the future delivering innovation, sustainability and economic growth in the coming years (Bocken et al., 2016).

3. SUMMARY

This chapter begins with an introduction and a literature review that places the "Product as a Service" (PaaS) business model in the larger framework of circular economy. It is an innovative economy model that aims to respond to all the environmental and resources issues of the current linear economy. Circular economy is based on the three principles of eliminating waste and pollution, using products and materials effectively, and restoring the ecosystem. These principles contradicted the dominant model of economic development by practicing the ever-use and ever-renewal of a resource, making resource waste near impossible. The PaaS model appears as one of the solutions that can help attain these objectives as it changes ownership from manufacturing products to delivering services that inspire producing long-lasting items which can be sustained and updated. The circular economy also brings forth newly emerging business models that can generate more value by means of repetitive or subscription-based revenues and improved customer interactions.

A brief discussion on the development and relevance of PaaS model is also included in the literature review. It underpinned servitization as a principal concept based on the shift towards business from pure product manufacturing and supply to providing products as well as services. The review also highlights how digitalization is an essential component for the advancement of PaaS by employing IoT, AI, and block chain technologies for effective PLM and services. However, there are significant challenges in the adoption of PaaS model. There is compelling evidence that substantial shifts in organizational strategies, culture, and capability are required, along with addressing cultural resistance and the need for supportive regulation. However, there are multiple benefits offered by PaaS, especially for industries where high initial investment is required or in industries where the technology becomes outdated in relatively short period of time like automobiles and technology industries. They have realized that access is way better than owning the product or service and as the technology advances, PaaS is set to be the model that will shape the new economy for a sustainable future.

Table 1. Benefits of the product-as-a-service model

For producers		For consumers		Sustainable development
Offer	Improved offer optimized for customer needs	Offer	Personalized offer	Longer use of resources
	New services and revenue streams		Availability of updated products	Best use due to multiple users
	Significant differentiation and individualization of products		Complementary services	Extended product life
Sales	Simple search for new customers	Expenses	Low initial investment	Ability to reuse and recycle
	Flexible pricing according to customer needs		Cost savings	More conscious consumption
Profitability and cash flow	Stable, predictable income	Operations	Improved customized service	
	Cost savings and increased profits			
Customer relationship	Close relationships with customers and increased customer loyalty			

(Authors Data)

Table 1 highlights all the advantages of the Product-as-a-Service (PaaS) model for the producer, consumer, and for sustainable development in society. The model holds great potential for producers, as it can open up opportunities for them by offering differentiation of product lines, additional sales and customer intimacy are targeted. It ultimately leads to guaranteeing producers a steady and reliable income stream. On the other hand, low initial investment, access to updated products and services as well as patronization of individually tailored services are notable benefits for the consumer. It also forms the basis of cost-saving strategy and improvement of service delivery. Discussing the PaaS model on the basis of sustainable development criteria, four benefits can be pointed out. Only tangible resources are used and the overall rate of their utility is multiplied. Simultaneously, while retaining the use, PaaS intends to sustain recycling, and ultimately reuse rates are optimized. It also pays close attention to what is being consumed and where it will end up at the conclusion of its life-cycle. The overview of these benefits incidentally shows how the PaaS model can achieve both economic and ecological preservation to form a key component of the circular economy plan.

Figure 1 illustrates the various business models of a circular economy, as compiled by the authors from multiple sources including Arenkov, Tsenzharik, and Vetrova (2019) and Accenture (2019). These models encapsulate the strategies and frameworks that businesses can adopt to shift towards a circular economy, emphasizing sustainability, resource efficiency, and the reduction of waste. The figure serves as a visual synthesis of the key components and practices that underpin a circular economy, providing a comprehensive insight into how different business models contribute to the sustainability goals. This visual representation is crucial for understanding the practical applications of circular economy principles in various industries.

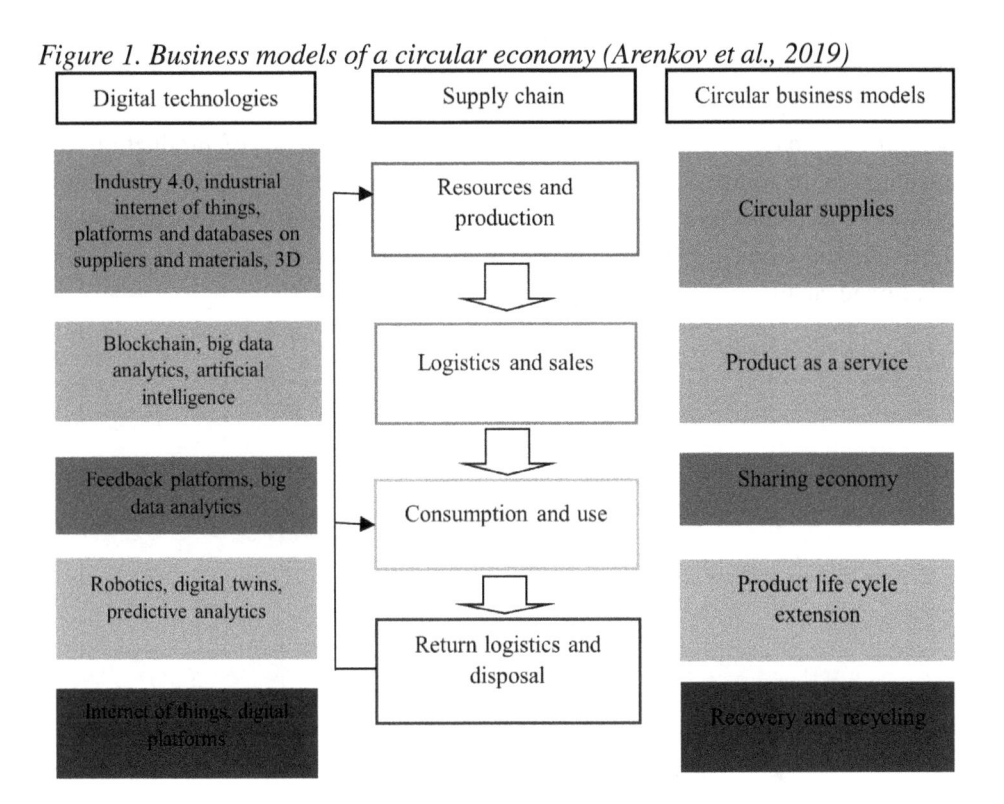

Figure 1. Business models of a circular economy (Arenkov et al., 2019)

Digital technologies	Supply chain	Circular business models
Industry 4.0, industrial internet of things, platforms and databases on suppliers and materials, 3D	Resources and production	Circular supplies
Blockchain, big data analytics, artificial intelligence	Logistics and sales	Product as a service
Feedback platforms, big data analytics	Consumption and use	Sharing economy
Robotics, digital twins, predictive analytics	Return logistics and disposal	Product life cycle extension
Internet of things, digital platforms		Recovery and recycling

4. CONCLUSION

The 'Product as a Service', also known as PaaS model can offer a viable theoretical structure for managing the shift towards a post-industrial and resource-scarce economy. Thus, empowering PaaS to regulate circular economy, entails the extension of product use, minimization of waste, and regeneration of natural resources. The integration of digital technologies like IoT, AI, and blockchain provides the opportunity to PaaS to progress to another level, enriching the company and solution partners with a more efficient way of creating, delivering, and managing value for clients and markets as well as reinforcing the structure needed to manage the full lifecycle of products.

Despite the challenges associated with implementing PaaS, such as the need for strategic shifts within organizations, cultural adaptation, and regulatory support, the benefits for businesses, consumers, and the environment are significant. As consumer preferences continue to evolve towards valuing access over ownership and as digitalization becomes more pervasive, PaaS is poised to play a crucial role in the future of sustainable business practices. The model not only offers businesses new opportunities for growth and differentiation but also contributes to the broader goals of sustainability by reducing the environmental impact of production and consumption. In conclusion, the PaaS model represents a vital pathway for achieving a more circular and resilient economy, and its adoption will be essential for addressing the global challenges of resource scarcity, environmental degradation, and climate change.

REFERENCES

Alghamdi, B., Potter, L. E., & Drew, S. (2021). Validation of architectural requirements for tackling cloud computing barriers: Cloud provider perspective. *Procedia Computer Science*, 181, 477–486. DOI: 10.1016/j.procs.2021.01.193

Anouar, A., Touhafi, A., & Tahiri, A. (2017). Towards an SOA architectural model for AAL-PaaS design and implimentation challenges. *International Journal of Advanced Computer Science and Applications*, 8(7). Advance online publication. DOI: 10.14569/IJACSA.2017.080708

Ardolino, M., Rapaccini, M., Saccani, N., Gaiardelli, P., Crespi, G., & Ruggeri, C. (2018). The role of digital technologies for the service transformation of industrial companies. *International Journal of Production Research*, 56(6), 2116–2132. DOI: 10.1080/00207543.2017.1324224

Arenkov, I., Tsenzharik, M., & Vetrova, M. (2019, September). Digital technologies in supply chain management. In *International Conference on Digital Technologies in Logistics and Infrastructure (ICDTLI 2019)* (pp. 448-453). Atlantis Press. https://doi.org/DOI: 10.2991/icdtli-19.2019.78

Belmonte-Ureña, L. J., Plaza-Úbeda, J. A., Vazquez-Brust, D., & Yakovleva, N. (2021). Circular economy, degrowth and green growth as pathways for research on sustainable development goals: A global analysis and future agenda. *Ecological Economics*, 185, 107050. DOI: 10.1016/j.ecolecon.2021.107050

Bocken, N. M., De Pauw, I., Bakker, C., & Van Der Grinten, B. (2016). Product design and business model strategies for a circular economy. *Journal of Industrial and Production Engineering*, 33(5), 308–320. DOI: 10.1080/21681015.2016.1172124

Brezet, H., & van Hemel, C. (Eds.). (1997). *Ecodesign: A promising approach to sustainable production and consumption* (1st ed.). U.N.E.P.

Che, H., Erdem, T., & Öncü, T. S. (2015). Consumer learning and evolution of consumer brand preferences. *Quantitative Marketing and Economics*, 13(3), 173–202. DOI: 10.1007/s11129-015-9158-x

Classen, M., Blum, C., Osterrieder, P., & Friedli, T. (2019, September). Everything as a service? Introducing the St. Gallen IGaaS management model. In J. Meierhofer, S.S. West (Ed.), *Proceedings of the 2nd Smart Services Summit* (pp. 61-65). Data Innovation Alliance.

Compagnoni, M., & Stadler, M. (2021). *Growth in a circular economy* (No. 145). [Working Papers, University of Tübingen]. UT Campus Repository. DOI: 10.15496/publikation-56495

Defra. (2013). *Waste Prevention Programme for England* [Policy Paper, Department for Environment, Food & Rural Affairs]. GOV.UK.

Firoiu, D., Ionescu, G. H., Băndoi, A., Florea, N. M., & Jianu, E. (2019). Achieving sustainable development goals (SDG): Implementation of the 2030 agenda in Romania. *Sustainability (Basel)*, 11(7), 2156. DOI: 10.3390/su11072156

Freet, D., Agrawal, R., John, S., & Walker, J. J. (2015, October). Cloud forensics challenges from a service model standpoint: IaaS, PaaS and SaaS. In *Proceedings of the 7th International Conference on Management of Computational and Collective intElligence in Digital EcoSystems* (pp. 148-155). ACM Digital Library. https://doi.org/DOI: 10.1145/2857218.2857253

Frosch, R. A., & Gallopoulos, N. E. (1989). Strategies for manufacturing. *Scientific American*, 261(3), 144–153. DOI: 10.1038/scientificamerican0989-144

Garcia-Gomez, S., Jimenez-Ganan, M., Taher, Y., Momm, C., Junker, F., Biro, J., Menychtas, A., Andrikopoulos, V., & Strauch, S. (2012). Challenges for the comprehensive management of cloud services in a PaaS framework. *Scalable Computing: Practice and Experience*, 13(3), 201–214. DOI: 10.1109/ISPA.2012.72

Gesvindr, D., & Buhnova, B. (2016). Performance challenges, current bad practices, and hints in PaaS cloud application design. *Performance Evaluation Review*, 43(4), 3–12. DOI: 10.1145/2897356.2897358

Giessmann, A., & Stanoevska-Slabeva, K. (2012). Business models of platform as a service (PaaS) providers: Current state and future directions. *Journal of Information Technology Theory and Application*, 13(4), 31.

Hernandez, R. J. (2019). Sustainable product-service systems and circular economies. *Sustainability (Basel)*, 11(19), 5383. DOI: 10.3390/su11195383

Ibn-Mohammed, T., Mustapha, K. B., Godsell, J., Adamu, Z., Babatunde, K. A., Akintade, D. D., Acquaye, A., Fujii, H., Ndiaye, M. M., Yamoah, F. A., & Koh, S. C. L. (2021). A critical analysis of the impacts of COVID-19 on the global economy and ecosystems and opportunities for circular economy strategies. *Resources, Conservation and Recycling*, 164, 105169. DOI: 10.1016/j.resconrec.2020.105169 PMID: 32982059

Ikundi, F., Islam, R., & White, P. (2017, October). Platform as a Service (PaaS) in public cloud: Challenges and mitigating strategy. In R. Deng, J. Weng, K. Ren, & V. Yegneswaran, (Eds.), *Security and Privacy in Communication Networks: 12th International Conference Proceedings* (pp. 296-304), [Lecture Notes of the Institute for Computer Sciences, Social Informatics and Telecommunications Engineering], Vol 198. Springer, Cham. https://doi.org/DOI: 10.1007/978-3-319-59608-2_17

Iqbal, K. M. J., Barykin, S. Y., Kharlamov, A. V., Kharlamova, T. L. V., & Khan, M. I. (2021). Innovative multivariate energy governance model for climate compatible development: The case of Pakistan. *Academy of Strategic Management Journal*, 20, 1–22.

Jensen, J. P. (2015). Routes for extending the lifetime of wind turbines. In T. Cooper, N. Braithwaite, M. Moreno, & G. Salvia (Eds.), *Product Lifetimes and The Environment: Conference Proceedings* (pp. 152-157). Nottingham Trent University.

Kharlamova, T. L., Kharlamov, A. V., & Antohina, Y. A. (2020). Influence of information technologies on the innovative development of the economic system. In I. V. Kovalev, A. A. Voroshilova, G. Herwig, U. Umbetov, A. S. Budagov, & Y. Y. Bocharova (Eds.), *Economic and Social Trends for Sustainability of Modern Society (ICEST 2020), Vol 90. European Proceedings of Social and Behavioural Sciences* (pp. 391-401). European Publisher. DOI: 10.15405/epsbs.2020.10.03.44

Khun, J. A. (2021). Relationship between macroambient factors, circular economy, and sustainability. In *No Poverty* [ENUNSDG Series]. (pp. 771–782). Springer., DOI: 10.1007/978-3-319-95714-2_31

Knill, C., Schulze, K., & Tosun, J. (2012). Regulatory policy outputs and impacts: Exploring a complex relationship. *Regulation & Governance*, 6(4), 427–444. DOI: 10.1111/j.1748-5991.2012.01150.x

Lampkin, N., Foster, C., & Padel, S. (1999). *The policy and regulatory environment for organic farming in Europe: Country reports*. Organic Farming in Europe: Economics and Policy, Vol. 2. Universität Hohenheim, Stuttgart-Hohenheim. https://doi.org/DOI: 10.52825/gjae.v50i7.1484

Li, N., Binti Mohd Ariffin, S. Z., & Gao, H. (2024). Optimizing Ecotourism in North Taihu Lake, Wuxi City, China: Integrating Back Propagation Neural Networks and Ant-Colony Algorithm for Sustainable Route Planning. *International Journal of Management Thinking*, 2(1), 1–15. DOI: 10.56868/ijmt.v2i1.53

Liu, H., Zhu, Q., Khoso, W. M., & Khoso, A. K. (2023). Spatial pattern and the development of green finance trends in China. *Renewable Energy*, 211, 370–378. DOI: 10.1016/j.renene.2023.05.014

Liu, Z., Liu, J., & Osmani, M. (2021). Integration of digital economy and circular economy: Current status and future directions. *Sustainability (Basel)*, 13(13), 7217. DOI: 10.3390/su13137217

Lüthi, S., & Prässler, T. (2011). Analyzing policy support instruments and regulatory risk factors for wind energy deployment—A developers' perspective. *Energy Policy*, 39(9), 4876–4892. DOI: 10.1016/j.enpol.2011.06.029

Matsumoto, O., Kawai, K., & Takeda, T. (2017). FUJITSU cloud service K5 PaaS digitalizes enterprise systems. *Fujitsu Scientific and Technical Journal*, 53(1), 17–24.

Nurulin, Y. R., Skvortsova, I., & Vinogradova, E. (2020). On the issue of the green energy markets development. In Arseniev, D., Overmeyer, L., Kälviäinen, H., & Katalinić, B. (Eds.), *Cyber-Physical Systems and Control* [Lecture Notes in Networks and Systems]. Vol. 95). Springer., DOI: 10.1007/978-3-030-34983-7_34

OECD. (2020). *The territorial impact of COVID-19: Managing the crisis across levels of government*. Organisation for Economic Co-operation and Development.

Osembe, L., & Padayachee, I. (2016). Perceptions on benefits and challenges of cloud computing technology adoption by IT SMEs: A case of Gauteng province. *Journal of Contemporary Management*, 13(1), 1255–1297.

Özkan, P., & Yücel, E. K. (2020). Linear economy to circular economy: Planned obsolescence to cradle-to-cradle product perspective. In *Handbook of research on entrepreneurship development and opportunities in circular economy* (pp. 61–86). IGI Global., DOI: 10.4018/978-1-7998-5116-5.ch004

Phaphoom, N., Wang, X., Samuel, S., Helmer, S., & Abrahamsson, P. (2015). A survey study on major technical barriers affecting the decision to adopt cloud services. *Journal of Systems and Software*, 103, 167–181. DOI: 10.1016/j.jss.2015.02.002

Pupentsova, S. V., Alekseeva, N. S., & Stroganova, O. A. (2020, February). Foreign and domestic experience in environmental planning and territory management. [IOP Publishing.]. *IOP Conference Series. Materials Science and Engineering*, 753(3), 032026. DOI: 10.1088/1757-899X/753/3/032026

Saygıner, C. (2023). Software as a service (SaaS) adoption as a disruptive technology: Understanding the challenges and the obstacles of non-SaaS adopters. *Uluslararası Yönetim İktisat ve İşletme Dergisi*, 19(3), 501–515. DOI: 10.17130/ijmeb.1249540

Sharma, V. S., & Santharam, A. (2013, December). Implementing a resilient application architecture for state management on a PaaS cloud. In *2013 IEEE 5th International Conference on Cloud Computing Technology and Science,* Vol. 1, (pp. 142-147). IEEE. DOI: 10.1109/CloudCom.2013.26

Silkina, G. (2019, April). From analogue to digital tools of business control: Succession and transformation. In *IOP Conference Series: Materials Science and Engineering,* Vol. 497, 012018. IOP Publishing. https://doi.org/DOI: 10.1088/1757-899X/497/1/012018

Sinkevičius Virginijus. (2020, June). *Commissioner speech at post-COVID green deal technology powered recovery in Europe.* European Commission, Brussels.

Taylor, C. M., Gallagher, E. A., Pollard, S. J., Rocks, S. A., Smith, H. M., Leinster, P., & Angus, A. J. (2019). Environmental regulation in transition: Policy officials' views of regulatory instruments and their mapping to environmental risks. *The Science of the Total Environment,* 646, 811–820. DOI: 10.1016/j.scitotenv.2018.07.217 PMID: 30064107

Tischner, U., Schmincke, E., Rubik, F., & Prösler, M. (2000). *How to do EcoDesign?: A guide for environmentally and economically sound design.* Verlag Form Publisher.

Tohanean, D., & Toma, S. G. (2024). The impact of cloud systems on enhancing organizational performance through innovative business models in the digitalization era. In *Proceedings of the International Conference on Business Excellence,* Vol. 18(1), (pp. 3568-3577). Sciendo. https://doi.org/DOI: 10.2478/picbe-2024-0289

Vetrova, M., & Ivanova, D. (2021). Closed Product Life Cycle as a Basis of the Circular Economy. [JBER]. *Journal of Business & Economics Research*, 5(4).

Vezzoli, C., & Manzini, E. (2008). *Design for environmental sustainability* (1st ed.). Springer Science and Business Media., DOI: 10.1007/978-1-4471-7364-9

Wortmann, F., Gebauer, H., Lamprecht, C., & Fleisch, E. (2024). Drive the PaaS transformation. In *Understanding Products as Services: How the Internet and AI are Transforming Product Companies* (pp. 57–67). Emerald Publishing Limited., DOI: 10.1108/978-1-83797-823-620241009

Wulf, F., Lindner, T., Strahringer, S., & Westner, M. (2021). IaaS, PaaS, or SaaS? The why of cloud computing delivery model selection: Vignettes on the post-adoption of cloud computing. In *Proceedings of the 54th Hawaii International Conference on System Sciences* (pp. 6285-6294). Ostbayerische Technische Hochschule Regensburg. https://doi.org/DOI: 10.24251/HICSS.2021.758

Chapter 2
Application of the Business Ecosystem Approach to the Transformation of Business Models in Industry:
The Example of the Light Industry Transformation

Sviatlana Hrytsevich
Belarusian National Technical University, Belarus

Muhammad Asad Tahir
University of Leicester, UK

ABSTRACT

Traditional industries that have dominated the market for decades are now facing new competitors that are changing the boundaries of industries and business models. These competitors are taking leadership positions due to the growth of digital technologies. One of the possible ways of successful functioning in the digital world can be the use of the concept of business ecosystems (BES), namely the introduction of a business ecosystem approach (BEA) to the formation of a new strategy for managing traditional business models. The article aims to reveal the relevance of the development of the BES concept and the use of BEA as a transitional element to the modern transformation of traditional industries. In this article, the author

DOI: 10.4018/979-8-3693-3423-2.ch002

presents key conclusions on the refinement and use of criteria characterizing the functioning of business ecosystems for the organization of transformational processes, considering digitalization and intercompany integration. The author examines the example of the transformation of the light industry (textile and clothing) in the Republic of Belarus.

NOVELTY STATEMENT

This chapter contributes significantly to the sustains a business ecosystem approach (BEA) that delivers outcomes which are up to par to the modern world needs and achieves the high-end goals for both the supplier and the consumer i.e., being advantageous to all stakeholders. The proposed approach offers a transition path to the traditional industries to upgrade themselves rather than demolishing these industries and relying solely on the digitilised new industries that came into being in the modern era.

1. INTRODUCTION

In the competitive struggle, traditional industries like manufacturing industry supports the emerging strategy of maintaining positions for years. Ironically, they are currently faced with the irrelevance of existing strategies. Expanding the influence of new high-tech on market participants of the global economy, demonstrating exponential growth in the short term, is a challenge to maintain a firm competitive position in traditional business. Typically, traditional industries are the primary source of employment in developing countries. They are at the stage of maturity or decline of their life cycle, and differ in the relative labor intensity of the production process. However, they retain the ability to innovate. Despite the challenges, traditional industries continue to be a significant contributor to national product creation and employment.

As they adapt their business models and seek new competitive advantages, the potential for transformation in these industries is a source of hope and optimism. The success stories of startup projects in high-tech industries serve as inspiration, influencing the revision and change of development strategies in long-established sectors and companies. Maintaining innovative activity in the traditional sector of the economy is a pressing need that requires new approaches to management. The changing business environment, influenced by automation, robotization, and digitalization processes, necessitates the explicit integration of various market participants. In this context, the creation of a business ecosystem (BES) emerges as a strategically important solution for increasing the business and innovation potential of both individual enterprises and entire industries.

In this chapter, the author defines the directions for changing the business model, using the example of one of the most important types of traditional manufacturing industry - the light industry (i.e., textile and clothing). The future dynamics have modified business model of this sector in the context of digitalization. The subject-matter discusses the development of innovative integrated solutions based on a combination of manufactured products and the possibility of providing industrial services that comprises of digital components in order to meet the ever-increasing needs of customers.

The experience of the development of the textile and clothing industry in the USA and the EU countries has presented opportunities for the transition of the so-called low-tech industries to the category of high-tech through cooperation. It encourages to achieve sustainable innovative solutions (Abbes et al., 2022; Beyers & Heinrichs, 2020; Todeschini et al., 2017). This transition was facilitated by the revision of development strategies towards the introduction of innovations, namely information and communication technologies in the design, and creation of fabrics, and products, as well as electronic trading platforms for the needs of customers.

It is of due consideration that the presence of platform business models within the framework of changes in international trade policy has allowed textile and clothing consortia of developed countries to create and sell products with exceptional characteristics and advantages. It has simultaneously solved the problems of textile waste and meeting high demands in the fashion industry (Ghoreishi & Happonen, 2022; Jensen & Whitfield, 2022). Participants of the BES in the textile and clothing industry platforms play a key role of a guide in the creation of new products and their transition from production to retail sales, allowing the use of digital design and ceaseless production processes without additional costs. It should be emphasized that data-driven product design allows the industry to achieve waste reduction goals, which is of significant concern in current environmental conditions.

2. LITERATURE REVIEW

Adner and Kapoor (2010) investigated the effectiveness of the management strategy in the business ecosystem. In the context of global semiconductor equipment for lithography, the study describes the architecture of business ecosystem (BES by pointing out the consequences of external innovation challenges depending not only on the scale of each of the participants but also on their location in the ecosystem relative to the leading company. Chen et al. (2022) clarified the concept of innovation ecosystem based on a sound theory. It allows us to present a conceptual model comprising of an innovative population and an innovative habitat in the innovation ecosystem, as well as the input and output activities of the subjects of innovation in the ecosystem. Jacobides et al. (2018) in their work focuses on supplementing the literature on BES in the direction of causes of ecosystem, and their differences from other forms of management. Along with this, the authors point out the importance of such a property as modularity which contributes to the emergence of an ecosystem that allows several separate but interdependent organizations to coordinate their actions without a complete hierarchical order.

Kapoor and Lee (2013) likewise consider firms in the context of their business ecosystems and explore how differences in the ways firms organize in relation to complementary activities affect their decision to invest in new technologies. Trinh et al. (2014) developed the topic of ecosystems in business by applying the game theory approach to the study of strategic behavior of firms in systems of joint value creation. The authors pay close attention to such important issues as cost optimization and its subsequent distribution among ecosystem participants.

Although some researchers criticize the concept of BES itself, the described empirical experience shows the importance of the transition to the use of a new theory in various sectors of industry, both high-tech and traditional (also known as low-tech). However, analytical work is still underway to create models that will allow evaluation of the effectiveness of this new form of interaction between economic entities in the future.

Furthermore, other studies were conducted under the necessary to clarify certain aspects. As Battistella et al. (2013) analyze BES models according to a methodology called "methodology of business ecosystem network analysis" (MOBENA). The authors illustrate its application via a case study conducted inside the Telecom Italia Future Centre, and in particular taking it as an example for the digital imaging ecosystem. Bithas et al. (2018) consider the development of innovative activities in service sector organizations within the framework of a two-dimensional model that displays the options ("roles") of a service system for positioning itself in a network of service systems. Creating joint value with an emphasis on innovative leadership and differentiation through partnerships along with making variations in the supplier-consumer relationship. Kim (2016) considers the activity of platforms as ecosystems of coexistence of suppliers and consumers, which can provide benefits to both sides. Based on these studies, it can be anticipated to build a platform model in BES within "12 different types of quality management strategies and revenue structure".

Radicic et al. (2020) explore seven EU regions to analyze government support programs for SMEs in the traditional manufacturing industry, which will consolidate and expand innovative ecosystems beyond the usual business partners by encouraging cooperation of SMEs with both knowledge providers from the private and public sectors. Wieninger et al. (2019) uses the example of the telecommunications industry to propose a new approach for the analysis of the BES by leading players, and assess the level of their control over the entire system.

The concept of BES was formed in context of revisiting economic thought in the 90s (Moore, 2006). It marked the beginning of an active study of complex economic dynamics. The theoretical core of the research was to analyse the concept of "ecosystem". Along with Moore, other scientists that contributed significantly are Teece D. (2016), Adner and Kapoor (2010), Rong et al. (2013), Williamson and De Meyer (2012) among others. In their study, they investigated the relation of participants in modern business that develop in a dynamic environment of network interactions. Subsequently, defining the purpose of the BES as a landscape for the integration of resources, the outcome of joint creation, and the receipt of values.

Several Russian scientists have also considered the study of the entire variety of business systems and the dynamics of their development. Research in this field of inter-firm interactions in the conditions of the formation of a post-industrial economy, emerging uncertainty, and the need for increasing stability of the values created are considered of paramount importance. This is endorsed in the publications of Ivanova et al. (2020). Vertakova et al. (2022) highlight the issues of digitalization of the economy, the development of transformation processes of inter-firm interaction, and describes various methodologies for assessing the transformation of traditional industries in modern conditions. The issues that arise in the formation of industrial ecosystem are discussed by Babkin et al. (2021). Gamidullaeva et al. (2021) provide thorough insight into the study of digital platforms as tools for the development of business ecosystems, and the description of intersectoral digital mechanisms of business relationships. Sheresheva et al. (2022) undertakes in their articles the elaboration of forms of interactions between participants of complex economic systems as the example of clusters, and establishes possible ways of identifying each of the participants using a network approach. Furthermore, Popov et al. (2021) explore the entrepreneurial and innovative ecosystems in publications and presents their analytical models.

This comprehensive literature elucidates that BES concept has changed the view from traditional inter-firm competition to a joint approach with simultaneous cooperation and competition between participants in economic systems. Therefore, the ecosystem business models are formed to collectively create value propositions that companies cannot create alone using only their resources. In other words, the ecosystem is determined by the structure through which partners interact to convey a value proposition to the end consumer (Adner, 2017). Ultimately, the BES concept emphasizes the process of the joint evolution of industrial system and their dynamic environment which offers new opportunities in conditions of uncertainty (Rong et al., 2013).

BES can be designed based on various values, methods, and technologies. However, the concept of BES is based on a basic set of principles inherent in any ecosystem, both in the biological environment and in the business environment. These principles include the principles of coevolution, dynamic transformation, continuous innovative development, competitive cooperation, coordination, self-adaptation, self-organization, joint value creation, modularity, multidimensionality, and scale (Gawer & Cusumano, 2014; Hein et al., 2019). Inter-company dynamic systems created in the context of the BES concept are subject to changes and adaptation considering digital transformation.

They involve the coordinated involvement of stakeholders in a complex network architecture of interaction to create common value. To simplify the complex architecture of inter-organizational interaction, it is intended to create platforms in ecosystem that allow direct and rapid exchange of data and information between participants. A digital platform in BES with a multi-level modular architecture can be used for a separate company, as well as serve as a component for other companies participating in the ecosystem (Hein et al., 2019; Jacobides & Lianos, 2021; Rong et al., 2021).

Modularity in the functioning of the platform facilitates the joint work of the BES participants, and the coordinating function of the platform indicates the joint work of the suppliers and partners hosted on it. In the process of ecosystem integration involving digital platforms, enterprises can reduce their labor and capital costs in software development by obtaining the necessary technologies, resources, and information from partners. A set of principles describing the conceptual boundaries of ecosystems in business underlie the emergence of such a complex category as the business ecosystem approach (BEA). The use of BEA to form a strategy for the transition to business ecosystem integration on the role model of foreign companies (Acs et al., 2017; Cavallo et al., 2021; Chen et al., 2022). It suggests the emergence of a new integrated approach to the management of enterprises. This will jointly create a value proposition describeing a new architecture of interaction and complements the indicators for evaluating the effectiveness of managing complex structures.

In BES, joint value creation occurs in a network of shared assets, when all participants create value simultaneously. Creating a common value involves directing resources to the development of innovations. On the basis of close networking, enterprises can get more resources and partners aimed at improving productivity through flexibility and reducing customer order fulfillment time (Rong et al., 2021). The development of networks is closely related to the emergence of digital platforms. The platforms allow the accumulation of network effects and the scaling of the business. A great number of participants of the BES increase the added value accumulated around the platform. Thereby, the structure of the BES is formed due to network effects. It should be noted that value propositions can provide benefits for end users not only in the form of tangible products, but also in the form of transactions, services, and new information (Trinh et al., 2014). The proposed approach fundamnetally describes the elements of the BES concept and the use of a business ecosystem approach to the management of traditional enterprises, which will allow changing development strategies and transforming business models of enterprises and entire industries in the modern digital era.

3. MATERIALS AND METHODS

The authors opted for the theoretical analysis. Taking into consideration all the aspects, this approach employed general scientific methods i.e, analysis, synthesis, generalization, and system analysis. It is imperative to anticipate the possibilities of applying the BES concept to management in traditional industries, utilizing methods of theoretical analysis of materials from reputable researchers as Cavallo et al. (2021), Jacobides et al., (2018), Wieninger et al. (2019). These derived materials were used to identify important properties that future participants of BES should meet, and the functional environment that should be sustained, in order to increase the intensification of development of their business processes during the period of active digitalization. These methods served as a conceptual basis for describing a set of criteria by which it is possible to assess the readiness of modernization of mature industries in both developed and developing countries.

The author argues that it is necessary to qualitatively assess the potential of traditional industries for the formation of a new form of organization of inter-company relations – BES. BES as a new method of doing business makes it possible to involve various firms (including small and medium-sized businesses) in the cooperative chains of large manufacturing enterprises, providing comprehensive support to the country's economy at the stage of development.

The materials entail changing the business model, by taking the light industry as an example. It prompts to redefine our future dynamics by integrating innovative solutions in the context of globally digitalized world. The traditional manufacturing industry won't cope with the strategically growing and all-encompassing models if they still retain redundant or declining life-cycles. It induces the need for robust innovative initiatives as assembled and put forward by the materials of this chapter. By encouraging sustainable business ecosystems, it anticipates the potential of providing industrial services that comprises of digital components, innovative solutions, and data-driven approach in order to meet the ever-increasing needs of customers.

The transformation experience inadvertently unleashes certain challenges to traditional manufacturing industry. The use of digital technologies and tools in production and business processes deems that it is high-time for the traditional industries to make a transition from low-tech industries to high-tech industry. As highlighted in the materials, this precept leads to an era of new cooperation and the integration of principles that delineates digital platforms to inter-play with the industry sectors unleashing an ecosystem with the creation of joint value and enables a differentiation of roles for service systems. It manifests unequivocally that the entire supply-chain is affected by the strategies towards design, production, and distribution of goods and services; dwelling on the significance of business ecosystem approach for the management of traditional enterprise or industry. With an inclusive approach of theoretical analysis of literature along side employing scientific method, makes this an impactful study to transform the empirical experience of participants by adoption of transformative business models throughout industry.

Consequently, the author concludes that for the development of light industry, it is necessary to use case examples from the industry of developed countries, reinforcing them in our country with products of the "smart" industry. Incorporating the empirical experience of foreign researchers along with the current framework of textile and clothing industry in addition to the theoretical analysis of materials derived from the works of reputable scientists clarifies concept of BES. The author of this chapter considers it imperative to identify several criteria that will allow the introduction of BEA in order to lead to the transformation of management models in the light industry.

It exemplifyes the Republic of Belarus, which belongs to the category of developing countries, as the role model. The use of the criteria of the BES concept will also make it possible to create an industry that is at a mature stage of its life cycle, with an increase in total value added through the expansion of value chains. This is made possible by including new participants who create and provide operations and business processes within the framework of digital transformation.

4. ANALYSIS AND RESULTS

The author carried out a theoretical analysis of the main criteria derived from the collected literature, and based its concluions on these characteristics of modern BES that had led foreign experts for successful implementation in their countries. In the context of various types of research on organization-leaders of Western and Asian ecosystems, the results prompted adopting those frame-works in our home country.

Analyzing the criteria and presenting explanations of the criteria within the framework of functioning BES, the author attempts to describe the signs. In compliance to it, will indicate the readiness of traditional industries to transform their business models into the direction of the creation of BES that adapts in resonance with the modern conditions of the digital business environment. The results are presented in table 1.

Table 1. Theoretical analysis of the possibilities of traditional industry transformation in BES

Main Criteria of Modern BES	Explanation of the selected criteria	Signs of Transformation
Dynamics	Dynamics within the ecosystem, which determines the nature of interaction of participants who create value in several ways. It presents a dynamic environment for the "life" of ecosystem.	It causes changes in the traditional boundaries of industries, complementing vertically integrated value chains with horizontal ones.
Coevolution	It creates the process of joint evolution on a mutually beneficial basis to enhance the value proposition with co-vision; co-design; co-create.	It leads to understanding the common ecosystem vision, and ecosystem thinking is necessary for the joint creativity of participants.
Integration	It expands cooperation with partners and competitors to share various types of resources, knowledge, along with the extension of cooperative ties.	It implies organization of inter-company communication, joint business processes, and infrastructure construction.
Coordination	It develops the existence of an active participant (platform) that allows you to distribute tasks, distribute risks and transaction costs, and define standards of interaction.	It induces the creation of a platform for the accumulation and organization of flows of different types of resources.
Networking	It creates a structure of interaction, the emergence of network effects that affect the scale of increasing the value of the total supply of ecosystem participants.	It brings revision of the value chain, and its expansion into a value-creation network with many stakeholders.
Knowledge Intensity	The exchange of skills and knowledge between partners is the main source of sustainable growth in various business areas to increase productivity and the uniqueness of the value proposition being created.	It sustains knowledge management throughout the life cycle of the created product/service.
Innovativeness	It increases the number of sources of innovative activity due to the emergence of new knowledge providers.	It maintains the ability to innovate, with commitment to the theory of open innovation for access to knowledge and resources of partners.
Digital maturity	It delivers digital development, with the use of advanced information technologies in the field of communications.	It prompts development of information and communication infrastructure, operational and secure access to data, automation, and digitalization of business processes.
Modularity	It creates a flexible modular architecture allowing participants to complement each other's activities outside of value chains.	It entails understanding the composition of value creating participants, choosing active and passive ones, and reviewing value creation networks.
Servitization	The integrated "product+service" system allows you to combine goods and services into complex integrated solutions, thereby providing higher added value for customers.	It urges at designing new indicators and describing a comprehensive contribution to an integrated value proposition aimed not only at the product but also at the service.

Table 1 highlights the main criteria for the development of BES approach. It elucidates how these criteria become the basis for the formation of strategic direction towards the transformation of traditional industries to the creation of a model of BES. Such a criterion as servitization, which allows the BES to jointly create an effective value proposition for customers, is currently becoming the main goal for the traditional industry to increase the added value created. The described set of criteria in this research are assessed in context of the light industry (i.e, the textile and clothing industry) of the Republic of Belarus.

The light industry in this country has a long history of being one of the most important sectors of the manufacturing industry during the Soviet Union, which has preserved the most important textile and clothing industries in the country since the 90s. During its new thirty-year history, the industry has faced several difficulties, such as outdated equipment, lack of raw materials, reduction in production volumes, changes in prices for thermal and electrical energy, etc. However, the ongoing socio-economic policy of the state successfully transformed traditional industry to model of BES. This was aimed at via maintaining domestic production, financing their modernization, gradual automation, switching to domestic raw materials and maintaining the high quality of manufactured products, supports the functioning of major large textile enterprises and the development of the private sewing sector. The author suggests applying a business ecosystem approach in a similar fashion as creating a future strategy for the development of enterprises in the light industry. The strategies opted to increase the performance and efficiency indicators and ultimately multiply the growth of total value added in the Republic of Belarus can be employed as a framework to transform other business models in industries.

The analysis of the readiness of the Republic of Belarus's light industry for transformation in the direction of applying the BES concept according to the proposed criteria is presented below. The author divided all the presented criteria into 2 groups according to the degree of readiness for transformation into BES. Group I indicates "there are prerequisites for changes in this direction", and Group II indicates that "there is still work to be done in this direction with the support of the institutional environment". The Group I criteria for the readiness of the analyzed industry included (1), (4), (5), (6), (7), and (8). The Group II criteria for readiness for transformation, which are still to be developed in the country, were attributed to (2), (3), (9), and (10).

According to the analysis carried out in the Republic of Belarus, out of 10 possible signs of changes in traditional business models in the direction of BES, 6 (i.e. more than half) have clear prerequisites for implementation. In this regard, under the influence of crisis events in the last few years, supply chains have been disrupted, which affected the activities of textile enterprises in the field of raw materials, resulting in a phased revision of value chains in favor of including participants (mainly suppliers) from agriculture and the chemical industry, allowing for uninterrupted supplies of their domestic raw materials (linen, synthetic yarns, etc).

Changes in value chains and revisions of manufactured products have become possible in the industry due to the adoption of new state programs to support innovative development, the creation of cluster initiatives, and the support of scientific organizations and institutes engaged in the creation of new production technologies and fabric compositions with consistently high quality recognised outside the country.

Attracting more stakeholders to the activities of the light industry, especially in the garment industry and fashion industry is reflected in the development of processes of subcontracting with representatives of small and medium-sized businesses. The initiator of such interactions was the Entrepreneurship Support Fund. The increase in interest in the development of the light industry in the Republic of Belarus is also caused by the country's participation in the EAEU and the active involvement of partners and their investments from other countries.

A key role in the prerequisites for the transformation of the traditional business model in the light industry is played by a high degree of development of information and communication infrastructure, internet coverage of the territory, as well as the development of information innovation projects in the created hi-tech park. Within the framework of the functioning light industry holdings (e.g. BELWEST), the first electronic platform has already been created and is being tested, connecting representatives of the business-to-business sector for joint activities to increase the value proposition.

The author points out that with the positive dynamics leading to the transformation of business models in traditional industries, some of the criteria constraining the transition to BES are still not mastered. For the future implementation of all elements of the BES concept in relation to the light industry, there is still work to be done on changing the business philosophy and management methods within the framework of understanding the ecosystem vision, developing standards for common value creation combining products and digital services, and a fair distribution of the total value created among all participants of the BES. We should expect a gradual revision of government programs and initiatives in favor of integrating business processes. This is an integral part of obtaining highly sustainable results in the creation, design, production, and implementation of modern innovative solutions, without which it is impossible to imagine industrial management in the digital age.

5. DISCUSSION

The author suggests discussing the use of criteria peculiar to modern BES in terms of their applicability. In order to prompt changes in traditional industries of various developing countries, the example of the light house in Republic of Belarus is opted. For the application of a viable Business Ecosystem Approach (BEA), this example is scrutinized to comprehend the framework of transition that can be taken for business models to initiate transformation and development from the prior state to the modern BES (Business Eco-System). While many studies focus on high-tech industry by taking the examples of the functioning of Western BES or the Asian context, one should not forget about improving established traditional industries that allow developing countries, including the Republic of Belarus, to transform the existing real sector of the economy. However, the transformation approach is under the influence of widespread digitalization trends of the modern world. Therefore, we can come up with a holistic strategy by integrating both, the theoretical conclusions obtained about the possibilities of the business ecosystem approach and the development of the BES concept in the future.

Furthermore, it can become the basis for making managerial decisions to improve inter-organizational interaction among enterprises that are at the stage of maturity of their life cycle, and striving while preserving their managerial traditions (which have been developing for decades). It ensues developing the main criteria for modern BES, that has been derived from the past literature in form of works by various scientists, data-analyts and researchers. This establishment of main criteria will determine the future of the enterprise or industry to cope with global development by a new paradigm shift in the dynamics that evolve the traditional boundaries of industries, entailing value-proposition by coevolution, extending inter-company communication for sharing resources and community ties, and creating platform with an ecosystem that promotes co-ordination, networking, knowledge intensity, productive innovativeness, and an infrastructure with digital maturity. Moreover, modular architecture along with integration of service-product goods helps the enterprise in crossing a millennial development spurt.

Now-a-days, there are various methodologies for assessing the transformation of traditional industries, especially for assessing the readiness of modernization of these individual enterprises and entire industries. The assessment, as discussed above, is partaken through a basic set of criteria to assess their readiness to ultimately emerge as a transformed business model with elements such as joint value creation and network of shared assets. This development of inter-firm interaction as elucidated by author, is up to par to meet the conditions of digitalization of economy, and leads to formation of a new form of organization within company as well as provide framework of conceptual basis of a business ecosystem approach.

In this era, the traditional industries are susceptible to declining life cycle. Though their capacity to innovate and subsequently transform according to modern-world needs still holds potential for the ultimate transition. It is to be stated that, as these business models adapt to the new competitive advantages, they will require to incorporate automation, robotization, and digital processes. This will integrate the various participants of the platform i.e., the supplier and the consumer, leading to the co-existence and delivering benefits to both sides. Digitization of economy has holistically changed the dynamics of business environments, assessing the need to fundamentally make a change in their management of infrastructure and incorporate development strategies.

Therefore, the incessant scope of urging the enterprises to evolve with the changing dynamics of the world, all the while retaining their substantial traditional characteristics by modifying itself into a new model and cater to the business ecosystem that harness the best strategies and delivers the best outcome for the businesss enterprise leads to an economic boom. With an inter-play of efforts that seeks sustainable business models, the advantages turn out to be multiplied in terms of benefits.

6. CONCLUSION AND RECOMMENDATIONS

This chapter infers to constitute a framework that establishes a direction of transformation of traditional industries to modern network structures of intercompany interactions under the overwhelming influence of digitalization of the world economy. The author drawing upon the findings of scientists and researchers by employing a theoretical analysis of the respective literature came forward with the considered possibilities in the form of criteria. This criterion is identified as a result of the theoretical and empirical development of the BES concept, in response to which business ecosystem approach is construed under the principles of assessing dynamics, co-evolution, integration, coordination, networking, knowledge intensity, innovativeness, digital maturity, modularity, and service. As such, these elements are used as the basis for assessing the signs of readiness for transition of economic structures of an enterprise or industry. The business ecosystem (BES) during the transition from long-established infrastructure to a modernized business model subsequently requires a new level of upgrading their management processes. In order to empirically derive this framework upon which the future business models can substantially be applied to, the example of the light industry of the Republic of Belarus is adopted. Therefore, along with the theoretical analysis of the selected criteria and assessment of opportunities for the transformation of industries, the comparative analysis brings forth the implementation policies and methods to not only digitalize the production and sales processes, but also accelerate the understanding of the importance of strengthening inter-organizational interaction.

In order to achieve this, we need to incorporate additional resources, investments, information for the creation of innovative technologies, products, and services that can bring additional value to consumers and increase the added value of enterprises. Holistically, traditional manufacturing industries are still at an early stage of relationship management in the emerging BES. Therefore, it is recommended that stakeholders should focus their efforts on finding new solutions, designed to optimize value chains for future successful integration.

REFERENCES

Abbes, N., Sejri, N., Xu, J., & Cheikhrouhou, M. (2022). New lean six sigma readiness assessment model using fuzzy logic: Case study within clothing industry. *Alexandria Engineering Journal*, 61(11), 9079–9094. DOI: 10.1016/j.aej.2022.02.047

Acs, Z. J., Stam, E., Audretsch, D. B., & O'Connor, A. (2017). The lineages of the entrepreneurial ecosystem approach. *Small Business Economics*, 49(1), 1–10. DOI: 10.1007/s11187-017-9864-8

Adner, R. (2017). Ecosystem as structure: An actionable construct for strategy. *Journal of Management*, 43(1), 39–58. DOI: 10.1177/0149206316678451

Adner, R., & Kapoor, R. (2010). Value creation in innovation ecosystems: How the structure of technological interdependence affects firm performance in new technology generations. *Strategic Management Journal*, 31(3), 306–333. DOI: 10.1002/smj.821

Babkin, A., Glukhov, V., Shkarupeta, E., Kharitonova, N., & Barabaner, H. (2021). Methodology for assessing industrial ecosystem maturity in the framework of digital technology implementation. *International Journal of Technology*, 12(7), 1397–1406. DOI: 10.14716/ijtech.v12i7.5390

Battistella, C., Colucci, K., De Toni, A. F., & Nonino, F. (2013). Methodology of business ecosystems network analysis: A case study in Telecom Italia Future Centre. *Technological Forecasting and Social Change*, 80(6), 1194–1210. DOI: 10.1016/j.techfore.2012.11.002

Beyers, F., & Heinrichs, H. (2020). Global partnerships for a textile transformation? A systematic literature review on inter- and transnational collaborative governance of the textile and clothing industry. *Journal of Cleaner Production*, 261, 121131. DOI: 10.1016/j.jclepro.2020.121131

Bithas, G., Kutsikos, K., Warr, A., & Sakas, D. (2018). Managing transformation within service systems networks: A system viability approach. *Systems Research and Behavioral Science*, 35(4), 469–484. DOI: 10.1002/sres.2543

Cavallo, A., Ghezzi, A., & Sanasi, S. (2021). Assessing entrepreneurial ecosystems through a strategic value network approach: Evidence from the San Francisco area. *Journal of Small Business and Enterprise Development*, 28(2), 261–276. DOI: 10.1108/JSBED-05-2019-0148

Chen, W. M., Wang, S. Y., & Wu, X. L. (2022). Concept refinement, factor symbiosis, and innovation activity efficiency analysis of innovation ecosystem. In E. G. Nepomuceno (Ed.), Mathematical Problems in Engineering, Vol. 2022(4), 1-5. DOI: 10.1155/2022/1942026

Clarysse, B., Wright, M., Bruneel, J., & Mahajan, A. (2014). Creating value in ecosystems: Crossing the chasm between knowledge and business ecosystems. *Research Policy*, 43(7), 1164–1176. DOI: 10.1016/j.respol.2014.04.014

Gamidullaeva, L., Tolstykh, T., Bystrov, A., Radaykin, A., & Shmeleva, N. (2021). Cross-sectoral digital platform as a tool for innovation ecosystem development. *Sustainability (Basel)*, 13(21), 11686. DOI: 10.3390/su132111686

Gawer, A., & Cusumano, M. A. (2014). Industry platforms and ecosystem innovation. *Journal of Product Innovation Management*, 31(3), 417–433. DOI: 10.1111/jpim.12105

Ghoreishi, M., & Happonen, A. (2022). The case of fabric and textile industry: The emerging role of digitalization, internet-of-things and industry 4.0 for circularity. In X. S. Yang, S. Sherratt, N. Dey, A. Joshi, (Eds), Lecture Notes in Networks and Systems: Vol. 216. Proceedings of Sixth International Congress on Information and Communication Technology (pp. 189–200). Springer Nature. DOI: 10.1007/978-981-16-1781-2_18

Hein, A., Weking, J., Schreieck, M., Wiesche, M., Böhm, M., & Krcmar, H. (2019). Value co-creation practices in business-to-business platform ecosystems. *Electronic Markets*, 29(3), 503–518. DOI: 10.1007/s12525-019-00337-y

Ivanova, I., Smorodinskaya, N., & Leydesdorff, L. (2020). On measuring complexity in a post-industrial economy: The ecosystem's approach. *Quality & Quantity*, 54(1), 197–212. DOI: 10.1007/s11135-019-00844-2

Jacobides, M. G., Cennamo, C., & Gawer, A. (2018). Towards a theory of ecosystems. *Strategic Management Journal*, 39(8), 2255–2276. DOI: 10.1002/smj.2904

Jacobides, M. G., & Lianos, I. (2021). Regulating platforms and ecosystems: An introduction. *Industrial and Corporate Change*, 30(5), 1131–1142. DOI: 10.1093/icc/dtab060

Jensen, F., & Whitfield, L. (2022). Leveraging participation in apparel global supply chains through green industrialization strategies: Implications for low-income countries. *Ecological Economics*, 194(1), 107331. DOI: 10.1016/j.ecolecon.2021.107331

Kapoor, R., & Lee, J. M. (2013). Coordinating and competing in ecosystems: How organizational forms shape new technology investments. *Strategic Management Journal*, 34(3), 274–296. DOI: 10.1002/smj.2010

Kim, J. (2016). The platform business model and business ecosystem: Quality management and revenue structures. *European Planning Studies*, 24(12), 2113–2132. DOI: 10.1080/09654313.2016.1251882

Moore, J. F. (2006). Business ecosystems and the view from the firm. *Antitrust Bulletin*, 51(1), 31–75. DOI: 10.1177/0003603X0605100103

Popov, E., Dolghenko, R., Simonova, V., & Chelak, I. (2021). Analytical model of innovation ecosystem development. In 1st Conference on Traditional and Renewable Energy Sources: Perspectives and Paradigms for the 21st Century (TRESP 2021): Vol. 250, 01004. EDP Sciences. DOI: 10.1051/e3sconf/202125001004

Radicic, D., Pugh, G., & Douglas, D. (2020). Promoting cooperation in innovation ecosystems: Evidence european traditional manufacturing SMEs. *Small Business Economics*, 54(1), 257–283. DOI: 10.1007/s11187-018-0088-3

Rong, K., Hu, G., Hou, J., Ma, R., & Shi, Y. (2013). Business ecosystem extension: Facilitating the technology substitution. *International Journal of Technology Management*, 63(3/4), 268–294. DOI: 10.1504/IJTM.2013.056901

Rong, K., Li, B., Peng, W., Zhou, D., & Shi, X. (2021). Sharing economy platforms: Creating shared value at a business ecosystem level. *Technological Forecasting and Social Change*, 169(2), 120804. DOI: 10.1016/j.techfore.2021.120804

Sheresheva, M. Y., Valitova, L. A., Sharko, E. R., & Buzulukova, E. V. (2022). Application of social network analysis to visualization and description of industrial clusters: A case of the textile industry. *Journal of Risk and Financial Management*, 15(3), 1–17. DOI: 10.3390/jrfm15030129

Teece, D. J. (2016). Business ecosystem. In Augier, M., & Teece, D. (Eds.), *The Palgrave Encyclopedia of Strategic Management* (pp. 1–4). Palgrave Macmillan., DOI: 10.1057/978-1-349-94848-2_724-1

Todeschini, B. V., Cortimiglia, M. N., de-Menezes, D. C., & Ghezzi, A. (2017). Innovative and sustainable business models in the fashion industry: Entrepreneurial drivers, opportunities, and challenges. *Business Horizons*, 60(6), 759–770. DOI: 10.1016/j.bushor.2017.07.003

Trinh, T. H., Liem, N. T., & Kachitvichyanukul, V. (2014). A game theory approach for value co-creation systems. *Production & Manufacturing Research*, 2(1), 253–265. DOI: 10.1080/21693277.2014.913124

Vertakova, Y., Klevtsova, M., & Zadimidchenko, A. (2022). Multiplicative methodology for assessing investment attractiveness and risk for industries. *Journal of Risk and Financial Management*, 15(10), 419. DOI: 10.3390/jrfm15100419

Wieninger, S., Gotzen, R., Gudergan, G., & Wenning, K. M. (2019). The strategic analysis of business ecosystems : New conception and practical application of a research approach. In 2019 IEEE International Conference on Engineering, Technology and Innovation (ICE/ITMC) [Conference Session]. Institute of Electrical and Electronics Engineers Inc. DOI: 10.1109/ICE.2019.8792657

Williamson, P. J., & De Meyer, A. (2012). Ecosystem advantage: How to successfully harness the power of partners. *California Management Review*, 55(1), 24–46. DOI: 10.1525/cmr.2012.55.1.24

Chapter 3
Desk Research Methods for Place Reputation Monitoring and Evaluation in the Digital Economy:
Place Reputation Monitoring of the Region

Valeriia V. Kulibanova
Institute for Regional Economic Studies, Russian Academy of Science, Russia

Tatiana R. Teor
Saint Petersburg Electrotechnical University "LETI", Russia

Irina A. Ilyina
Saint Petersburg Electrotechnical University "LETI", Russia

ABSTRACT

A positive place reputation is the important competitive advantage, which allows it to attract resources and to facilitate mutually beneficial partnership with stakeholder groups. The main task is not only to identify the mechanism of the place reputation formation, but to analyze the factors influencing this process and their real-time dynamics. Now the information obtained by monitoring social networks, electronic media, and other looks more reliable and unbiased. Desk research has recently come to the fore, as it allows a fairly rapid analysis of large amounts of information. Another trend is a shift of focus from quantitative indicators' study

DOI: 10.4018/979-8-3693-3423-2.ch003

to the identification of qualitative ones. The article reveals methods of conducting desk research using various digital platforms. On the example of St. Petersburg, the authors analyzed possibilities of monitoring the place reputational capital, allowing to assess the reputation characteristic expressed quantitatively, and a qualitative characteristic in the form of an attitude reflection towards the region expressed in publications' tone.

1. INTRODUCTION

Social capital and reputation, as its most crucial component in the modern world, are the most important intangible assets not only for a specific person or company, but also for a place. The reputation of any object, including the place, is what people think about it based on their own experience or the opinion of experts. The reputation of any object, including the place, is what people think about it based on their own experience or the opinion of experts. Trust in territories is based on their reputation. A number of works have convincingly proved that a high level of confidence, based on the place's reputation, greatly simplifies communication, which not only attracts a large number of tourists and highly qualified specialists but also contributes to the inflow of foreign direct investment (Morgan and Pritchard 2014; Metaxas 2010; Ingenhoff et al. 2018; Bell 2016; Braun et al. 2018; Foroudi et al. 2016).

Considering the facts mentioned above, it can be argued that the future of the place today largely depends on the existing reputation and systematic work on it. Shirvani Dastgerdi and De Luca point out that such work should be carried out continuously: "Strengthening the reputation of the city through branding strategy is a dynamic and continuous process that should be considered as a long-term and public policy at all levels. This process forms in a complex and multidimensional context including economic, cultural, political, media and urban planning, in which various stakeholders participate through a variety of ideas, motives and goals" (Shirvani Dastgerdi and De Luca, 2019). However, the question immediately arises of how to track the change in reputation. Since it is regular monitoring will make it possible to judge the effectiveness of the reputation management mechanisms and, if necessary, adjust the tools used. Attempts to measure the influence of reputation on the performance of companies have been undertaken for quite a long time (Boyd, Bergh, and Ketchen 2010; Jin and Drozdenko 2010; Shin et al. 2015), but scientists have paid attention to measuring a similar indicator for the place relatively recently (Delgado García and De Quevedo Puente 2016; Ali et al. 2021). Therefore, first of all, it is necessary to answer the question about the relevant methods of analysis and assessment of the place's reputation. The modern world is becoming more complex.

New industries and spheres of activity appear new players are gaining strength; discoveries are constantly being made that quickly change the existing order of things.

For reputation monitoring, digitalization is a very important trend. Its constant deepening, spurred by the COVID-19 pandemic, has dramatically increased the importance of information circulating in the online environment. A number of studies on the impact of online reputation on the financial success of tourist attractions have convincingly demonstrated this relationship. It has been proven that the more conventional "stars" we see on Yelp or TripAdvisor, the more impressive the economic results of such properties will be (Luca 2011; Kim, Li, and Brymer 2016; Xie, Zhang, and Zhang 2014; Sayfuddin and Chen 2021). The virtual space has become the most important platform for the formation of a favorable attitude towards the place. That is why, when forming the reputation of the region, it is necessary to read the "digital footprint", which forms the media image of the place and directly affects the increase of its competitiveness. In addition, modern technology is helping to take a fresh look at the traditional methods of marketing research, including research related to reputation. If earlier the highest degree of trust was caused by a field survey, now more and more attention is paid to desk research (Bozhuk et al. 2019; Krasnov et al. 2019). New opportunities associated with the analysis of big data made it possible to add high reliability and relevance to the previously noted advantages of secondary information – low cost and low labor intensity, which has always been worthy of primary information.

At the same time, field surveys, especially polls, which previously were the main focus in the analysis of place brand image, now raise more and more questions from researchers. In particular, the "rand image data – as currently collected in consumer surveys – is not a valid source of market information" and suggest using associative methods as an alternative (Dolnicar and Grün 2013). Wilson and Joye point out that response bias can occur during surveys, which occurs when participants answer survey or interview questions systematically while in a certain perspective (Wilson and Joye 2020). If we turn to Luhmann's position (Luhmann 1996), the shortcomings of polling methods associated with the socio-psychological aspects of broadcasting the approved opinion are obvious. Therefore, good results are obtained by clarifying opinions and deepening understanding of stakeholders' perceptions of the reputation of a particular region in social media. Quantitative and qualitative characteristics of publications obtained using semantic and content analysis can serve as indicators here.

Finally, the pandemic has already led to significant changes in traditional views of marketing and marketing tools for both professional marketers and consumers (He and Harris 2020; Mazzoleni, Turchetti, and Ambrosino 2020). Therefore, at present, the question of the ethics of collecting primary data will become acute. The principles of voluntary participation, confidentiality, anonymity, dignity, and safety may not always be observed when conducting field research. Compliance with the latter principle has become especially urgent at the present time in connection with the increased threat of the COVID-19 pandemic. In this context, desk research comes to the fore. The focus is on secondary information, which can be found online. The purpose of this study is to develop recommendations on the use of modern methods for place reputation monitoring in the online environment. The object of the study is the regions of the Russian Federation; the subject of the study is the place's reputation as the most important intangible asset. Based on the object, subject, and purpose of the study, the following tasks have been formulated: to consider the main directions of place reputation formation; to identify the criteria for monitoring and describe the measurement procedures used in the processing of secondary information; to consider the practical application of the proposed author's methodology on the example of St. Petersburg.

2. LITERATURE REVIEW

The literature on place reputation monitoring and evaluation has evolved significantly over the years, with a growing emphasis on digital technologies, data analytics, and the dynamic nature of reputation management. In particular, this review will focus on two key areas: (1) The Conceptualization of Place Reputation and Its Importance and (2) Methods and Approaches for Monitoring Place Reputation in the Digital Age. These two areas provide a comprehensive understanding of how place reputation is shaped, monitored, and evaluated in the context of modern technological and economic shifts, particularly within the framework of the digital economy.

1. Place Reputation and Social Capital in the Digital Age

Having a good place reputation which is a part of social capital is now considered an important non-physical asset for regions cities in the global economy (Gandini & Gandini, 2016). Social capital can be defined as the sum of actual or potential resources available in a society, in terms of concrete networks, reliance and common conventions with a view of realizing collective advantage (Liu et al., 2023). Reputation thus being an essential component of social capital impacts perception and subsequent decision in matters concerning, investment, tourism and immigration. Positive image can help in lowering the transaction costs, and can attract external resources besides helping in improving the economic and social status of the place (Morgan & Pritchard, 2014). In the past, reputation management for places has was mainly connected with branding concepts. Destination branding is a deliberate effort of communicating and managing a place's image that set it apart from other places (Kavaratzis & Ashworth, 2005). These results show that brand image positively influences tourist arrivals, skilled labor migration and FDI inflows in cities and regions. However, in modern world place reputation has gone beyond simple branding which involves a range of mechanisms that are determined by the interactivity through social networks and media coverage.

Therefore, the digital economy has influenced the formation and perception of place reputation vastly. Thus, new communication channels of social networks Facebook, Twitter, Instagram, and sites for sharing opinions about places, such as TripAdvisor, Yelp, etc., have emerged. From the perspective of Ferlander, (2003), social media is characterized by two-way communication through which people can share their experiences and opinions regarding cities and regions, and therefore, the overall reputation of such locations. In line with the above argument, several research works have pointed to the effects of OR on the quantitative value of cities and tourist sites. For instance, Luca (2011) showed that in the context of restaurants, higher rating on Yelp lead to increased revenues. Likewise, Xie et al. (2014) posited that rating is direct in driving hotel performance, where rated hotels record higher occupancy and revenue. Kim, Li, and Khoso et al. (2022) posit that online reputation influences tourism, thereby boosting local region and city economies.

The COVID-19 pandemic has further intensified the importance of digital reputation management. As travel restrictions limited physical mobility, the virtual image of places became a crucial factor in sustaining tourism and investment flows. Sayfuddin and Chen (2021) noted that destinations with a strong online presence were better able to recover from the pandemic's economic impacts. Therefore, digital tools and platforms have become indispensable for managing and monitoring the reputation of places, particularly in times of crisis.

2. Methods for Monitoring and Evaluating Place Reputation

Because place reputation has been seen as central to the success of LSPs, scholars and practitioners have proposed different ways of tracking it. Some of the conventional tools employed in the collection of data with regard to how people perceive places include field surveys and interviews. Nonetheless, these methods are also not without their drawbacks such as response bias, involving high costs, and sheer time consumption (Dolnicar & Grün, 2013). Furthermore, the theoretical account provided by Luhmann (1996) pointed to socio-psychological problems to surveys arguing that they can neither reveal the limits of opinions and attitudes properly. Consequently, there is a trend towards the use of desk research approaches where quantitative data analysis and web technologies are more effective and cheaper than others. Desk research is the process of gathering secondary information, which is easily accessible from sites like webpages, social media platforms, and online databases among others. This approach enables the researchers to collect a large amount of data with limited resources (Bozhuk et al., 2019). Krasnov et al. (2019) also note that the big data approach provides reliability and relevance to the reputation analysis, making it a suitable tool for assessing place reputation.

Content analysis is one of the most popular techniques that is employed when conducting desk based research on reputation. This involves coding content of media, blogs, social media, and other online publications to determine the perception citizens and the media have for a place and how it is depicted in cyberspace (Krippendorff, 2018). When it comes to analyzing the tone and sentiment of a given area or city researchers can collect key words and phrases in the content. For instance, comparing bits of information based on whether they form positive, negative or a neutral perception of a place may be helpful in assessing its image (Wilson & Joye, 2020).

Another popular method is social media monitoring which means identification of mentions of a place in Twitter, Facebook and Instagram. This method enables/one to monitor the perception that people hold about a specific place at a particular time with regard to residents, tourists, and investors. They stressed that the analysis of reputation status on social media should be performed continuously as well as capture dynamic changes of the opinion. With the help of such tools as sentiment analysis, machine learning, NLP, social media monitoring has turned to be an effective tool for reputation management.

Besides, content analysis and social media monitoring, big data analytics become the powerful tool to assess place reputation. There is a large amount of data produced on the internet and this means that there are chances of finding out how some things are related in ways which otherwise would not be discovered if the data was analyzed on a smaller scale. For instance, Jin and Drozdenko (2010) analyzed big data to establish the connection between a place's online image and the corresponding economic outcomes; in such research, positive online sentiment appeared to yield higher investment levels. Shin et al., (2015) also discussed the future reputation trend forecasting capabilities of big data based on current trend of public sentiments. Even though the digital tools have proven to be instrumental in the monitoring of place reputation, there are issues accompanying their use. The two main issues that have been deemed questionable under the context of ethical considerations especially when incorporating social media data for research include data privacy and consent which has been discussed by He & Harris 2020. Further, accuracy of the online data may be influenced by fake reviews, bots, and other forms of configurations. Hence, there is a need to handle a lot of care and use a lot of research and quality methods while dealing with online reputation data.

3. MATERIALS AND METHODS

To monitor place reputation effectively in the digital economy, a wide range of secondary data resources can be leveraged, including general-purpose web search engines, social networks, blogs, forums, websites, electronic media, and digital versions of traditional print media. Monitoring place reputation in the online space begins with a comprehensive audit. This audit involves understanding what information is already available online, assessing the nature of discussions surrounding the place, and analyzing the sentiment or emotional tone of the messages. The goal is to determine the current state of reputation, which helps in identifying areas for improvement or action. This process can be carried out through either manual or automated methods.

1. Manual Monitoring of Place Reputation

Manual monitoring involves systematically searching for relevant publications and discussions on various platforms such as city directories, review websites, forums, and social media channels. This process mimics the actions of a typical user but in a more extensive and targeted manner. The information gathered manually often includes user reviews, expert opinions, and feedback from various online sources, providing insights into how a place is perceived across multiple dimensions. One of the most critical sources of secondary data for reputation monitoring is the set of city and territory ratings. The position of a place within these ratings often indicates its overall reputation—whether positive or negative. Established expert organizations create these ratings using publicly recognized methodologies, allowing for a credible and standardized assessment. One of the global leaders in reputation measurement is the Reputation Institute, which conducts annual studies such as Country RepTrak® and City RepTrak®. These studies assess the reputations of major countries and cities worldwide, based on a variety of factors including governance, culture, and economic prospects. In the context of regional analysis, country-specific reputation evaluations are also valuable. For example, RAEX Analytics, a leading rating agency in Russia, provides credit ratings and assesses the investment attractiveness of Russian regions. Such studies offer essential insights into how a region's reputation impacts its economic and social development.

The reputation of a place is often based on various parameters, which include:
- Investment Attractiveness: This refers to the region's ability to attract foreign and domestic investments. Investment portals, such as the one provided by the Government of St. Petersburg, offer crucial data for

assessing the region's appeal to investors (see, for example, https://www.gov.spb.ru/gov/otrasl/invest/statistic/development/).
- Tourist Attractiveness: Platforms like TripAdvisor are pivotal in assessing the region's reputation as a tourist destination, offering reviews, ratings, and travel-related feedback from both domestic and international visitors.
- Credit Ratings: Financial health and the ability to meet debt obligations play a significant role in reputation. International credit rating agencies such as Fitch Ratings and national agencies like RAEX provide critical credit rating information that influences perceptions of stability and investment potential (https://www.fitchratings.com, https://raexpert.ru).
- Innovation Ratings: Innovation is a key factor in modern economic development. Rankings like the Innovation Cities™ Index 2021, which lists the world's top 100 most innovative cities, are useful indicators of a city's standing in the global innovation landscape (https://www.innovation-cities.com).

2. Automated Monitoring of Place Reputation

Automated tools provide a more efficient way to monitor place reputation, especially given the vast amount of data available online. These tools can rapidly collect, organize, and analyze data, offering real-time insights. Some of the most commonly used automated tools include:
- Google Alerts: A free tool that monitors new online content related to specific keywords and sends notifications when new mentions appear (https://www.google.ru/alerts).
- Babkee: This tool analyzes the volume, frequency, and sentiment of mentions across various platforms, including social networks and forums. It also provides insights into the demographics of the people discussing the place (http://www.babkee.ru).
- Medialogia: A professional platform that monitors and analyzes media content across Russia and neighboring countries. It offers advanced analytical tools for tracking the reputation of a place based on media mentions (https://mlg.ru).
- Brand Analytics: A powerful tool for monitoring and analyzing brand mentions in social media, blogs, and forums. It offers sentiment analysis, allowing users to assess the emotional tone of discussions about a place (https://br-analytics.ru).
- YouScan: A social media listening platform that uses artificial intelligence (AI) to track brand mentions and analyze sentiment across vari-

ous digital platforms. This tool provides insights into public opinion and identifies trends in real time (https://youscan.io/ru).
- IQBuzz: Another popular tool for tracking and analyzing media mentions, offering both free and paid services. It monitors social media and news websites, providing sentiment analysis and demographic data (https://iqbuzz.pro).
- These tools allow for comprehensive reputation monitoring, which is particularly valuable for businesses, government agencies, and other stakeholders who rely on timely and accurate information. By analyzing the data obtained from these platforms, users can adjust their strategies to manage and improve the region's reputation effectively.

3. Accessibility of Reputation Monitoring Tools

One of the challenges in monitoring place reputation is the cost associated with many of these tools. Automated monitoring platforms often provide premium services, making it difficult for small companies or individuals to access them. In such cases, manual monitoring or free services like Google Alerts may be the only viable options. This study, therefore, focuses on secondary data available through open sources or low-cost subscription services, ensuring that the reputation monitoring process remains accessible and affordable. In conclusion, place reputation monitoring is an essential part of regional management in the digital economy. By leveraging both manual and automated tools, it is possible to assess public opinion, identify key reputation drivers, and adjust strategies to enhance the region's image. This study applies these methods to the case of St. Petersburg, demonstrating how the reputation of a place can be systematically monitored and evaluated to inform policy and strategy development.

4. RESULTS

As mentioned above, the most important source of information on the place reputation is the rating system. One of the basic parts of any rating is factors related to the quality of life: security, the development of political and legal institutions, the socio-ecological and economic environment, and transport infrastructure (Akhmetshin et al. 2020). For example, in the Ranking of federal subjects by the quality of life in Russia, which is represented by IIA «Rossiya Segodnya» (https://ria.ru), such parameters as safety, environment, accessibility of stores, leisure and sporting activities, cleanliness, quality of life and cost of services, public transport, children's infrastructure, the work of public utilities, and others' good attitudes were evaluated. Seventy indicators were analyzed while providing the rating, which are grouped into 11 groups, characterizing the main aspects of quality of life in the region: income level of residents, employment and labor market, housing conditions, safety of residence, demographic situation, environmental and climatic conditions, health and education, provision of social infrastructure, economic development, the level of small business development, land use and development of transport infrastructure (see Table 1).

Table 1. Ranking of federal subjects with the highest quality of life in Russia

No.	Federal subject	Rating points in 2020	Rating position in 2019
1	Moscow	82,164	1
2	St. Petersburg	80,634	2
3	Moscow region	76,068	3
4	Tatarstan	66,624	4
5	Belgorod region	64,769	5
6	Krasnodar Territory	63,714	6
7	Leningrad region	61,600	8
8	Voronezh region	61,046	7
9	Khanty-Mansi Autonomous Area	60,523	10
10	Kaliningrad region	59,253	9

Note. Adapted from https://ria.ru/20200217/1564483827.html accessed on 2021/09/20.

However, there are significant differences in data provided by the rating agency «National Credit Ratings» (NCR), where regions were not ranked according to traditional indicators of economic potential or investment activity but according to a number of indicators, the criterion for the calculation of which was Retail turnover per capita (as a demand indicator), housing affordability, levels of employment and savings, teachers' and physicians' prosperity. St. Petersburg took first place in the ranking, and its superiority over Moscow is due primarily to the capital's low score in housing affordability (Table 2).

According to the classification of cities and city rankings from the Index of Innovative Cities at the beginning of 2021 (Table 3), St. Petersburg is not in the top 100, while Moscow has risen ten places to 34th place compared to 2019 and 18 cities are included in the ranking of the Global Cities 500 itself. This ranking compared infrastructure development, network market, and cultural values in the five years before the COVID-19 pandemic and the year after. As it can be seen from Table 2, many large regional centres are not included in the ranking table. This is most likely due to a lack of interest in developing the place brand. People form their opinions about what is happening based on information received from various media, which is why the media's sphere of influence extends far beyond the boundaries of a particular territory. At the same time, in recent years, a clear priority has been given to electronic sources of information (Internet publications, social networks). Therefore, when analyzing a place's reputation and its competitive attractiveness, it is very important what kind of information reaches the media.

Table 2. The 2020 best federal subject's rankings for quality of life in Russia

Indicator	St. Petersburg	Moscow	Belgorod region	Moscow region	Voronezh region
Total score	7	6.10	5.64	5.61	5.31
Retail turnover	7.0	7.0	5.4	7.0	6.4
Average wage ratio in the region	1.0	1.1	3	2.8	2.7
Ratio of outstanding loans to individuals to average wage in the region	6.0	6.3	3	1.5	4.2
Proportion of employed with formal labor income	7.0	6.4	4.4	4.9	4.3
The ratio of bank deposits per capita to average wages	7.0	7.0	4.5	6.3	6.8
Cost of fixed assets in education, health, culture and sport	6.9	7.0	6.3	4.7	1.6

Note. Adapted from https://www.rbc.ru/economics/21/07/2020/5f0ece439a79470d37b66efcaccessedon2021/09/20.

A significant indicator in reputation analysis is the Media Index, which indicates the communication and PR activity's effectiveness/ineffectiveness of a region. In other words, it is an indicator of the level of representation of the object in the media for a selected period. To calculate the index, all mentions are summed up, taking into account the role of the object and "weight" of the publication. The weight of a publication is calculated by its citation rate in social media (Table 4). The media index is based on the Medialogy media database, which includes more than 66,000 sources: TV, radio, newspapers, magazines, news agencies, and Internet media. When calculating the rating, references to Russian regions in the context of the "May Decree No. 204" of the President of the Russian Federation, aimed, among other things, at improving the standard of living of citizens, were taken into account. The calculation of the media index is based on taking into account three main factors: the citation rate of the object and its speakers, the visibility of the object, and the tone of the messages in relation to the object. In the analysis of the place's reputation, the qualitative indicator will be the activity of diffuse groups of stakeholders, implemented in the publications in the media and social media according to certain headings. The rubrics serve as indicators in assessing the region's information background in the economic, administrative-political, and technological spheres and in the sphere of social development of the territory.

Table 3. Russia's cities in the world's 500 towns of the year 2021

No.	City	2021 Ranking	Change 2019	Cultural Assets	Human Infrastructure	Networked Markets	Pre-COVID 5-Year Average Ranking	Post-COVID year Result
1	Moscow	34	+4	18	14	16	44	+10
2	St. Petersburg	121	-16	17	13	13	81	-40
3	Kazan	366	+27	14	11	11	333	-33
4	Yekaterinburg	385	+31	13	11	11	349	-36
5	Volgograd	401	+43	13	11	11	419	+18
6	Kaliningrad	404	+33	13	11	11	391	-13
7	Novosibirsk	406	-1	13	11	11	365	-41
8	Samara	421	+19	13	10	11	396	-25
9	Nizhny Novgorod	423	-2	12	11	11	376	-47
10	Rostov-na-Donu	425	-6	12	11	11	381	-44
11	Vladivostok	428	+19	12	11	11	446	-11
12	Krasnoyarsk	437	+1	12	11	10	393	-44
13	Omsk	439	+10	12	10	11	418	-21
14	Saratov	448	+15	12	10	10	424	-24
15	Perm	450	-9	12	10	10	410	-40
16	Tomsk	452	+8	12	10	10	426	-26
17	Orenburg	454	+19	12	10	10	450	-4
18	Izhevsk	455	+27	12	10	10	451	-4
19								

Note. Adapted from https://www.innovation-cities.com/index-2019-global-city-rankings/18842/ accessed on 2021/09/20.

Table 4. Media imdex of Russian regions as of July 2021

No.	Federal subject	Media Index
1	Altai Territory	11 525.00
2	Republic of Bashkortostan	18 999.20
3	Volgograd region	15 487.30
4	Irkutsk region	13 407.60
5	Krasnodar Territory	26 598.50
6	Krasnoyarsk Territory	19 846.90
7	Republic of Crimea	12 261.00
8	Leningrad region	21 968.60
9	Moscow region	25 168.40

continued on following page

Table 4. Continued

No.	Federal subject	Media Index
10	Nizhni Novgorod region	17 367.20
11	Novosibirsk region	16 931.90
12	Omsk region	14 892.70
13	Primorsky region	28 216.30
14	Samara region	23 944.30
15	St. Petersburg	14 638.60
16	Sverdlovsk Region	14 070.70
17	Stavropol Territory	13 306.40
18	Republic of Tatarstan	16 226.50
19	Tver region	18 001.60
20	Udmurtian Republic	11 795.50

Note. Adapted from https://www.mlg.ru/ratings/research/8634/ accessed on 2021/09/20.

It should be noted that different monitoring systems do not give equal ranking indicators. Thus, Figure 1 shows the indicators of media activity of the regions by the number of publications. The analysis is carried out on the platform of the monitoring company press index https://pressindex.ru. This rating includes the regions rather than individual cities, which on the one hand does not allow to compare the level of a particular city. For example, if in the media presence rating the Leningrad region is in 8th place and the Moscow region in 9th place, when comparing regional centers, the picture will be different.

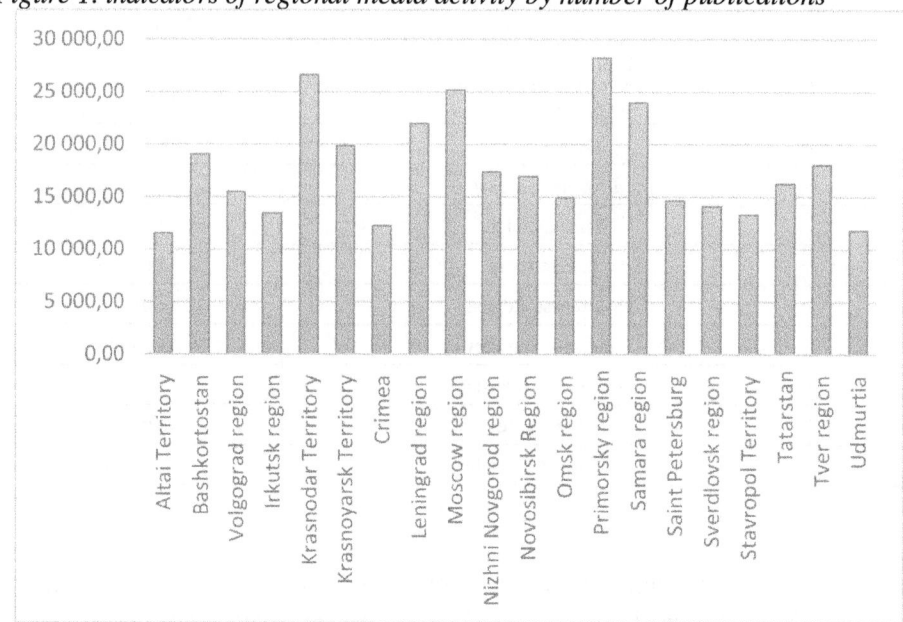

Figure 1. indicators of regional media activity by number of publications

But the number of publications does not give a complete picture of the state of affairs in the region. For example, an analysis of the media activity in the St. Petersburg region indicates an increase in the number of publications in general (see Figure 2).

Figure 2. Tonality of publications about the St. Petersburg region for the period January-August 2018–2021

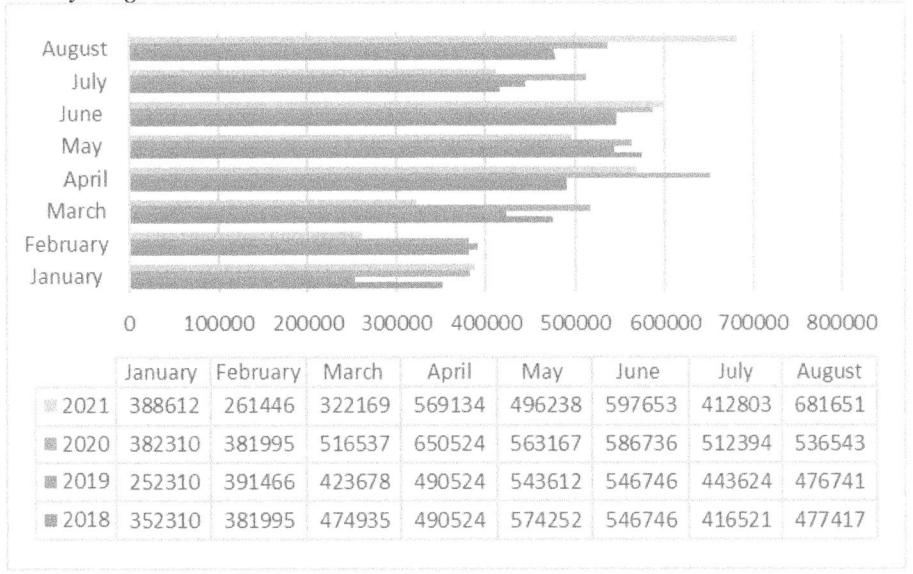

	January	February	March	April	May	June	July	August
2021	388612	261446	322169	569134	496238	597653	412803	681651
2020	382310	381995	516537	650524	563167	586736	512394	536543
2019	252310	391466	423678	490524	543612	546746	443624	476741
2018	352310	381995	474935	490524	574252	546746	416521	477417

Figure 3 shows the evolution of the tone of publications for the period January-August 2018-2021, which shows an apparent increase in negative publications for the year 2021. This is despite the fact that the region has entered all the top rankings and is in the lead. At the same time, the negativity is mainly focused on changes in the cultural and tourism spheres (news stories related to the cancellation of events and complications in access to museums), which is a significant indicator of the region's development. Thus, when monitoring reputation, it is essential to consider both quantitative and qualitative indicators. To sum up, it should be noted that while 5 years ago, the ratings and monitoring indicators were based on the public image presented by the media, in the last 2–3 years, social media reputation has become of greater interest, as it is a source of knowledge about the real state of affairs in the regions, where data is obtained from a group of consumer stakeholders.

Figure 3. The tone of the publications about St. Petersburg for the period January-August 2018–2021

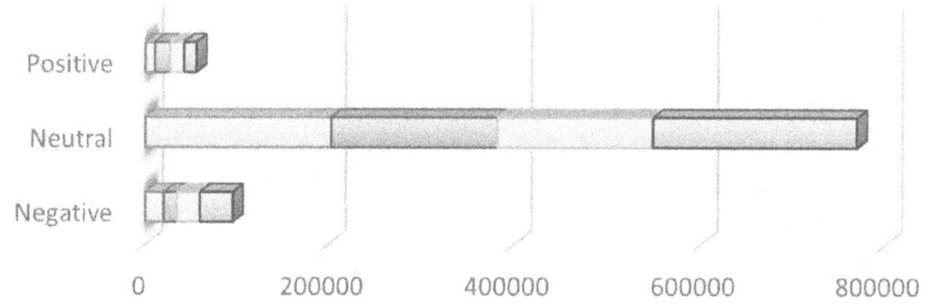

	Negative	Neutral	Positive
2018	19842	202695	10047
2019	14368	179332	16549
2020	24397	166537	14834
2021	36761	220654	13210

5. DISCUSSION

This chapter's findings underscore that place reputation management is not an easy task, especially in the era of media in which presence and social media activity shape reputations. The implications of the study present evidence showing that although coverage by conventional media is still relevant, social network sites offer more timely and stakeholder-oriented perception of a region's reputation. St. Petersburg example shows that the place can be objectively ranked high in national and international rankings, however, increasing negative connotations in the social construction of the place, especially in connection with elements of post-industrial culture and tourism. However, one of the most valuable discoveries of the results is the need for the integration of quantitative and qualitative methods in place reputation measurement. Public sentiment exposure in this case is a function of media activity; however, the number of publications (Figure 1) presents a limited picture of the regional media activity. For instance, media activity with regards to St. Petersburg has risen gradually from 2018 up to 2021, however, the trend of the articles has had a more negative inclination particularly in 2021 (Figures 2 and 3). This suggest that though the coverage has increased in terms of the number, the tone of the conversations has shifted from positive to negative. This negative change in the perceived image was observed especially in the sectors that have been most impacted by the

COVID-19 outbreak including the tourism and cultural events where restrictions and cancellation of events have been negative indicators despite the general positive position on economic and infrastructural development in the region.

This change of focus demands that place reputation monitoring needs to go beyond things that are easily quantifiable, such as the number of published articles or appearances in the media. The coverage of an ML/AL firm reveals quantitative and qualitative aspects of reputation: the impact of tone and content. More tone negativity in this global city, however this city ranks high in most of the indexes, this indicates the media and PR management of St Petersburg does not properly respond to the public sentiment especially the tourists and cultural areas. Such a situation where there are high ranks, but negative sentiments has the potential of damaging the region's future potentiality as a tourist attraction, an investment destination and habitation unless adequate measures are being implemented. The research also reveals that social media is a crucial factor which affects the views of the population. As mentioned in the results, traditional media are no longer a good representation of real-time public opinion as embodied by social media. This is particularly important in light of a reputation concern since social media brings the 'consumer stakeholders', who define a region based on the experiences into direct contact. Thus, the process of introducing analysis of social media data to the system of reputation management is the shift in the way regions operate. Despite the fact that, five years earlier, media coverage inexorably dictated public opinion, the Internet, particularly the extensive use of social networks such as Twitter, Instagram and Facebook, has become a powerful tool in managing organizations' reputation.

Moreover, that place investment attractiveness, the level of retail turnover, and employment indicators cannot be excluded from a general evaluation of place reputation is also evident. While these indices afford a reasonable starting point for developing reputational rankings, the results indicate that the influence on attitudinal reactions is occasionally overridden by the salient needs, for example, service quality and convenient access to cultural resources. In St. Petersburg, the region that had achieved high rankings in economic performance of infrastructural development, relevant stakeholders' concerns emerged on the unfavorable impact on the cultural and tourism sector. This implies that the stakeholders, people in the community and the tourists, focus on the short-term, in this case the economically tangible touch with the community, aspects of development than the overall economy, aspects of development that predictably concern those tourists with experiences that are clearly defined by the tangible aspects of the society.

6. CONCLUSION

This study highlights the evolving complexities of place reputation monitoring in the digital age, particularly through the case analysis of St. Petersburg. As regions and cities increasingly compete for investment, tourism, and talent, reputation becomes a critical intangible asset. The findings emphasize the need for a comprehensive approach to reputation monitoring that incorporates both quantitative and qualitative metrics, using a combination of traditional media analysis and real-time social media data. The case of St. Petersburg underscores the importance of balancing strong economic performance and infrastructural development with the need to address public sentiment, particularly in sectors such as culture and tourism. Despite the region's high rankings in national and global indices, negative perceptions in the media and social media—particularly regarding the impact of the COVID-19 pandemic—demonstrated the need for a more responsive reputation management strategy. This reinforces the idea that economic metrics alone are insufficient for maintaining a positive place reputation; stakeholder experiences and immediate concerns play a critical role in shaping public opinion. The growing influence of social media as a source of real-time public sentiment adds another layer of complexity to reputation management. Social media platforms provide a direct and unfiltered view of stakeholder opinions, which can rapidly shape or shift the reputation of a region. Therefore, future strategies for managing and enhancing place reputation should integrate social media analytics alongside traditional media monitoring to capture a more comprehensive view of public perception.

REFERENCES

Akhmetshin, E., Ilyina, I., Kulibanova, V., & Teor, T. (2020). The Methods of Identification and Analysis of Key Indicators Affecting the Place Reputation in the Modern Information and Communication Space. *IOP Conference Series. Materials Science and Engineering*, 940(October), 012100. DOI: 10.1088/1757-899X/940/1/012100

Ali, T., Marc, B., Omar, B., Soulaimane, K., & Larbi, S. (2021). Exploring Destination's Negative e-Reputation Using Aspect Based Sentiment Analysis Approach: Case of Marrakech Destination on TripAdvisor. *Tourism Management Perspectives*, 40(October), 100892. DOI: 10.1016/j.tmp.2021.100892

Bell, F. (2016). Looking beyond Place Branding: The Emergence of Place Reputation. *Journal of Place Management and Development*, 9(3), 247–254. DOI: 10.1108/JPMD-08-2016-0055

Boyd, B. K., Bergh, D. D., & Ketchen, D. J. Jr. (2010). Reconsidering the Reputation—Performance Relationship: A Resource-Based View. *Journal of Management*, 36(3), 588–609. DOI: 10.1177/0149206308328507

Bozhuk, S. G., Maslova, T. D., Pletneva, N. A., & Evdokimov, K. V. (2019). Improvement of the Consumers' Satisfaction Research Technology in the Digital Environment. *IOP Conference Series. Materials Science and Engineering*, 666(1), 012055. DOI: 10.1088/1757-899X/666/1/012055

Braun, E., Eshuis, J., Klijn, E. H., & Zenker, S. (2018). Improving Place Reputation: Do an Open Place Brand Process and an Identity-Image Match Pay Off? *Cities (London, England)*, 80, 22–28. DOI: 10.1016/j.cities.2017.06.010

Darazi, M. A., Khoso, A. K., & Mahesar, K. A. (2023). INVESTIGATING THE EFFECTS OF ESL TEACHERS' FEEDBACK ON ESL UNDERGRADUATE STUDENTS' LEVEL OF MOTIVATION, ACADEMIC PERFORMANCE, AND SATISFACTION: MEDIATING ROLE OF STUDENTS' MOTIVATION. *Pakistan Journal of Educational Research*, 6(2).

Dolnicar, S., & Grün, B. (2013). Validly Measuring Destination Image in Survey Studies. *Journal of Travel Research*, 52(1), 3–14. DOI: 10.1177/0047287512457267

Ferlander, S. (2003). The internet, social capital and local community.

Foroudi, P., Gupta, S., Kitchen, P., Foroudi, M. M., & Nguyen, B. (2016). A Framework of Place Branding, Place Image, and Place Reputation. *Qualitative Market Research*, 19(2), 241–264. DOI: 10.1108/QMR-02-2016-0020

Gandini, A., & Gandini, A. (2016). Reputation, the Social Capital of a Digital Society. *The Reputation Economy: Understanding Knowledge Work in Digital Society*, 27-43. DOI: 10.1057/978-1-137-56107-7_3

García, D., Bautista, J., & Esther, D. Q. P. (2016). The Complex Link of City Reputation and City Performance. Results for FsQCA Analysis. *Journal of Business Research*, 69(8), 2830–2839. DOI: 10.1016/j.jbusres.2015.12.052

He, H., & Harris, L. (2020). The Impact of COVID-19 Pandemic on Corporate Social Responsibility and Marketing Philosophy. *Journal of Business Research*, 116(August), 176–182. DOI: 10.1016/j.jbusres.2020.05.030 PMID: 32457556

Ingenhoff, D., Buhmann, A., White, C., Zhang, T., & Kiousis, S. (2018). Reputation Spillover: Corporate Crises' Effects on Country Reputation. *Journal of Communication Management (London)*, 22(1), 96–112. DOI: 10.1108/JCOM-08-2017-0081

Jin, K. G., & Drozdenko, R. G. (2010). Relationships among Perceived Organizational Core Values, Corporate Social Responsibility, Ethics, and Organizational Performance Outcomes: An Empirical Study of Information Technology Professionals. *Journal of Business Ethics*, 92(3), 341–359. DOI: 10.1007/s10551-009-0158-1

Khoso, A. K., Darazi, M. A., Mahesar, K. A., Memon, M. A., & Nawaz, F. (2022). The impact of ESL teachers' emotional intelligence on ESL Students academic engagement, reading and writing proficiency: Mediating role of ESL students motivation. *Int. J. Early Childhood Spec. Educ*, 14, 3267–3280.

Kim, W. G., Li, J. J., & Brymer, R. A. (2016). The Impact of Social Media Reviews on Restaurant Performance: The Moderating Role of Excellence Certificate. *International Journal of Hospitality Management*, 55(May), 41–51. DOI: 10.1016/j.ijhm.2016.03.001

Krasnov, A., Chargaziya, G., Griffith, R., & Draganov, M. (2019). Dynamic and Static Elements of a Consumer's Digital Portrait and Methods of Their Studying. *IOP Conference Series. Materials Science and Engineering*, 497(April), 012123. DOI: 10.1088/1757-899X/497/1/012123

Kulibanova, V. V., & Teor, T. R. (2017). Identifying Key Stakeholder Groups for Implementing a Place Branding Policy in Saint Petersburg. *Baltic Region*, 9(3), 99–115. DOI: 10.5922/2079-8555-2017-3-7

Liu, H., Zhu, Q., Khoso, W. M., & Khoso, A. K. (2023). Spatial pattern and the development of green finance trends in China. *Renewable Energy*, 211, 370–378. DOI: 10.1016/j.renene.2023.05.014

Luca, M. 2011. "Reviews, Reputation, and Revenue: The Case of Yelp.Com." *SSRN Electronic Journal*. https://doi.org/DOI: 10.2139/ssrn.1928601

Luhmann, N. (1996). *Social Systems*. Stanford University Press.

Mazzoleni, S., Turchetti, G., & Ambrosino, N. (2020). The COVID-19 Outbreak: From 'Black Swan' to Global Challenges and Opportunities. *Pulmonology*, 26(3), 117–118. DOI: 10.1016/j.pulmoe.2020.03.002 PMID: 32291202

Metaxas, T. (2010). Place Marketing, Place Branding and Foreign Direct Investments: Defining Their Relationship in the Frame of Local Economic Development Process. *Place Branding and Public Diplomacy*, 6(3), 228–243. DOI: 10.1057/pb.2010.22

Morgan, N., & Pritchard, A. (2014). Destination Reputations and Brands: Communication Challenges. *Journal of Destination Marketing & Management*, 3(1), 1. DOI: 10.1016/j.jdmm.2014.02.001

Sayfuddin, A. T. M., & Chen, Y. (2021). The Signaling and Reputational Effects of Customer Ratings on Hotel Revenues: Evidence from TripAdvisor. *International Journal of Hospitality Management*, 99(October), 103065. DOI: 10.1016/j.ijhm.2021.103065

Shin, Y., Sung, S. Y., Choi, J. N., & Kim, M. S. (2015). Top Management Ethical Leadership and Firm Performance: Mediating Role of Ethical and Procedural Justice Climate. *Journal of Business Ethics*, 129(1), 43–57. DOI: 10.1007/s10551-014-2144-5

Shirvani Dastgerdi, A., & De Luca, G. (2019). Strengthening the City's Reputation in the Age of Cities: An Insight in the City Branding Theory. *City, Territory and Architecture*, 6(1), 2. DOI: 10.1186/s40410-019-0101-4

Wilson, J. H., & Joye, S. W. 2020. "Research Designs and Variables." In *Research Methods and Statistics: An Integrated Approach*, 40–72. 2455 Teller Road, Thousand Oaks California 91320: SAGE Publications, Inc. https://doi.org/DOI: 10.4135/9781071802717.n3

Xie, K. L., Zhang, Z., & Zhang, Z. (2014). The Business Value of Online Consumer Reviews and Management Response to Hotel Performance. *International Journal of Hospitality Management*, 43(October), 1–12. DOI: 10.1016/j.ijhm.2014.07.007

Chapter 4
Digital Technologies in Ensuring Sustainable Development of Oil and Gas Territories in the Russian Federation

Danil P. Egorov
St. Petersburg State University, Russia

Anatoliy I. Chistobaev
St. Petersburg State University, Russia

Kirill P. Egorov
St. Petersburg State University, Russia

ABSTRACT

The introduction of digital technologies in oil and gas industry helps to reduce the anthropogenic burden on the environment and ensures sustainable development of territories in Russia. The article assesses the conditions for the introduction of digital technologies in the oil and gas business, identifies barriers to the spread of innovative methods for making managerial decisions, including everyday issues of people's lives and operating oil and gas industry facilities. Based on the statistical analysis, authors of the article identified the main causes of emergency situations in the oil and gas industry and developed a list of recommendations for its prevention or elimination of consequences. The study is shown that when using digital technologies the costs of the management process are reduced, bureaucratic delays in the preparation of the necessary documents for the population are eliminated; the

DOI: 10.4018/979-8-3693-3423-2.ch004

Copyright © 2025, IGI Global. Copying or distributing in print or electronic forms without written permission of IGI Global is prohibited.

opportunities for members of society to make managerial decisions are realized. The institutional imperfections and other factors were identified as the main barriers for innovation.

NOVELTY STATEMENT

The chapter presents a novel exploration of the integration of digital technologies in fostering sustainable development within the oil and gas territories of the Russian Federation. It uniquely addresses the intersection of digital innovation and environmental stewardship in the context of the oil and gas industry, emphasizing how digital transformation can drive sustainable practices. The chapter also offers fresh insights into the potential of digital tools to mitigate environmental impacts while enhancing operational efficiency and resilience in oil and gas operations. This work contributes to the evolving discourse on sustainable development in resource-intensive industries through a digital lens.

1. INTRODUCTION

The oil and gas industry is the leading branch of the domestic industry, which largely determines the foundations of the country's economic development. In modern realities, there is a tendency for greening industries all over the world, including the oil and gas industry. However, the problematic situation in the raw materials markets, macroeconomic and geopolitical instability caused by Western sanctions, and with the objective deterioration of the mining and geological characteristics of the explored mineral resource base are factors that complicate the implementation of large projects. As a result of working in a tight time frame with the operation of outdated equipment, the accident rate increases, which, in turn, negatively affects the environmental aspects of the activity (Egorov 2021).

Currently, active work is being carried out at the legislative level aimed at intensifying the development of the hydrocarbon-rich territories of the Russian north, in this regard, conditions are being created that are attractive for business communities, as well as measures to guarantee the protection of the natural environment from possible negative impacts (Solodovnikov et al., 2016). At the same time, according to the author, the implemented environmental policy lacks a strategic vision of the problem that would contribute to the sustainable development of the territories. One of the ways to solve this problem is the improvement and introduction of the latest digital technologies in the fields of the energy industry and environmental quality control in the territories of the oil and gas field (Krotov et al., 2019). According to this measure, it will be possible to maintain (preserve) the natural state of the environment: landscape and biological diversity.

Based on the above, the purpose of this study is to assess the role of digital technologies in maintaining the sustainable development of oil and gas territories in Russia. To achieve this goal, the authors had to solve the following tasks: to analyze the impact of the oil and gas industry on the sustainable development of the country and regions; to identify obstacles to the spread of digital technologies and determine ways to eliminate them; to identify regional differences in the introduction of digital technologies; to show the role of digital technologies in the prevention of accidents at oil and gas industry facilities.

1.1 The Impact of the Oil and Gas Industry on Sustainable Development in Russia

Consequently, Russia relies on it to cater for its growing economic needs such as the Gross Domestic Product, employment and export earnings. On the one hand, this industry has significant impact on pollution of the environment: greenhouse effect, destruction of the local flora and fauna, water pollution (Filimonova et al., 2020). The effect of oil and gas production on environment is worst in existing areas vulnerable to environmental degradation such as the Russian north. The problem of economic rent and its relation to environmental impact is the key concern at the core of sustainable development of these territories (Eder et al., 2017). The impact of the oil and gas in the environment is not a single dimensional issue, but instead, it is complex. It encompasses the CO_2 and CH_4 emissions that occur during extraction, transportation and refining of the fuels, the emissions that arise from the infrastructure development that is required to support such fuels like deforestation and soil erosion (Sergi & Berezin, 2018). These problems are worsened by the ageing infrastructures in the industry; it results in a higher accident rate and environmental spills that can cause long-term negative impacts on local ecosystems and communities (Liu et al., 2023). For instance, oil can also pollute water which affects the lives of animals, and makes the affected piece of land unsuitable for cultivation or settlement.

Nonetheless, the Russia government has embarked on several policies that can reduce the impacts of the oil and natural gas sector on environment. These are the tightening of legislation on emissions and waste disposal, and the provision of stimulus for companies to employ technologies and practices that are less damaging to the environment (Tyaglov et al., 2021). However, those measures are often compensated by the economic and geopolitical factors, for example, sanctions and fluctuations of oil prices that make officials put off or reduce the scope of environmental projects (Cherepovitsyn et al., 2021). Also, the absence of fair distribution of resources as well as technological development all over the territory of the Russian Federation provokes sharp regional distinctions in the establishment of the usage of affordances. Some regions e. g. Western Siberia are better connected with technology infrastructure and are ready to embrace digital technologies for management of environment while others like Russian Far East are not ready due to low investment and lack expertise (Romasheva & Dmitrieva, 2021). It is therefore important to argue that there is the need to come up with sustainable development models that will adequately capture the context of the various regions.

Thus, one can claim that, on the one hand, the oil and gas industry plays an essential role in the Russian economy, and, on the other hand, its adverse effects on the environment complicate the process of sustainable development (Filimonova et al., 2020). These challenges can only be fought with a complex strategy aimed at the legal changes, technological advancements, and international coalition. IT institutionalization represents a positive opportunity for improving the industry's environmental management as well as guaranteeing the environmental stability of Russian oil and gas regions.

1.2 The Role of Digital Technologies in Enhancing Environmental Sustainability in the Oil and Gas Sector

Mobile and information technologies as well as other forms of technological advancement are able to dramatically influence the landscape of the oil and gas industry through enhancement of outcomes, reduction of costs and adverse environmental effects (Anaba et al., 2024). In a broader framework of the Russian oil and gas industry, the use of digital technologies is especially important because of the evolution of the ageing facilities and the growth of the technical and geographic challenges for extracting the resources. AI, IoT solutions, and blockchain can significantly improve environmental efficiency because they allow monitoring, predicting problems with systems, and making explicit reports (Daneeva et al., 2020). Digital technologies can and do, for instance enable the enhancement of operations productivity, thus minimizing the impact of oil and gas activities on the environment. For instance, AI will help in improving the analytics of the drilling process so as to determine the correct pathways of drilling and the energy that ought to be used in the process (Esiri et al., 2024). Environmental parameters, as provided by IoT sensors or drones for instance, will be able to alert for any instances of leakages, spills or any other environmental disamenities. These technologies reduce the possibility of damaging the environment and they also reduce the costs incurred in an event of an accident or failure to obey environmental laws Anaba et al., 2024).

Other emerging field that greatly benefit from assimilation of digital technologies is the field of predictive maintenance. It identifies that the traditional breakdowns calendar maintenance approach involves regular, rigid intervals that results in either over-maintenance or under maintenance: the wastes resources and can lead to equipment failure, or risk an environmental accident (Romasheva & Dmitrieva, 2021). Applying artificial intelligence and machine learning to the machinery helps firms to forecast when certain machinery is likely to develop faults and so fix it at the appropriate time instead of at frequent intervals thus increasing the lifespan of the machinery and reducing on probabilities of accident (Liu et al., 2023). Besides increasing safety, the method has benefits that make the industry more sustainable by minimizing wastage of resources. This is particularly true in the case of the oil and gas sector, where this new technology, which involves the use of blocks of chained data, guarantees the efficiency of interactions and can also help increase the level of transparency of business processes. When using blockchain to support the supply chain, it is possible to provide an impartial and permanent record of transactions and, thus, eliminate the possibility of fraud and guarantee compliance with the environment standards (Arinze et al., 2024). For instance, blockchain can be applied to monitor the genesis of oil and gas products and guaranteeing that they are derived from socially appropriate activities and all the rules and regulations made regarding the commodity are followed to the latter.

Still, there are some challenges that could be linked to the use of the digital technologies in Russian oil and gas industry. These are inadequate capital investment in new technologies, legal restraints and scarcity of personnel who can help in the implementation and management of these technologies (Lukyanova, 2020). Also, the geopolitical and economic risk in the region contributes to the lack of adoption of digital tools (Krotov, Karatabanov & Zan, 2019). Mitigating these hurdles will call for collaboration among the government, industry and academia to foster a conducive environment for digital advancement. digital technologies is an effective solution for increasing the indicator of environmental friendliness of the Russian oil and gas sector. Through the optimization of the processes involved, real-time monitoring, and transparency, these technologies have the potentiality of reducing the adverse effects of oil and gas activities on the environment and the promotion of the sustainable nature of the industry. Nonetheless, achieving the full utilization of digital technologies will entail surmounting of numerous challenges such as investment hurdles and skills gap.

2. LITERATURE REVIEW

2.1. The impact of digital technologies on the implementation of the concept of sustainable development.

For the past few years, the application of digital technologies in the SDG implementation process has gained significant attention, especially in the oil and gas sector (Pigola et al. 2021). The utilization of sophisticated technologies including big data analytics, artificial intelligence and blockchain platforms have opened new ways through which the firms can certainly monitor their environmental impacts and effectively control them. For example, the use of predictive analytics in AI can improve the productivity of resource extraction activities, meaning that there are fewer losses and the processes take less time, which entails a lesser impact on the environment (Alieksieienko et al., 2022). So, the application of blockchain technology can help to track material and its adherence to environmental regulation in the supply chain (Kottmeyer, 2021). These technological developments do not only enhance productivity of corporations but also enable them to be in sync with global sustainability goals thus making them more appealing to investors and consumers who are now more socially responsible.

In addition, digitization of environmental management activities has promoted company' increased responsibility and the oil and gas industry accountability when reporting on their environmental impact. Such transparency is critical for sustaining public confidence particularly seeing that environmental and social governance (ESG) standards are today more than ever under specific focus by the public (Borodina et al., 2023). For example, satellite remote-sensing and the Internet of Things (IoT) sensors enable the real-time monitoring of environmental factors and give companies a chance to act on potential threats affecting the environment promptly and report the rightful data to the regulating authorities (Marcovecchio et al., 2019). Moreover, the application of these technologies into the corporate sustainability concept helps the company to respond to the social issues resulting from industrial developments in the course of achieving the corporate goals in the affected areas. They are able to engage with stakeholders using digital platforms and in so doing, attend to issues to do with environment and the development of sustainable solutions (Ershova et al., 2020).

Sustainable development is a set of measures aimed at meeting current human needs while preserving the environment and resources, that is, without compromising the ability of future generations to meet their own needs. Sustainable development is possible when three main components are balanced: economic growth, social responsibility, and environmental balance (Korobitsyn 2016). The basis of the Russian economy is the oil and gas industry, a branch of production that includes the production, processing, and transportation of hydrocarbon raw materials (oil and natural gas). The development of digital technologies has made a significant contribution for the sustainable development of territories, including in the Russian Federation. Thanks to social media and universal openness, a person can create information content that is instantly distributed on the network. Thus, major environmental disasters or acts of social inequality cannot be hidden from the masses and receive maximum publicity (Kraus et al. 2021). A comparison of the information publicity of accidents at fuel and energy complex facilities during Soviet period and over the past couple of years looks indicative in this regard. Despite the desire of regional governments to conceal information about accidents, the response of emergency response services and public outcry from the oil spill in Norilsk (May 2020) (Yakutseni and Solovyev 2020) and the fire on the Ob River (March 2021) (Moskovchenko et al., 2020) was more intense than from the explosions of the reactor at the Chornobyl nuclear power plant (April 1986) and the gas pipeline near the railway near Ufa (June 1989) (Magdich et al., 2019).

Thus, the awareness of people all over the world has led to the fact that sustainable development has become a part of our lives and an agenda not only for international organizations and countries but also for companies. The Sustainable Development Goals declared by the UN are increasingly being implemented in the development strategies of companies, including Russian ones. As an example, we can cite social programs implemented by domestic oil and gas campaigns. These include the "social production" at the Samatlor field (Mitrova and Melnikov 2019) and the system of compensation payments to representatives of tiny indigenous peoples of the north. In addition, we should not forget about the record compensation for environmental damage in the amount of 146.2 billion rubles, which MMC Norilsk Nickel paid. Thus, the sustainable business development agenda, which began with large companies, increases companies' social, economic, and environmental responsibility through direct investments in the medical and social spheres to improve the quality of life through social programs for local Indigenous communities affected by the extractive industry. Recent studies show that adherence to the principles of sustainable development in companies makes employees more loyal. In a changing world, companies need to be more aware in order to be chosen by young employees (Sulyandziga 2019).

2.2. The Role of Digital Technologies in the Prevention of Emergencies

The use of new digital solutions for the prevention of emergencies in the oil and gas sector is gradually changing the industry's strategies in the sphere of safety and environmental management. A couple of significant strides created in this field include the use of predictive maintenance that is the application of big data analysis and artificial intelligence algorithms with a view of identifying that a particular piece of equipment is likely to fail (Sulyandziga 2019). These systems consist of data gathering sensors placed on pipelines or drills, and by constant data analysis, these systems can identify potential problems including corrosion, pressure fluctuations, as well as mechanical deterioration (Moskovchenko et al., 2020). Besides reducing the probability of accidents, it proactively enables intervention before a given disaster and its subsequent environmental effects, can happen. Furthermore, these capabilities apply to the predictive qualities of these systems, the applications of which are enhanced and become more accurate when exposed to larger datasets (Korobitsyn, 2016).

Moreover, the importance of the remote monitoring and control systems concerning the prevention of the emergencies has significantly risen. Thanks to IoT it is now possible to observe the process of oil and gas production in time including the incredibly distant and hard to reach ones (Kottmeyer, 2021). IoT sensors can be used to measure a number of operational parameters including pressure temperature and flow rates and send this data to IoT control centers where the information is sometimes processed and acted upon through the use of other IoT systems or by human operatives. This real-time monitoring enables the quick identification of any anomaly since the process is in progress and takes instant decisions that are critical for avoiding occurrences that could cause harm to the environment (Liu et al., 2019). For instance, if there is a consequent drop in pipeline pressure that signals leakage, then turnoff processes can commence or adjustments can be made to redirect the flow this drastically curtails the possibility of polluting the environment (Arinze et al., 2024).

And, finally, the 'intelligent fields', or 'smart fields' is the highest level of oil and gas Industry digitalization that provides a total solution for constant emergencies and environmental control. These smart fields use AI, IoT, robotics for making each aspect of oil and gas production connected, integrated, and 'smart,' that is, capable of its own self-regulation (Kozlovtseva et al., 2021). For instance, intelligent drilling systems can accommodate changes in the regimes of drilling parameters that are responsive to subsurface environment and therefore minimize incidents such as blowouts or any other form of incidents of drilling (Lyshchikova et al., 2019). Furthermore, smart fields can also forecast and control the effects on the environment by constantly measuring and assessing emissions, waste, and any other related factors and subsequently, regulating and responding to such threats. The application of these systems contributes to increasing operational safety; at the same time, it contributes to the achievement of the sustainable development objectives by reducing the oil and gas industry's negative impact on the environment (Schulz et al., 2020).

As noted earlier, in recent years, a series of major accidents have occurred in oil and gas industry enterprises in Russia, which acted as prerequisites for reforming the implemented environmental policy and searching for new solutions in the field of nature protection (Makarov 2016). Having considered some examples of emergencies (Moskovchenko et al., 2020) on pipelines (the Ufa disaster, a fire on the Ob River), we can confidently identify standard features and, based on them, formulate a primary mechanism that allows maintaining the safety of the ecological framework of territories in the vicinity of main linear objects. The leading indicator is the level of the current pressure in the pipe: its increase may indicate the formation of intra-pipe deposits that can lead to damage to the main line; a fall, as a rule, means a leak that has already occurred and is a signal for emergency measures. In this regard, all pipeline systems should be equipped with pressure monitoring and monitoring systems, and highly qualified specialists should monitor the indicators around the clock. It is also necessary to introduce an intellectual knowledge base system that provides a comprehensive account of the experience of implementing similar projects. It will allow expanding approaches to predicting the risk of the linear part of the main pipelines and ensuring their safety (Schipachev and Dmitrieva 2021; Bianco et al., 2021). At the same time, strict compliance with the established design requirements is necessary, and maintenance and reconstruction of objects should occur regularly, according to the specified frequency.

However, the most significant impact on the environment during the development of hydrocarbon deposits occurs during the construction of wells. And since it is easier to prevent an accident than to eliminate its consequences, it is necessary to constantly modernize the anti-blowout equipment (Tarantola et al. 2018) and improve the systems of control and measuring devices of drilling rigs and geological and technical research stations (Chernikov et al. 2020). There may be gas, water, and oil in the drilled formations. The gas penetrates the well through cracks and pores. Timely detection of gases at the wellhead is one of the main tasks for the prevention of gas and oil occurrences (Lyubimov et al., 2018). Modern gas analysis equipment is a product of digital technologies designed to automate control systems for technological processes and emissions. The installation of gas analysis equipment on the drilling rig is carried out in accordance with the current regulatory documents. In contrast, in practice the regulatory documents for the equipment of geological and technical research stations are wholly ignored (Pichtel 2016). The highest form of digitalization in the oil and gas industry at the moment is the concept of "intelligent fields" or the "smart field" system implementation. Intelligent fields combine artificial intelligence, intelligent drilling, and machine learning, which together can help oil and gas companies reduce costs, increase production, and reduce the level of anthropogenic load (Dmitrievskiy et al. 2020; Iliinskij et al. 2020).

2.3. Barriers to Promoting Digital Technology in the Oil and Gas Business

Currently, there is a reduction in the flow rates of oil and gas wells due to the depletion of field reserves. In recent years, the maintenance of production volumes has been achieved by drilling new wells: new fields are being put into operation, and horizontal drilling through productive formations is actively used in old fields to increase the return coefficient, and work is underway to create cluster wells. In conditions of increased market competition, Russian oil and gas companies are focused on reducing production costs. During periods of falling global energy prices, the most used cost-saving tools are staff reduction and restructuring (Mitrova and Melnikov 2019). Under current circumstances, one of the main directions for maintaining production is to increase the pace of production drilling. However, as practice shows, production profitability decreases. In this regard, company and government heads face the question of choosing between reducing production volumes or profitability (Egorov 2021).

The use of digital technologies and the introduction of breakthrough technologies based on them to increase the efficiency of the development of oil and gas are associated with large-scale investments. The existing potential may not be realized if the domestic oil and gas sector does not actively increase investments in digital technologies in the future. Thus, the main problem of the Russian oil and gas sector is the need for significant long-term costs, which are investments in innovations, including digital solutions (Olisaeva 2019). However, not all companies are ready to take risks and prefer instant profit to long-term prospects, exposing the objects of operation to an increased risk of emergencies. The development of digital technologies requires the introduction of appropriate technical standards. Such standards exist in Russia, but they have inherent disadvantages that reduce the effectiveness of management. For example, 3G coverage is absent not only in a large area of Siberia and the Russian Far East but also in the European part of Russia; this is especially true for 4G (Chistobaev and Kulakovskiy 2020).

There are also a number of factors that complicate the implementation of the concept of "intelligent deposits". In the Russian Federation, all Russian offshore projects, including the Sakhalin projects, are intellectual (Olisaeva 2019). However, the offshore project is characterized by territorial compactness: all infrastructure facilities are localized within the area of the drilling platform. The implementation of this concept in large continental fields is complicated by the remoteness of drilling sites and the material costs associated with their maintenance in relation to digital systems. In addition, the current political situation has a significant impact on the development of domestic digital systems. In 2018, one of the largest companies, Oracle, specializing in database management systems and enterprise resource management (ERP), notified its Russian partners about the termination of providing services and technologies for deep-sea and Arctic offshore exploration and production projects (Rouissi 2020). As a result, they were subject to sanctions that applied to both new projects and renewal (extension) for all the major players in the sector who had already signed contracts. In this regard, the goal of import substitution should be to create favorable conditions for the emergence of domestic digital solutions and increase their competitiveness (Olisaeva 2019).

The lack of adequate human resources as a result of the inability to produce enough skilled professional who can effectively put in place and manage the enhanced systems is another factor the slows down the use of digital technology among the Russian oil and Gas industry (Øien et al., 2020). The digitalization of the oil and gas industry is a rather heavy investment but it is also necessary to introduce a capable workforce that can work with digital technologies like data analysis, artificial intelligence, and Internet of Things. Nevertheless, even at the current, the educational and training process in Russia has not developed pace with the growth in these technologies to close the skills gap that limits the correct implementation of these solutions (Haouel & Nemeslaki, 2024). This is further compounded by the centralized geographical distribution of oil and gas operations predominantly in difficult-to-reach areas such as Siberia and Russian Far East the access to education and training and development opportunity is restricted (Su et al., 2022). Thus, the speed of finding talented people who would be able to contribute to companies' digitalization and salary competition hinders the fast rate of industry's digitalization.

Furthermore, there still exist serious problems associated with the old equipment in many oil and gas fields of Russia that hampers the digital technologies integration. The technology of many apparatuses and infrastructures apply in the industry was developed and erected in the 1980s in the least, considerably earlier than recent digital advancements (Roberts et al., 2021). These ageing structures when this has to be done involves extensive modification or replacement of the existing systems by the new digital technologies which may be expensive. This challenge is further compounded by the fact that many companies are currently, very to invest in such upgrades due to the perceived risks and uncertainties of digitalization especially because of volatile energy markets as well as geopolitical instabilities (Goyal et al., 2020). Failure to upgrade ageing fixed assets with new ones fit for the digital era not only hinders most of the advantages of digitalization but also renders the operations at a higher risk of either sub-optimization or actual disasters, making it more difficult for the countries involved to come close to achieving their sustainable development objectives.

2.4. Assessment of Digital Technology Usage Levels

Despite the significant financial opportunities of the Russian oil and gas sector, the industry as a whole is quite conservative, which hinders the development of new technologies. Meanwhile, the introduction of digital technologies requires prompt decision-making and new approaches to the organization of staff work. Over the past 2–3 years, a number of new digital technologies, if not passed into the category of familiar and routine, then at least have received more than isolated examples of application. We are talking about the use of drones for scanning the terrain and construction control, remote assistants based on augmented reality technologies, and video analytics to ensure safety at production facilities. Companies have experience in implementing these technologies (Fernandez-Vidal et al. 2022).

Despite this, the use of expensive analytical equipment at remote production facilities requires the involvement and constant presence of qualified personnel. An alternative for using qualified personnel is the robotization of remote production facilities and their equipment with fully automatic equipment. The oil and gas industry has all the prerequisites for a wider use of robots: the shift of production to hard-to-reach regions with difficult climatic conditions, the increasing complexity of production, and the growing shortage of personnel of working specialties (Litvinenko 2020). There are already examples of the introduction of sparsely populated technologies in the oil and gas sector in the world; for example, the autonomous platform (Norway) is the first utterly deserted production platform that requires only 1–2 visits per year for maintenance (Cordes et al. 2016). Nevertheless, in the Russian Federation, in the field of industrial robotics, there is a lack of investment and production capacity, as well as a lack of specialists, research, and design bureaus. So, over the past five years, only about five thousand robots have been introduced in Russia.

An important problem, which rises in the attempt of evaluating the current level of digital technology implementation in the Russian oil and gas sector, is the disparities in the companies' digital technologies utilization (Arinze et al., 2024). Thus, large companies as market leaders are often ahead, leaving behind smaller firms and those function in less developed economic areas or regions (Kozlovtseva et al., 2021). This is partly so, because, practicing digital technologies can be very expensive especially for firms that cannot afford to invest large sums of money that perhaps larger firms can afford. Further to this, the extremities in climate of the Russian territory especially Siberia and Russian Far East compounding the challenges –there are high stiffening factors regarding installation and maintenance of digital platforms in such territories respectively (Lyshchikova et al., 2019). Therefore, although new technologies of digitization could have a profound impact on efficiency and safety of the oil and gas sector, the actual application of these technologies remains weak and patchy.

Another aspect of the usage of digital technology that needs to be dealt with is the challenge of linking data management systems and the capacity to process the big amounts of data provided by the majority of contemporary digital tools. The data originates from several sources such as seismic surveys, drilling and well completion, and production data (Schipachev and Dmitrieva 2021). However, it has emerged that many companies are just starting to collect or already have vast arrays of valuable data, but they are unable to efficiently manage it or make good use of it, primarily due to their lack of better data analysis that a lot of the time is intertwined with poor integration of multiple digital platforms. For instance, drones and remote sensors are capable of offering real-time information on the state of infrastructure as well as the environment but this information is hardly utilized owing to the absence of suitable analytical tools or profound knowledge on how to use it propound-making (Liu et al., et al., 2023). With missing or inadequate data management, and analysis, this gap thus poses a major challenge to how the industry can harness digital technologies fully.

In addition, the existing legal requirements in Russia also influence the rate and degree of digitization of the technology in the oil and gas industry. Legal frameworks may be slow in embracing the new technologies, and this puts uncertainties and barriers to firms that would prefer to use new digital technologies (Arinze et al., 2024). For instance, there is a limited awareness about the use of drones or any autonomous system within the industrial facilities hence such industries might hesitate to adopt these systems due to legal aspects that are involved. Also, lack of an established normative base for digital systems integration implies that most companies may encounter challenges when trying to integrate various digital tools and platforms; this would make it challenging to adopt efficient digital strategies for the business (Iliinskij et al. 2020). Managing these regulatory issues is critical to promoting the enabling digital environment that will be needed for sustaining the competiveness of the upstream oil & gas industry going forward.

3. DISCUSSION

Information technology in the Russian oil and gas sector: opportunities and threats There are some prospects for using digital technologies in the Russian oil and gas sector and, at the same time, there are some risks also. As discussed in the literature review on the impacts of digitization, more specifically artificial intelligence, IoT and robotics, are likely to bring about changes that would improve operation efficiency, safety and minimize the impacts on the environment. It means that the better solution for the growth of the sector is to stimulate investments and develop corresponding policies that will help the quintessential industry transform digitally without leaving the weaker links of the financial chain. Furthermore, the current legal environment of Russia prescribes other challenges. Since new forms of technology are developed before new laws are formulated to govern them, there is always some ambiguity in the operation of most modern industries including the use of autonomous systems and drones. This lack of regulation can delay the integration of new technologies because companies are not sure on the legal frameworks that are available to support them and protect them from risks. In addition, the absence of a clear pattern to the integration of digital systems aggravates these problems since it is not easy to reach the degree of digital integration that requires companies. The topic also emphasizes the lack of skill in the work force a factor, which the discussion also examines. That is why, as digitalization continues to expand throughout the oil and gas sector, the requirement for qualified professionals who can operate arising digital technologies effectively would continue to rise. But without such effort this talent remains untapped and the industry lags behind the current global trend toward digitalization and sustainability. Thus, it is recommended that investments in IT systems, changes in legislation, and development of employees should embrace the Russian O&G sector in order to grasp the potential of digitalization and remain sustainable in the future.

4. CONCLUSION

This chapter has given a detailed insight to the executive together with an analysis of the integration of digital technologies within the Russian oil and gas sector but importantly, how crucial they are in enhancing sustainable development. Technologies like AI, IoT, robotics etc are excellent opportunities for increasing overall productivity, decreasing the negative consequences to the environment and raising safety standards in the industry. However, the Implementation of these technologies is not without some difficulties. Due to inherent conservatism of the sector, high costs of digitalization, and regulatory uncertainties, as well as the lack of qualified specialists, the level of digitalization remains relatively low. Such challenges have to be resolved with the help of specific investments, qualitative changes in legislation, and the preparation of a competent staff if the industry is to reap the bulk of the potential gains of digitization.

The chapter also discusses the need for the support of the environment for digitalization, especially for the Russian oil and gas fields characterized by their large extent and geographic distribution. If such endeavors are to overcome the aforementioned obstacles and Russian oil and gas industry is to enhance its digital readiness, the sector can not only enhance its competitiveness, but it could also bring more added value to the international sustainable development objective. And as the industry undergoes more Darwinian twists and turns in the new world of energy, it is the ability to adopt the digital technologies that will be central to the future viability and prosperity of the industry. Chapter 9 therefore posits that complex and planned efforts are the proper way to exploiting the value of digital transformation in the oil and gas industry while also bringing them in line with environmental and economic objectives.

REFERENCES

Alieksieienko, T. F., Kryshtanovych, S., Noskova, M., Burdun, V., & Semenenko, A. (2022). The use of modern digital technologies for the development of the educational environment in the system for ensuring the sustainable development of the region. *International Journal of Sustainable Development and Planning*, 8(17), 2427–2434. DOI: 10.18280/ijsdp.170810

Anaba, D. C., Kess-Momoh, A. J., & Ayodeji, S. A. (2024). Digital transformation in oil and gas production: Enhancing efficiency and reducing costs. *International Journal of Management & Entrepreneurship Research*, 6(7), 2153–2161. DOI: 10.51594/ijmer.v6i7.1263

Arinze, C. A., Ajala, O. A., Okoye, C. C., Ofodile, O. C., & Daraojimba, A. I. (2024). Evaluating the integration of advanced IT solutions for emission reduction in the oil and gas sector. *Engineering Science & Technology Journal*, 5(3), 639–652. DOI: 10.51594/estj.v5i3.862

Bianco, I., Ilin, I., & Iliinsky, A. (2021). Digital technology risk reduction mechanisms to enhance ecological and human safety in the northern sea route for oil and gas companies. In *E3S Web of Conferences* (Vol. 258, p. 06047). EDP Sciences. DOI: 10.1051/e3sconf/202125806047

Borodina, M., Idrisov, H., Kapustina, D., Zhildikbayeva, A., Fedorov, A., Denisova, D., Gerasimova, E., & Solovyanenko, N. (2023). State Regulation of Digital Technologies for Sustainable Development and Territorial Planning. *International Journal of Sustainable Development and Planning*, 18(5), 1615–1624. DOI: 10.18280/ijsdp.180533

Cherepovitsyn, A., Rutenko, E., & Solovyova, V. (2021). Sustainable development of oil and gas resources: A system of environmental, socio-economic, and innovation indicators. *Journal of Marine Science and Engineering*, 9(11), 1307. DOI: 10.3390/jmse9111307

Chernikov, A. D., Eremin, N. A., Stolyarov, V. E., Sboev, A. G., Semenova-Chashchina, O. K., & Fitsner, L. K. (2020). Application of Artificial Intelligence Methods for Identifying and Predicting Complications in the Construction of Oil and Gas Wells: Problems and Solutions. *Georesursy*, 22(3), 87–96. DOI: 10.18599/grs.2020.3.87-96

Chistobaev, A., & Kulakovskiy, E. (2020, September). Digital technologies in ensuring local government in the Russian Federation. []. IOP Publishing.]. *IOP Conference Series. Materials Science and Engineering*, 940(1), 012039. DOI: 10.1088/1757-899X/940/1/012039

Cordes, E. E., Jones, D. O., Schlacher, T. A., Amon, D. J., Bernardino, A. F., Brooke, S., Carney, R., DeLeo, D. M., Dunlop, K. M., Escobar-Briones, E. G., Gates, A. R., Génio, L., Gobin, J., Henry, L.-A., Herrera, S., Hoyt, S., Joye, M., Kark, S., Mestre, N. C., & Witte, U. (2016). Environmental impacts of the deep-water oil and gas industry: A review to guide management strategies. *Frontiers in Environmental Science*, 4, 58. DOI: 10.3389/fenvs.2016.00058

Daneeva, Y., Glebova, A., Daneev, O., & Zvonova, E. (2020, August). Digital transformation of oil and gas companies: energy transition. In *Russian Conference on Digital Economy and Knowledge Management (RuDEcK 2020)* (pp. 199-205). Atlantis Press. DOI: 10.2991/aebmr.k.200730.037

Dmitrievsky, A. N., Eremin, N. A., Filippova, D. S., & Safarova, E. A. (2020). Digital oil and gas complex of Russia. Georesursy= Georesources, Special issue, 32-35.

Eder, L. V., Filimonova, I. V., Provornaya, I. V., Komarova, A. V., & Nikitenko, S. M. (2017). New directions for sustainable development of oil and gas industry of Russia: Innovative strategies, regional smart specializations, public-private partnership. *International Multidisciplinary Scientific GeoConference: SGEM, 17*(1.5), 365-372.

Egorov, D. P. (2021). A CLASSIFICATION OF OIL AND GAS-BEARING TERRITORIES AND A QUALITATIVE ASSESSMENT OF THE RUSSIAN OIL AND GAS INDUSTRY. International Research Journal, 2021(5107).

Ershova, I., Obukhova, A., & Belyaeva, O. (2020). Implementation of innovative digital technologies in the world. *Economic Annals-XXI/Ekonomičnij Časopis-XXI, 186*.

Esiri, A. E., Babayeju, O. A., & Ekemezie, I. O. (2024). Implementing sustainable practices in oil and gas operations to minimize environmental footprint.

Fernandez-Vidal, F., Gonzalez, R., Gasco, J., & Llopis, J. (2022). Digitalization and corporate transformation: The case of European oil & gas firms. *Technological Forecasting and Social Change*, 174, 121293. DOI: 10.1016/j.techfore.2021.121293

Filimonova, I., Komarova, A., Nemov, V., & Provornaya, I. (2020). Sustainable development of Russian energy sector: Hydrocarbons of Eastern Siberia. *International Multidisciplinary Scientific GeoConference: SGEM, 20*(1.2), 777-783.

Filimonova, I. V., Komarova, A. V., Provornaya, I. V., Dzyuba, Y. A., & Link, A. E. (2020). Efficiency of oil companies in Russia in the context of energy and sustainable development. *Energy Reports*, 6, 498–504. DOI: 10.1016/j.egyr.2020.09.027

Goyal, R., Pokhriyal, D. S., Nechully, D. S., & Gupta, D. S. (2020). Management Barriers in Implementation of Integrated Operations Solutions in Indian Upstream Companies. *International Journal of Management*, 11(10).

Haouel, C., & Nemeslaki, A. (2024). Digital transformation in oil and gas industry: Opportunities and challenges. *Periodica Polytechnica Social and Management Sciences*, 32(1), 1–16. DOI: 10.3311/PPso.20830

Haouel, C., & Nemeslaki, A. (2024). Digital transformation in oil and gas industry: Opportunities and challenges. *Periodica Polytechnica Social and Management Sciences*, 32(1), 1–16. DOI: 10.3311/PPso.20830

Iliinskij, A., Afanasiev, M., Wei, T. X., Ishel, B., & Metkin, D. (2020, November). Organizational and management model of smart field technology on the arctic shelf. In Proceedings of the International Scientific Conference-Digital Transformation on Manufacturing, Infrastructure and Service (pp. 1-5). DOI: 10.1145/3446434.3446510

Khoso, A. K., Darazi, M. A., Mahesar, K. A., Memon, M. A., & Nawaz, F. (2022). The impact of ESL teachers' emotional intelligence on ESL Students academic engagement, reading and writing proficiency: Mediating role of ESL students motivation. *Int. J. Early Childhood Spec. Educ*, 14, 3267–3280.

Korobitsyn, B. A. (2016). Regional resilience of the Ural Federal District in economic shocks and crises: medico-demographic and environmental aspects. Ekonomika Regiona= Economy of Regions, (3), 790-801. .DOI: 10.17059/2016-3-15

Kottmeyer, B. (2021). Digitisation and sustainable development: The opportunities and risks of using digital technologies for the implementation of a circular economy. *Journal of Entrepreneurship and Innovation in Emerging Economies*, 7(1), 17–23. DOI: 10.1177/2393957520967799

Kozlovtseva, V., Demianchuk, M., Koval, V., Hordopolov, V., & Atstaja, D. (2021). Ensuring sustainable development of enterprises in the conditions of digital transformations. In *E3S Web of Conferences* (No. 280, p. 02002).

Kraus, S., Jones, P., Kailer, N., Weinmann, A., Chaparro-Banegas, N., & Roig-Tierno, N. (2021). Digital transformation: An overview of the current state of the art of research. *SAGE Open*, 11(3), 21582440211047576. DOI: 10.1177/21582440211047576

Krotov, A. V., Karatabanov, R. A., & Zan, V. M. (2019, November). Problems and prospects of competitiveness of the territories of Greater Altai in the context of sustainable developmen. In *International Conference on Sustainable Development of Cross-Border Regions: Economic, Social and Security Challenges (ICSDCBR 2019)* (pp. 257-260). Atlantis Press. DOI: 10.2991/icsdcbr-19.2019.54

Litvinenko, V. S. (2020). Digital economy as a factor in the technological development of the mineral sector. *Natural Resources Research*, 29(3), 1521–1541. DOI: 10.1007/s11053-019-09568-4

Liu, H., Zhu, Q., Khoso, W. M., & Khoso, A. K. (2023). Spatial pattern and the development of green finance trends in China. *Renewable Energy*, 211, 370–378. DOI: 10.1016/j.renene.2023.05.014

Lyshchikova, J. V., Stryabkova, E. A., Glotova, A. S., & Dobrodomova, T. N. (2019). The'Smart Region'concept: The implementation of digital technology. *Journal of Advanced Research in Law and Economics*, 10(4 (42)), 1338–1345. DOI: 10.14505//jarle.v10.4(42).34

Lyubimov, I. L., Lysyuk, M. V., & Gvozdeva, M. A. (2018). Atlas of economic complexity, Russian regional pages. *Voprosy Ekonomiki*, (6), 71–91. Advance online publication. DOI: 10.32609/0042-8736-2018-6-71-91

Magdich, I. A., Petrov, V. P., & Pyatibrat, A. O. (2019). Analiz sanitarnykh i bezvozvratnykh poter'v zavisimosti ot kharaktera i usloviy chrezvychaynykh situatsiy na zheleznoy doroge= Analysis of Sanitary and Irreparable Losses Depending on the Nature and Conditions of Railway Emergencies. *Medical-biological and Socio-psychological Problems of Safety in Emergency Situations*, 1, 72–80. DOI: 10.25016/2541-7487-2019-0-1-72-80

Makarov, I. A. (2016). Russia's participation in international environmental cooperation. *Strategic Analysis*, 40(6), 536–546. DOI: 10.1080/09700161.2016.1224062

Marcovecchio, I., Thinyane, M., Estevez, E., & Janowski, T. (2019). Digital government as implementation means for sustainable development goals. [IJPADA]. *International Journal of Public Administration in the Digital Age*, 6(3), 1–22. DOI: 10.4018/IJPADA.2019070101

Minakir, P. A. (2020). Political value of expectations in economy. Spatial Economics= Prostranstvennaya Ekonomika, (3), 7-23. .DOI: 10.14530/se.2020.3.007-023

Mitrova, T., & Melnikov, Y. (2019). Energy transition in Russia. *Energy Transitions*, 3(1-2), 73–80. DOI: 10.1007/s41825-019-00016-8

Moskovchenko, D. V., Babushkin, A. G., & Yurtaev, A. A. (2020). The impact of the Russian oil industry on surface water quality (a case study of the Agan River catchment, West Siberia). *Environmental Earth Sciences*, 79(14), 355. DOI: 10.1007/s12665-020-09097-x

Øien, K., Hauge, S., & Grøtan, T. O. (2020). Barrier Management Digitalization in the Oil and Gas Industry-Status and Challenges. In *e-proceedings of the 30th European Safety and Reliability Conference and 15th Probabilistic Safety Assessment and Management Conference (ESREL2020 PSAM15)*. Research Publishing Services.

Olisaeva, A. V. (2019). Technological development of Russia: HR policy, digital transformation, Industry 4.0. *Planning and Teaching Engineering Staff for the Industrial and Economic Complex of the Region*, 86-89. Advance online publication. DOI: 10.17816/PTES26310

Pichtel, J. (2016). Oil and gas production wastewater: Soil contamination and pollution prevention. *Applied and Environmental Soil Science*, 2016(1), 2707989. DOI: 10.1155/2016/2707989

Pigola, A., da Costa, P. R., Carvalho, L. C., Silva, L. F. D., Kniess, C. T., & Maccari, E. A. (2021). Artificial intelligence-driven digital technologies to the implementation of the sustainable development goals: A perspective from Brazil and Portugal. *Sustainability (Basel)*, 13(24), 13669. DOI: 10.3390/su132413669

Roberts, R., Flin, R., Millar, D., & Corradi, L. (2021). Psychological factors influencing technology adoption: A case study from the oil and gas industry. *Technovation*, 102, 102219. DOI: 10.1016/j.technovation.2020.102219

Romasheva, N., & Dmitrieva, D. (2021). Energy resources exploitation in the russian arctic: Challenges and prospects for the sustainable development of the ecosystem. *Energies*, 14(24), 8300. DOI: 10.3390/en14248300

Rouissi, C. (2020). The influence of the enterprise resource Planning (ERP) on management controllers: A study in the Tunisian context. *International Journal of Business and Management*, 15(4), 25–35. DOI: 10.5539/ijbm.v15n4p25

Schipachev, A. M., & Dmitrieva, A. S. (2021). Application of the resonant energy separation effect at natural gas reduction points in order to improve the energy efficiency of the gas distribution system. Записки Горного института, 248, 253-259. .DOI: 10.31897/PMI.2021.2.9

Schulz, K. A., Gstrein, O. J., & Zwitter, A. J. (2020). Exploring the governance and implementation of sustainable development initiatives through blockchain technology. *Futures*, 122, 102611. DOI: 10.1016/j.futures.2020.102611

Sergi, B. S., & Berezin, A. (2018). Oil and gas industry's technological and sustainable development: Where does Russia stand? In *Exploring the future of Russia's economy and markets: Towards sustainable economic development* (pp. 161–182). Emerald Publishing Limited. DOI: 10.1108/978-1-78769-397-520181009

Smirnova, O., & Ponomaryova, A. (2021). Assessment of the Industrial Development Rates of Russian Regions in the Context of the Digitalization of the Economy. In Digital Transformation in Industry: Trends, Management, Strategies (pp. 283-290). Springer International Publishing. DOI: 10.1007/978-3-030-73261-5_26

Solodovnikov, A. Yu., Chistobaev, A. I., & Semenova, Z. A. (2016). Influence of Oil and Gas Facilities of Western Siberia on the Fauna. *Geography and Natural Resources*, (4). Advance online publication. DOI: 10.21782/GiPR0206-1619-2016-4(48-54)

Su, J., Yao, S., & Liu, H. (2022). Data governance facilitate digital transformation of oil and gas industry. *Frontiers in Earth Science (Lausanne)*, 10, 861091. DOI: 10.3389/feart.2022.861091

Sulyandziga, L. (2019). Indigenous peoples and extractive industry encounters: Benefit-sharing agreements in Russian Arctic. *Polar Science*, 21, 68–74. DOI: 10.1016/j.polar.2018.12.002

TARANTOLA, S., ROSSOTTI, A., CONTINI, P., & CONTINI, S. (2018). A guide to the equipment, methods and procedures for the prevention of risks, emergency response and mitigation of the consequences of accidents: Part I.

Tyaglov, S. G., Sheveleva, A. V., Rodionova, N. D., & Guseva, T. B. (2021, March). Contribution of Russian oil and gas companies to the implementation of the sustainable development goal of combating climate change. []. IOP Publishing.]. *IOP Conference Series. Earth and Environmental Science*, 666(2), 022007. DOI: 10.1088/1755-1315/666/2/022007

Yakutseni, S. P., & Solov'ev, I. A. (2020). Calculation of environmental damage as a result of an accident at a fuel depot in Norilsk. Geograficheskaya sreda i zhivye sistemy, (4), 48-56. https://doi.org/DOI: 10.18384/2712-7621-2020-4-48-56

Chapter 5
Digital Transformation Features of Mining Industry Enterprises:
Mining 4.0 Technologies Development Trends

Alisa Torosyan
ITMO University, Russia

Olga Tcukanova
ITMO University, Russia

Natalia Alekseeva
Peter the Great St. Petersburg Polytechnic University, Russia

Abdul Hameed Qureshi
Pakistan Telecommunication Authority, Islamabad, Pakistan

Shafiya Qadeer Memon
Department of Software Engineering, UET Jamshoro, Pakistan

ABSTRACT

The concept of the 4th industrial revolution is becoming a strategic driver of sustainability, success, and competitiveness in the modern mining sector. An important factor is that the industry in question has its own features in operational and production processes, which are changing due to the development and implementation of digital technologies. The purpose of this article is to determine the features of the digital transformation of enterprises in the mining industry and to study the role of

DOI: 10.4018/979-8-3693-3423-2.ch005

Copyright © 2025, IGI Global. Copying or distributing in print or electronic forms without written permission of IGI Global is prohibited.

digital technologies implemented in companies. To conduct the analysis, the method of content analysis of scientific studies and articles related to digital transformation in the mining sector was used. The results of the analysis are the formulation of the main directions for the development of digital technologies for mining enterprises, the definition of their role in the mining industry under consideration, and clarification of the features of digital transformation. The results of the study can be used in subsequent work aimed at further emplementing modern digital technologies.

1. INTRODUCTION

People engage in multiple digital activities in their day to day lives and many industries go through digitization. Digitalization and its core and key part is the digitalization of the company and members of the company systems, business processes, and business models (Bertayeva et al., 2019). The creation and improvement of information technologies presuppose changes in the activity of enterprises and companies as the suppliers of goods and services utilizing new forms of manufacturing and delivering goods and services, as well as the general transformation of the economy at large (Barnewold & Lottermoser, 2020). The further evolution of Information Technology, the wants and needs of Industry 4. The objective of obtaining profit is now equaled to 0 and the need for sustainable development set new tasks for enterprises. In the recent past, mining companies have been very keen on deploying technologies that enable cost-saving measures in mining, and enhance production capacity as well as being capable of initiating mining projects that were uneconomical previously (Bertayeva et al. 2019). Thereby, it is noteworthy to state that mining is a technology leader and an adopter of disruptive technologies simultaneously (Young & Rogers, 2019). In general, digital transformation is connected with the digital positioning, supply chain, management, creation of agency, or behavior of a firm. There is an increase in the connection of companies' economic activity with the technologies of big data collection and analysis, artificial intelligence, virtual and augmented reality, and 3D printing. In recent research, the analysis of digitalization showed that many firms adopt digital technologies in all the exploration and mining phases (Zhironkina & Zhironkin, 2023). The process associated with the life cycle of deposits include the exploration and development of deposits, mining, production and lastly, completion. Depends on the thickness of the mineral deposits, they are mined in three methods, namely open-pit mining, underground mining, or a combination of the two (Zhironkin & Ezdina, 2023).

The mining industry is one of the most relevant fields in the development of modeling as many processes and operations are directly m-empirical and generate a vast amount of data for quantitative analysis, which can be suitable for the application of digital intelligence (Kagan et al., 2021). Specific emphasis should be placed on the development of equipment used by mining-related enterprises. Over the course of industrial growth in the world, it has evolved with each new technological stage, updating. Modern technology has enabled exploration companies to examine what is in the ore, sample it, and identify whether a deposit should be developed (OECD, 2019, p. 8). Before proceeding with the analysis, it is necessary to note the mining equipment, its availability, and operational safety as potential risks that appear in the processes of mining companies (Paschke et al., 2020). The digital technologies that define Industry 4. 0 and that are integrated into mining companies are called Mining 4. Concerning the concept of Mining 4. 0 thus embraces the knowledge of relations between technologies and therefore seeks to provide a more holistic view to the changes taking place in the mining industry through technology adoption. The main directions for the development of Mining 4. 0 are Internet of Things, Mechatronics which is a combination of mechanics, electronics, control and information technology, Telematics which covers telecommunication systems, Distribution which is a part of an enterprise logistics (Gao et al. 2019). Automation is viewed as a way to enhance the flow of coal and eliminate certain adverse circumstances for miners as well. They identified what digital technologies are mostly reported in the mining scientific literature based on the following findings (Sánchez & Hartlieb, 2020).

Figure 1. The ten most relevant digital technologies for the mining industry are rated by PF (Barnewold and Lottermoser 2020)

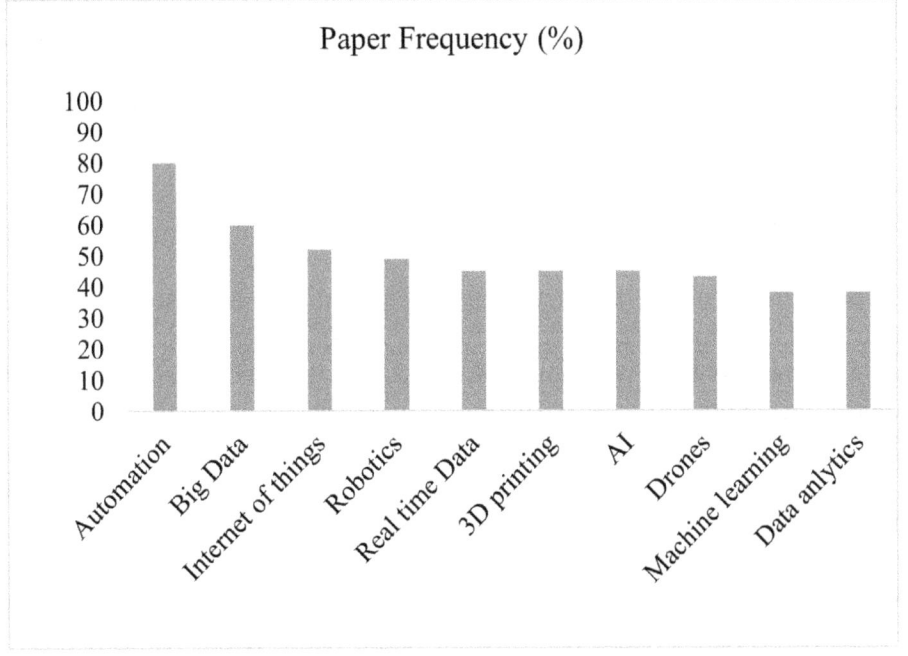

In this regard, it is imperative and relevant to note that digital transformation exerts a very positive effect on enterprise value (Hushko et al., 2021). Innovation represents one of the most effective strategies to generate value at every state of the resource price cycle (Choros et al. 2022). This forms the base on which a consolidated automated system that consist of a lot of critical information and analytical data can be developed. In the relation to the general consideration of data, it should be stated that all the data is transferred to computer systems and becomes more accessible (Pereira et al., 2022). In the case of the mining industry, the benefits of using databases are the possibility of the automation of different processes, which would entail the optimization of resource consumption and personnel (Stenin 2021). For instance, one of the target of big data for the mining industry is to minimize the occurrences of mining disruptions, optimize the processes, minimize costs and maximize gains.

Assessment and management of risks is an important element in organization of production and other processes in mining organizations. If risk management is not done well this may result in losses in production and at some points the consequences are severe to personnel and problems to environment (Domingues et al., 2017). These are the following characteristics of this industry: some companies develop the assets and deposits situated in the difficult access areas with low metal content in the ore, which adds the costs of the companies (Zhukova 2019). However, company size plays the most essential role in identifying whether a firm will innovate, since market barriers push definite smaller companies to engage in non-technological innovation, join forces with other national companies and institutions, and appeal to public funding (Yamashita and Fujii 2022). Mining 4. 0 technologies enhance human abilities as well as enhance perception via sensors in for example, wears during the periods of continued, unbroken perception necessary in challenging working environments.

1.1 The Evolution of Mining 4.0 Technologies

Traditionally, mining companies employed manpower and used a number of mechanical instruments (Zhironkin & Ezdina, 2023). Though there were enhancements in productivity by mechanization, effects and drawbacks were seen in terms of accuracy, control and, safety of the workers. Mining 4. 0 is founded based on the concept of Industry 4. 0 and, where digitally connected systems and solutions can play a part in industrial workflows (Sishi & Telukdarie, 2020). These technologies such as automation, IoT, big data, and AI have started defining the operations of the mining companies right from exploration through to extraction and others. Arising from these technologies, mining enterprises are now in a position to track their operations in real time, make decisions based on predictive information, and employ best practices that will increase efficiency in order to obtain the maximum levels of production without compromising on efficiency (Zhironkina & Zhironkin, 2023).

For example, smart sensors in mining for instance are used to measure and capture data on the state of equipment, the surrounding environment, and possible failure. This saves time, ensure safety, and improves the management of the assets (Bartnitzki, 2017). In the same manner, AI-based analytics highlight geological conditions and adapt the drilling and extraction procedures which in turn, minimizes expenses and harm to environment. Another significant component of Mining 4.0 is the use of self-driving cars as well as automated equipment's (Ulewicz et al., 2017). These machines are self-automated and work in dangerous environments to avoid endangering human life while at the same time cutting cost and increasing accuracy. Autonomous vehicle is becoming a preferred investment for mines, to enhance productivity and safety in fleets used for ore transportation, drilling and material movement.

Another important aspect of mining also involves the utilization of Robotics and Drones for inspection, surveying and monitoring purposes (Bertayeva et al., 2019). Drones can maneuver through restricted areas and collect quality images, thus enable operators to make the right decisions. In the same context, robots are employed in underground mining especially in repetitive tasks or in areas that pose major threats to human beings. The use of Block chain technology is another trend that is emerging in Mining 4. There has been 0 progress for supply chain transparency and traceability (Pałaka et al., 2020). Block chain allows keeping the records of transactions and material flows, which is especially valuable for such industries as precious metals and rare earth minerals, the proper sourcing of which is a concern. Smart mines therefore symbolize the integration of all these technologies where everything in the mining value chain from the exploration of the resources to extraction and transportation is automated and controlled in real time. That, on the one hand, contributes to operation optimization, and on the other hand, enables mining enterprises to decrease the level of adverse environmental impact, including waste and energy consumption as well as emissions (Zhironkin et al., 2022).

1.1 Key Trends in Digital Transformation of the Mining Industry

Several key trends have emerged in the digital transformation of the mining industry, each of which is helping reshape traditional mining processes (Lazarenko et al., 2021). These trends are critical in understanding the direction of technological development and the future of Mining 4.0. The current advancement in technology is also changing how a mining industry's workers perform their duties. In the past, mining was characterized by the employment of many people who spent most time in dangerous terrains (Barnewold & Lottermoser, 2020). However, mining 4.

0 is changing this by cutting on the use of manpower especially in risky sections, and enhancing safety. This change is best exemplified by automation and robotics where such process as drilling, blasting, handling of materials, etc. are all done by the machines. This not only increases effectiveness of operations, but it also greatly reduces the probabilities of occurrence of accidents. For instance, the traditional smart or a fully autonomous car can traverse underground mines where the triumphant dangers including leakage in poisonous gases, collapse of mines or poisonous gas and fumes can cost many human lives (Mottaeva & Gordeyeva, 2024). Indeed, the emergence of the digital workforce requires one to look for new human capital. There is a growing trend where mining firms are spending on training of employees in the manner of developing their competencies when it comes to utilizing digital technologies such as artificial intelligence systems, IoT, data, and analytic tools. This shift is changing the traditional interpersonal relationships at the mining sites in that there is increasing need of data scientists, automation specialists and AI specialists. Thus, the training and education of the workforce are important factors in the process of digital transformation of mining enterprises (Paschke et al., 2020).

Applications including virtual reality (VR) and augmented reality (AR) are being applied in improving training and safety of workers. Workers can be trained what they have to do in a particular contract by use of virtual reality simulations and this way, they will be exposed to real life activities but they are safe (Bertayeva et al., 2029). While, AR technologies are used to present the real-time information and necessary instructions to the workers in the field where they need to make decisions and do not want to encounter any risks. Also wearables in mining are present and even becoming more popular. IoT sensors integrated into wearables can also check the workers' health and the environmental conditions and inform the workers and their managers about possible hazards. This real-time monitoring can assist in avoiding mishaps in addition to improving security measures in risky locations hence Mining 4. 0 not only about the optimization of its own operations but about organizational safety as well (Gao et al., 2019). The changes of the mining industry through the implementation of Mining 4. This paper finds that 0 technologies are transformative in determining how mining enterprises are being run. Automation, IoT, AI, as well as other smart technologies can improve mining industry's functioning, decrease negative effects on the environment, and increase the safety of workers. Some of these trends include sustainability, transformation of workforce whereby miners are more skills, and safety measures, transcription of mining practices to meet better practice (Dragičević & Bošnjak, 2019). About the future of mining, it is necessary to mention that Mining 4. 0 technologies will prove to be central in the overall planning of the future of the healthcare sector. Businesses that successfully adopt these technologies will be at an advantage over their counterparts and part of the effort to bring about positive changes in mining that will have minimum negative impacts on the environment and the society. Mining 4. 0 has presented the future of mining and the development trends of mining digital transformation. 0 are establishing the framework of the new paradigm of mining innovation.

2. LITERATURE REVIEW

The mining industry has always played a significant role in the overall structure of world industrialization and with the introduction of Mining 4. 0, meaning that the sector is at a very early stage of the digitalization process. Contemporary publications explore the impact of digitization initiatives, including automation, artificial intelligence (AI), and the Internet of Things (IoT), on aspects of mining including, but not limited to, productivity, environmental impacts, and safety of employees and contractors (Sánchez & Hartlieb, 2020). This paper aims at presenting and reviewing the most relevant literature related to the Mining 4. 0 technologies and how these can be applied in improving the first sustainable dimension and the second operational dimension.

2.1 The Impact of Digital Transformation on Mining Efficiency and Productivity

According to the literature review, the digital revolution has been instrumental in enhancing efficiency and productivity within the mining industry (SAVAS, 2022). The use of advanced technology solutions like autonomous vehicles, smart sensors, and real-time data analytics has improved decision-making and resource utilization processes in mining companies. For instance, the sensors in the Internet of Things can always track the performance of the machinery and the conditions of the environment, making it possible for the mining operators to determine when equipment is likely to fail and when to maintain the assets. Liu et al. (2023) notes that the use of IoT in predictive maintenance has improved operations in mines by reducing operational downtimes and thus increasing the operating efficiency and decreasing operating costs. Likewise, AI-imbued systems can scan data to identify issues and suggest ways to augment processes, thereby improving operations' results (Khoso et al., 2022). Autonomous mining vehicles can be considered as another important aspect in the context of Mining 4. 0 However, it means that efficiency increases as it is not necessary to use people in dangerous conditions. Research conducted by Zhang et al. (2019), show how AHS is useful in enhancing the overall mining production, given that such vehicles do not require rest or time to adapt to various weather conditions. They are most commonly used in underground mining, where human intervention could be dangerous due to the conditions. The use of AI drilling and blasting systems has also aided in increasing recovery rates as patterns are designed by AI to be efficient in avoiding unnecessary drilling and maximizing the amount of resource that can be recovered (Kalenov & Kukushkin, 2021). In addition to the concepts of autonomous systems, one of the advanced technologies is digital twin technology that can increase the efficiency of operations. Digital twins are models that represent physical mining assets or processes and can be used to test the impact of changes before applying them in the field. Based on Gupta et al. (2022), mining digital twin technology has been effective in improving decision-making as the use of models provides simulations of extraction best practices that are energy-efficient in the process. That is why the combination of digital twins, in conjunction with real-time data analysis, allows miners to make effective decisions in terms of resource allocation and organizational productivity. This technological advancement has led to efficiency, in doing so some researchers are advising against the overdependence on the use of these technologies (Sánchez & Hartlieb, 2020). This is especially true as most mining enterprises consider digitization as a critical success factor in the conduct of their operations and in the overall management of their resources. Mitigating these risks thus call for a harmonized approach to digital change whereby technological initiatives are combined with well-coordinated risk management frameworks.

2.2 Environmental Sustainability and Digital Transformation in Mining

The effects mining has on the environment is a concern that has been in focus for quite some time now, and it has been known that this practice causes destruction of habitats, pollution and depletion of resources (Zhou, 2024). However, mining digitalization, especially through Mining 4. 0 technologies are opening up new avenues for improvement of the sustainability of mining operations. In this way, through application of digital technologies, the mining enterprise can conform to strict environmental requirements as well as minimize its impact on environment. Some of the previous works have attempted at examining how the utilization of different digital technologies can help mine in a way that is environmentally friendly (Xu et al., 2022). For example, Liu et al. (2019) show that for monitoring the energy usage and emissions, IoT-based systems aid the mining industries to detect inefficiencies in near real time and further, to manage their climate change impact. This particular technology is useful where energy is likely to be used in large quantities for instance in extracting and processing minerals, a slight saving in the amount of energy used has huge impacts in averting the emission of greenhouse gases.

In the same way, artificial intelligence, and machine learning are now used to improve the use of techniques of extraction with very many impacts of mining on the environment negated. Feroz et al. (2021) highlighted that the application of the AI algorithm enhances the drilling and blasting efficiency of processes thus reducing the overall loss of ores, thereby increasing the recovery in totality, which attributes the profitability factor in modelling waste rock and its effects depict the excesses of drilling too much or blasting (Liu et al., 2023). Therefore, extraction resource is crucial where balanced information submission could be deployed to improve on the resource-mining activities while reducing the impacts on the environment.

Another way in which the mining industries are availing digital transformation in the enhancement of the environmental sustainability is in water management (Akhtar et al., 2024). This paper seeks to conduct a policy analysis on mining following major policy issues on water policy, water legislation, water use, and its importance to the mining industry. Nonetheless, mining activities cause high water usage for both mining and the subsequent processing as well as polluting water resources especially in the arid zones. Liu et al (2019) also further noted that, IoT and AI are some of the technologies that are being deployed to monitor the usage of water in the mining processes in real time hence helping mining companies to cut on their water usage as well as meeting the legal requirements on the use of water.

The circular economy has also found application in mining and digitalization being a critical driver of the concept of circular economy has also been embraced in the mining industry (Lazarenko et al., 2021). The practice involves the use of products in order to reuse, recycle and or recover, thus minimizing the flow of waste towards sustainability and the environment. Xu et al. (2023) pointed out that block chain can be used to monitor the life cycle of mineral materials and guarantee the recycling and reuse of waste materials. The way materials are acquired and what happens thereafter is well followed in the system using block chain to add the credibility of the company. Specifically discussing about emissions reductions, electrification and hybridization of automobile fleet is emerging in Mining 4.0 operations. Diesel-operated equipment has been one of the major causes of greenhouse gas emissions in mining; however, the use of electric cars and renewable energy is slowly negating the effect (Kunkel & Matthess, 2020). Using solar and wind power, Silva et al. assert that dependency on fossil fuels in mining has been minimized and carbon emissions mitigated (Shvedina, 2020). These developments coupled with the use of mining vehicles that are powered by electricity means that mining is now environmentally friendlier.

While numerous scholars look into the positive impact of digital transformation on the environment there are equally a number of scholars that argue of the difficulties involved concerning implementation of such programs mainly in the developing world. Following the arguments by Maroufkhani et al., (2022), it is clear that, even though digital technologies present tremendous sustainability benefits, they come with considerable capital and human capital costs of implementation. As the resources are scarce in many regions, the application of Mining 4. 0 technologies may cause lower speed resulting to the likely hood of the technologies effect on the environment. In addition to this, issues like accuracy of data used, compatibility of equipment's as well as the regulatory environments pose other problems that has to be solved in support of digitization of mining enterprises (Van Hau et al., 2022). The use of digital technologies is however associated with the following limitations; Research including that conducted by Zhironkin and Ezdina, (2023), note that technology fixes can in fact cannot solve the problems in the mining industries. In its place we need systemic change which attributes accordant weight to digital transitions, policy changes, corporate responsibility, and societal involvement. For instance, in mining industries, there is need to collaborate with individuals, groups or organizations within the exercising territories to prevent the destruction of environment or inadmissible impairing of native people's lives.

3. METHODS

3.1 Materials and Methods

This section details the methods used to explore the digital transformation of the mining industry, focusing on the adoption of Mining 4.0 technologies. The methodology combines both qualitative and quantitative approaches, relying on data from The Future of Jobs Report 2020 by Lund et al. (2021) and a thorough content analysis of various sources related to digital transformation trends in mining. These approaches help to assess the integration of technologies such as artificial intelligence (AI), big data, cloud computing, Internet of Things (IoT), and robotics in mining operations, providing a holistic understanding of how these technologies are transforming the industry.

3.2 Survey Data Analysis

A key component of the study is the analysis of survey data from The Future of Jobs Report 2020, which highlights the technologies expected to be adopted by mining companies by 2025. The report provides detailed information on the percentage of companies that plan to integrate specific digital technologies into their operations. For example, 76% of the surveyed companies expressed an intention to adopt AI, while 90% anticipated using big data analytics, IoT, and non-humanoid robots by 2025. Other significant technologies expected to be integrated include cloud computing (87%), augmented and virtual reality (57%), and block chain (50%). These survey results form the foundation of the quantitative aspect of this study, allowing for an in-depth examination of how mining companies are preparing for digital transformation. Each technology's adoption potential is evaluated in relation to its impact on operational efficiency, sustainability, and workforce management. The results also offer insights into the different rates of adoption for various technologies, revealing which digital tools are likely to have the greatest influence on mining operations in the near future. The analysis further considers how these technologies can be leveraged to address current challenges in mining, such as environmental impact, operational efficiency, and safety improvements.

3.3 Content Analysis

The content analysis method is used to explore the broader context of digital transformation in the mining industry. This qualitative approach involved a systematic review of a wide range of sources, including peer-reviewed journals, industry reports, and case studies, all of which provide insights into the role of Mining 4.0 technologies in modernizing mining operations. The analysis focused on identifying key trends and themes, particularly regarding how technologies like AI, big data, IoT, and automation are reshaping traditional mining processes.

Through content analysis, the study highlights how digital technologies are being implemented across different phases of mining, from exploration to production and supply chain management. The literature reveals that digital solutions are being used to enhance operational efficiency by optimizing resource extraction, improving equipment monitoring, and enabling predictive maintenance. Additionally, the integration of AI and IoT technologies is helping mining companies reduce energy consumption and waste, contributing to more sustainable and environmentally friendly operations. Content analysis also uncovers the growing importance of real-time data analytics, which allows companies to make more informed decisions, thereby improving productivity and reducing operational risks. The review further identifies significant advancements in safety protocols, particularly through the use of autonomous vehicles, drones, and robotics, which reduce human exposure to dangerous mining environments. By removing workers from hazardous conditions and enabling remote monitoring, these technologies are playing a critical role in transforming the way safety is managed in the mining industry. Moreover, content analysis reveals that the workforce is undergoing a transformation, with an increased demand for digital skills as automation and AI technologies become more prevalent in mining operations.

3.4 Generalization Method

The generalization method was employed to synthesize the findings from both the survey data and content analysis. This approach enabled the study to draw broader conclusions about the role of digital technologies in mining, integrating insights from multiple sources to provide a comprehensive understanding of the key trends in Mining 4.0. The generalization process involved summarizing the data and identifying common themes related to the adoption of specific technologies across different mining contexts. For example, technologies such as AI, big data, and IoT were consistently highlighted across various studies as key drivers of operational efficiency and sustainability. The generalization method also allowed for the identification of critical areas where these technologies are having the most significant impact, such as in energy management, resource optimization, and safety improvements. Additionally, the method helped to categorize technologies based on their potential to enhance productivity, environmental sustainability, and workforce safety, offering a structured framework for analyzing the impact of digital transformation on the mining industry.

4. RESULTS

4.1 Technology Adoption in the Mining Industry

In this section, they review on the use of various digital innovations in the mining sector particularly on the future trends up to the year 2025. The information used for this purpose is extracted from The Future of Jobs Report 2020 (Lund et al., 2021) which reveals the level of mining organizations' appetence towards various technological applications including Artificial Intelligence (AI), Big Data Analytics, Internet of Things (IoT), and others. For the purpose of the enhanced understanding, the presented data has been presented broken down by the company size, namely, large, medium and small-scaled companies as well as by the regions identified as Region 1, Region 2 and Region 3. The analysis done here using quantitative descriptive research method provides fundamental understanding on the advancement of digital transformation in mining industry. This helps us to determine the prevalence and future outlook for adopting these technologies at the industry level while also understanding which types of companies and geographic locations are ahead of others in leverage these technologies. Table 1 also gives the level of adoption of the various technologies by small, medium and large companies and companies in different regions. These findings show that big firms' technology uptake ratios are significantly higher than smaller firms across all categories, primarily AI, Big Data Analytics, and Cloud Computing, as the clients' uptake ratios are over 80%. Small firms, on the other hand, are slow to adopt technology – especially in emerging fields including Robotics and Quantum Computing. Table 4 shown below gives the distribution of the percentage of mining firms using each technology by the size of firms and geographical locations. There are also noticeable gaps that compare different regions based on Technology Adoption Matrix where Region 1 excel in Big Data Analytics & IoT while Region 3 being slow in Robotics & Quantum Computing. With these differences it is understood that the implementation of the digital transformation strategy has to consider the size of the company and the technological environment of the region.

Table 1. Detailed technology adoption in mining by company size and region

Technology	Percentage (%)	Large Companies (%)	Medium Companies (%)	Small Companies (%)	Region 1 (%)	Region 2 (%)	Region 3 (%)
3D and 4D Printing	48	55	45	35	50	46	40
Artificial Intelligence	76	80	74	60	78	72	65
Augmented & Virtual Reality	57	65	55	40	60	55	50
Big Data Analytics	90	92	88	75	93	88	80
Biotechnology	16	18	15	10	17	15	12
Cloud Computing	87	90	85	70	88	85	78
Distributed Ledger Technology	50	52	48	30	53	48	45
E-commerce & Digital Trade	62	67	60	50	63	60	55
Encryption & Cybersecurity	83	88	80	65	85	80	75
Internet of Things	90	92	88	80	93	88	82
New Materials	37	40	35	25	38	35	30
Power Storage & Generation	57	60	55	45	58	55	50
Quantum Computing	29	32	25	20	28	25	22
Robots (Humanoid)	15	20	12	10	18	12	10
Robots (Non-Humanoid)	90	93	85	75	92	85	80
Text, Image & Voice Processing	76	78	74	65	75	74	70

Figure 2 thus depicts the types and levels of usage of digital technologies for large, medium, and small-sized firms. It gives insights into the degree of digitalization that has occurred within the mining industry, where several primary mining players are found to be at an advanced level of digital implementation than others, especially in the use of emerging technologies such as Artificial Intelligence, Big Data Analytics, and Cloud Computing. Smaller enterprises feature notably lower levels of technology integration, even for sophisticated tools like Robotics and Quantum Computing. The following diagram offers an understandable graphical display of the correlation between the company size and rate of technology adoption in the industry.

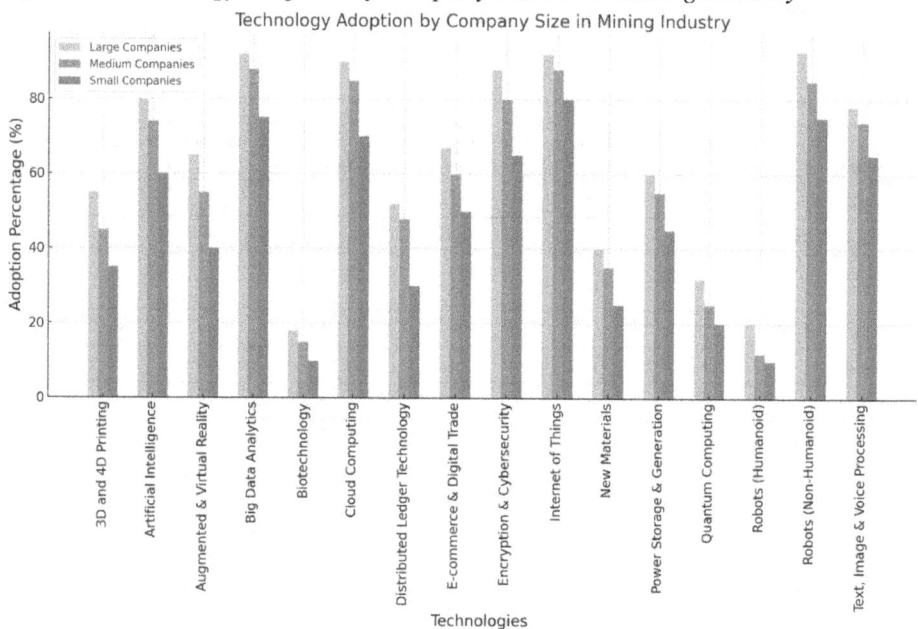

Figure 2. Technology adoption by company size in the mining industry

4.2 Content Analysis

The two articles were analyzed based on their content in a bid to understand whether there was an adoption of Mining 4. When a number of the 0 technologies are considered several trends, challenges and opportunities that are inherent in the mining industry can be identified. AI, Big Data Analytics, IoT and automation are some of the advanced technologies reviewed here that are revolutional different stages of mining; from exploration to the process stage. However, the study shows that factors limiting the development of wind power are lack of funds, poor infrastructure and lack of personnel, especially for small players and in less developed areas of the country. Table 4 below provides an overview of the findings developed in the organizational operational area, the adopted technologies and the main problems related to each.

Table 2. Application of digital technologies in mining companies

Activities of mining companies	Digital technologies	Problems of digital technologies implementation
exploration	– 3D modelling – specialized software and programs – digital twins – artificial intelligence – big data analytics – machine learning – drones	– insufficient funding – shortage of qualified personnel – insufficient development of infrastructure for the implementation and application of technologies – complexity of work conditions
design	– 3D modelling – specialized software and programs – artificial intelligence – big data analytics – machine learning – drones	– insufficient funding – shortage of qualified personnel – insufficient information base – complexity of work conditions
mining	– automated mining equipment – specialized software and programs – onboard robotic system – remote control – artificial intelligence – big data analytics – machine learning	– insufficient funding – shortage of qualified personnel – access difficulty to deposits – complexity of work conditions – insufficient development of infrastructure for the implementation and application of technologies – non-compliance with the health and safety policy
transportation of minerals	– autonomous (unmanned) loading dump trucks and excavators – specialized software and programs – artificial intelligence – big data analytics – machine learning	– insufficient funding – shortage of qualified personnel – insufficient development of infrastructure for the implementation and application of technologies – non-compliance with the health and safety policy
processing of minerals	– Internet of things – robots – specialized software and programs for artificial intelligence – big data analytics – machine learning – remote control	– insufficient funding – shortage of qualified personnel – non-compliance with the health and safety policy

continued on following page

Table 2. Continued

Activities of mining companies	Digital technologies	Problems of digital technologies implementation
procurement activities	– digital platforms – specialized software and programs for artificial intelligence – big data analytics – machine learning	– insufficient funding – lack of digital technologies for minor players in the market – information security threat – unwillingness to automate business processes among staff
operational process management	– specialized software and programs – remote control – robotization of hr processes – artificial intelligence – big data analytics – machine learning – use of VR in staff training	– insufficient funding – insufficient prioritization of processes for digitalisation – inconsistency between departments – information security threat – unwillingness of the staff to automate business processes

The analysis of Mining 4.0 technologies across various operational areas of the mining industry reveals both significant advancements and persistent challenges. While digital transformation offers clear benefits in terms of operational efficiency, safety, and sustainability, many companies face hurdles in implementing these technologies fully. The table below summarizes the findings, presenting the key technologies adopted in each operational area and the major challenges associated with their implementation. This summary offers a comprehensive view of how Mining 4.0 is reshaping the industry and the obstacles that must be addressed for broader adoption.

Table 3. Summary of findings on Mining 4.0 technologies

Operational Area	Adopted Technologies	Key Challenges
Exploration	3D modeling, digital twins, drones, AI, Big Data Analytics	Insufficient funding, shortage of qualified personnel, infrastructure gaps, complexity of conditions
Design and Planning	3D modeling, AI, digital twins, machine learning	Lack of information base, funding constraints, personnel shortages
Mining Operations	Automated equipment, remote control systems, robotics, AI	Difficult access to deposits, health and safety compliance, infrastructure development issues
Mineral Processing	IoT devices, robots, AI, Big Data Analytics, remote control systems	Funding limitations, personnel shortages, non-compliance with safety policies
Transportation	Autonomous trucks, specialized software, AI, machine learning	Lack of infrastructure, personnel shortages, health and safety challenges
Procurement and Process Management	Digital platforms, AI, machine learning, Big Data Analytics	Information security risks, staff reluctance to automate processes, interdepartmental inconsistencies

4.3 Comparative Analysis of Technology Adoption in the Mining Industry

The comparative analysis focuses on the implementation of Mining 4. 0 technologies by the company's size and geographical location. Thereby, we can identify whether the large companies, or some regions are ahead of others in terms of embracing technologies like AI, Big Data Analytics, IoTs, automations, etc.

1. Comparison by Company Size

The information also shows the trend in the uptake of the Mining 4. No difference was observed in the number of technologies implemented based on the size of the company. The utilization of these technologies is more profound in large firms as their uptake usually stands at above 80% for technologies like AI, Big Data Analytics, and Cloud Computing. Looking at the medium-sized companies, they are gradually advancing, yet not in a pace with bigger ones, with the scores varying from 55% to 88%. On the other hand, small companies demonstrate the lowest levels of adoption, especially in the Robotics and Quantum Computing technologies whereby adoption is below 40%. This has worldview implies that large organizations are typically endowed with more capital and hence they are well placed to incorporate new technologies that would enhance safe and efficient operation. Larger companies on the other hand do not have issues with funding and skilled personnel as they do with the small firms which have a smaller appetite for changing their operations to digital. These disparities suggest that smaller firms require special assistance through political backlash or partnership plan to gain and implement these technologies at a higher rate.

Table 4. Comparative analysis of technology adoption by company size

Technology	Large Companies (%)	Medium Companies (%)	Small Companies (%)
3D and 4D Printing	55	45	35
Artificial Intelligence	80	74	60
Augmented & Virtual Reality	65	55	40
Big Data Analytics	92	88	75
Biotechnology	18	15	10
Cloud Computing	90	85	70
Distributed Ledger Technology	52	48	30

continued on following page

Table 4. Continued

Technology	Large Companies (%)	Medium Companies (%)	Small Companies (%)
E-commerce & Digital Trade	67	60	50
Encryption & Cybersecurity	88	80	65
Internet of Things	92	88	80
New Materials	40	35	25
Power Storage & Generation	60	55	45
Quantum Computing	32	25	20
Robots (Humanoid)	20	12	10
Robots (Non-Humanoid)	93	85	75
Text, Image & Voice Processing	78	74	65

2. Comparison by Geographic Region

Geographical positions are also very influential in the application of Mining 4. Region 1 companies are those based in technologically advanced areas and report the highest percentages of adoption with percentages over 85% for AI, Big Data Analytics, and IoT. Region 3, on the other hand, has a much lower adoption rate, especially in new technologies such as Quantum Computing and Humanoid Robotics where adoption stands at less than 30%. Such a difference could be due to the disparities in the technological environment, government, and the market environment in different regions. Companies in developed areas with proper regulations have the ability to adapt to new technologies, but companies in the developing regions suffer from infrastructure weakness and restricted resources.

Table 5. Comparative analysis of technology adoption by region

Technology	Region 1 (%)	Region 2 (%)	Region 3 (%)
3D and 4D Printing	50	46	40
Artificial Intelligence	78	72	65
Augmented & Virtual Reality	60	55	50
Big Data Analytics	93	88	80
Biotechnology	17	15	12
Cloud Computing	88	85	78
Distributed Ledger Technology	53	48	45

continued on following page

Table 5. Continued

Technology	Region 1 (%)	Region 2 (%)	Region 3 (%)
E-commerce & Digital Trade	63	60	55
Encryption & Cybersecurity	85	80	75
Internet of Things	93	88	82
New Materials	38	35	30
Power Storage & Generation	58	55	50
Quantum Computing	28	25	22
Robots (Humanoid)	18	12	10
Robots (Non-Humanoid)	92	85	80
Text, Image & Voice Processing	75	74	70

The comparative analysis suggests for Australia's, Canada's and Peru's mines that their adoption of Mining 4. It cannot be said that there is homogeneity about the industry in terms of adoption of 0 technologies. Big and medium firms more importantly those operating within technologically developed areas are the most aggressive in the adoption of AI, Big Data, IoT, and other technologies. On the other hand, small businesses and firms based in developing countries have issues that put them off to expand digitally, including the following; lack of capital, inadequate infrastructure, and human capital constrain. This discussion emphasizes the need to augment a support to smaller organizations and regions that have not developed the positive sides of Mining 4. 0 technologies are not dominant from the viewpoint of the dispersion throughout the industry. Through mitigating the above issues, the mining industry would be in a position to see that the Advance of the Digital Transformation Brought About Positive Change that can benefit every stakeholder.

5. DISCUSSION

Based on the results of this study, it is clear that the concept of Mining 4. The calculation delivers 0 technologies in relation to the mining industry, where there is remarkable development and stagnation at the same time. Digital technology trends like AI, Big Data Analytics, IoT, and automation are also widely implemented, especially at large firms. They are automating and optimizing processes that include reducing downtime, real-time tracking, and making decisions based on gathered data. The increased use of Big Data Analytics and AI, with the expectation that more than 90% of large companies will implement them by 2025, proves just how essential these tools have become for mining enterprises that are struggling to maximize productivity with their resources while minimizing the amount of downtime that a mine experiences. As for smaller companies, the picture looks quite different. Small mining companies have been found to have impacted the ability to implement enhanced technologies as experienced by large-scale mining firms. For technologies like robotics, cloud computing, and quantum computing, the adoption rate is significantly less, even in small firms, below 40%. The likely cause for this may be attributed to issues such as inadequate funding for health facilities, the availability of health facilities' infrastructure, and a lack of adequate trained human resource. Small firms simply do not have the money to purchase expensive digital applications and also, do not have the staff to support and operate such systems, and as a result, there is a large technology gap within the field.

Geographic region is another key element that determine the level of adoption of Mining 4. 0 technologies. Region 1 industries are farther technologically advanced as compared to the industries of Region 3 and hence the AI, IoT, and other such tools and technologies are more practiced in Region 1 as compared to Region 3, robotics and quantum computing for example. Thus, the readiness regarding the infrastructural and technological environment of each region considerably influences the speed and efficiency of companies' adoption of these tools. Advanced digital economies within various geographies are well placed to leverage Mining 4. 0 The ones in the more developed region have it easy in terms of embracing change and catching up with what is current in the market.

These findings also echo the rising concern of sustainability of the mining industry. Large miners are now exploring smart technologies in a bid to address contentious issues such as energy concerns with the extraction processes. Smart devices and tools of analytics facilitate to control the emissions in real-time, enhance the use of energy, and adopt environmentally friendly measures. However, the implementation of sustainability-centered technologies that include the energy systems and power storage system differ from firm to firm and country to country. While some larger firms have made significant strides in decarburization efforts, smaller companies and those in less developed regions are lagging behind, largely due to the same barriers that hinder overall technology adoption: HIV/AIDS, lack of funding, infrastructure and skilled personnel, inadequate infrastructure and investment and human resource constraints. Safety is another of those areas that, despite having been initially seen as a technical challenge that would not require the hiring of executives with a business background, have emerged as a strategic issue that is critical to any mining company. Technology is seeing with 0 technologies as having the biggest impact. The use of robotics, automated vehicles and unmanned aerial vehicles have made it possible to eliminate human personnel from working in dangerous places thereby lowering the dangers involved in mining. Especially, the use of autonomous vehicles can be considered as one of the key driving forces that has notably changed the industry and provides better security and more efficient material delivery in hazardous or hard-to-reach areas. However, the configuration of analyzing the data reduces it to the discovery that health and safety compliance, particularly in feeble firms, still constrains optimum use of such safety-improvement technologies.

6. CONCLUSION

These are in line with the findings in the analysis of Mining 4. ZERO technologies depict apparent changes in the mining sector that involves the utilization of integrated opportunities like Artificial intelligence (AI), Big data & analytics, internet of things (IOT), and automation. These technologies are greatly increasing operational effectiveness, safety, and/or eco-friendliness and are much useful to big players in the industries who have big bucks to spend for such advancements' funding. Implementations of such technologies are currently being spearheaded by major mining firms whereby the firms stand to benefit through better decision making, reduced time to repair equipment and increased productivity. However, this change is not consistent throughout the industry; the change is gradual and depends on a number of factors such as the level of competition within the business. Newer and smaller firms and those situated in the less developed areas experience significant difficulties in implementing Mining 4. 0 technologies. Some of the challenges that hamper progress includes Lack of adequate funding, poor infrastructure, and an acute shortage of skilled personnel which hinders their ability to fully adopt digital solutions. The end-product is a further disconnection in terms of digital adoption that might open up new avenues for increasing disparity in the film making industry where established players extend their lead while the 'have nots' trail behind.

Thus, geographic distribution can be considered one of the primary factors that define the tempo of the digital transition. Administrative centers with developed technological settings provide fertile ground for implementing the new technologies while administrative centers in the developing countries inertia to implement the advanced technologies. This is why more support in terms of infrastructure, training and funding needs to be given to help the smaller and regionally disadvantaged firms to remain abreast with industry developments. Other than increasing the efficiency in operations, mining 4. None of the 0 technologies are being seen as useful in the push for sustainability. With real-time data analytics, IoT, and AI, organizations are in a position to minimize the use of resources thus lowering the emission of greenhouse gases besides enhancing energy efficiency throughout the industrial processes. But the improvement of sustainable practices is still not homogeneous – small businesses and specific regions remain behind in the implementation of green technologies, thanks to the same problems as in other digital initiatives.

REFERENCES

Akhtar, S., Tian, H., Alsedrah, I. T., Anwar, A., & Bashir, S. (2024). Green mining in China: Fintech's contribution to enhancing innovation performance aimed at sustainable and digital transformation in the mining sector. *Resources Policy*, 92, 104968. DOI: 10.1016/j.resourpol.2024.104968

Barnewold, L., & Lottermoser, B. G. (2020). Identification of digital technologies and digitalisation trends in the mining industry. *International Journal of Mining Science and Technology*, 30(6), 747–757. DOI: 10.1016/j.ijmst.2020.07.003

Barnewold, L., & Lottermoser, B. G. (2020). Identification of digital technologies and digitalisation trends in the mining industry. *International Journal of Mining Science and Technology*, 30(6), 747–757. DOI: 10.1016/j.ijmst.2020.07.003

Barnewold, L., & Lottermoser, B. G. (2020). Identification of digital technologies and digitalisation trends in the mining industry. *International Journal of Mining Science and Technology*, 30(6), 747–757. DOI: 10.1016/j.ijmst.2020.07.003

Bartnitzki, T. (2017). Mining 4.0: Importance of Industry 4.0 for the raw materials sector. *Artificial Intelligence*, 2(1), 25–31.

Bertayeva, K., Panaedova, G., Natocheeva, N., & Belyanchikova, T. (2019). Industry 4.0 in the mining industry: Global trends and innovative development. In *E3S Web of conferences* (Vol. 135, p. 04026). EDP Sciences.

Bertayeva, K., Panaedova, G., Natocheeva, N., & Belyanchikova, T. (2019). Industry 4.0 in the mining industry: Global trends and innovative development. In *E3S Web of conferences* (Vol. 135, p. 04026). EDP Sciences.

Bertayeva, K., Panaedova, G., Natocheeva, N., & Belyanchikova, T. (2019). Industry 4.0 in the mining industry: Global trends and innovative development. In *E3S Web of conferences* (Vol. 135, p. 04026). EDP Sciences.

Choros, K. A., Job, A. T., Edgar, M. L., Austin, K. J., & McAree, P. R. (2022). Can Hyperspectral Imaging and Neural Network Classification Be Used for Ore Grade Discrimination at the Point of Excavation? *Sensors (Basel)*, 22(7), 2687. DOI: 10.3390/s22072687 PMID: 35408301

Domingues, M. S., Baptista, A. L., & Diogo, M. T. (2017). Engineering complex systems applied to risk management in the mining industry. *International Journal of Mining Science and Technology*, 27(4), 611–616. DOI: 10.1016/j.ijmst.2017.05.007

Dragičević, Z., & Bošnjak, S. (2019). Digital transformation in the mining enterprise: The empirical study. *Mining and Metallurgy Engineering Bor*, (1-2), 73–90. DOI: 10.5937/mmeb1902073D

Feroz, A. K., Zo, H., & Chiravuri, A. (2021). Digital transformation and environmental sustainability: A review and research agenda. *Sustainability (Basel)*, 13(3), 1530. DOI: 10.3390/su13031530

Gao, S., Hakanen, E., Töytäri, P., & Rajala, R. (2019). Digital transformation in asset-intensive businesses: Lessons learned from the metals and mining industry.

Gao, S., Hakanen, E., Töytäri, P., & Rajala, R. (2019). Digital transformation in asset-intensive businesses: Lessons learned from the metals and mining industry.

Gupta, C., Jindal, P., & Malhotra, R. K. (2022, November). A study of increasing adoption trends of digital technologies-An evidence from Indian banking. In *AIP Conference Proceedings* (Vol. 2481, No. 1). AIP Publishing. DOI: 10.1063/5.0104572

Hushko, S., Botelho, J. M., Maksymova, I., Slusarenko, K., & Kulishov, V. (2021). Sustainable development of global mineral resources market in Industry 4.0 context. []. IOP Publishing.]. *IOP Conference Series. Earth and Environmental Science*, 628(1), 012025. DOI: 10.1088/1755-1315/628/1/012025

Kagan, E. S., Goosen, E. V., Pakhomova, E. O., & Goosen, O. K. (2021, July). Industry 4.0. and an upgrade of the business models of large mining companies. []. IOP Publishing.]. *IOP Conference Series. Earth and Environmental Science*, 823(1), 012057. DOI: 10.1088/1755-1315/823/1/012057

Kalenov, O., & Kukushkin, S. (2021). Digital transformation of mining enterprises. In *E3S Web of Conferences* (Vol. 278, p. 01015). EDP Sciences.

Kunkel, S., & Matthess, M. (2020). Digital transformation and environmental sustainability in industry: Putting expectations in Asian and African policies into perspective. *Environmental Science & Policy*, 112, 318–329. DOI: 10.1016/j.envsci.2020.06.022

Lazarenko, Y., Garafonova, O., Marhasova, V., & Tkalenko, N. (2021). Digital transformation in the mining sector: Exploring global technology trends and managerial issues. In *E3S Web of Conferences* (Vol. 315, p. 04006). EDP Sciences. DOI: 10.1051/e3sconf/202131504006

Lazarenko, Y., Garafonova, O., Marhasova, V., & Tkalenko, N. (2021). Digital transformation in the mining sector: Exploring global technology trends and managerial issues. In *E3S Web of Conferences* (Vol. 315, p. 04006). EDP Sciences. DOI: 10.1051/e3sconf/202131504006

Liu, R., Gailhofer, P., Gensch, C. O., Köhler, A., Wolff, F., Monteforte, M., ... & Williams, R. (2019). Impacts of the digital transformation on the environment and sustainability. *Issue Paper under Task, 3*.

Lund, S., Madgavkar, A., Manyika, J., Smit, S., Ellingrud, K., Meaney, M., & Robinson, O. (2021). The future of work after COVID-19. *McKinsey global institute, 18*.

Maroufkhani, P., Desouza, K. C., Perrons, R. K., & Iranmanesh, M. (2022). Digital transformation in the resource and energy sectors: A systematic review. *Resources Policy*, 76, 102622. DOI: 10.1016/j.resourpol.2022.102622

Mottaeva, A., & Gordeyeva, Y. (2024). Sustainable development of the mining industry in the context of digital transformation. In *E3S Web of Conferences* (Vol. 531, p. 01032). EDP Sciences. DOI: 10.1051/e3sconf/202453101032

Pałaka, D., Paczesny, B., Gurdziel, M., & Wieloch, W. (2020). Industry 4.0 in development of new technologies for underground mining. In *E3S Web of Conferences* (Vol. 174, p. 01002). EDP Sciences.

Paschke, M., Lebedeva, O., Shabalov, M., & Ivanova, D. (2020). Economic and legal aspects of digital transformation in mining industry. In *Advances in raw material industries for sustainable development goals* (pp. 492–500). CRC Press. DOI: 10.1201/9781003164395-61

Paschke, M., Lebedeva, O., Shabalov, M., & Ivanova, D. (2020). Economic and legal aspects of digital transformation in mining industry. In *Advances in raw material industries for sustainable development goals* (pp. 492–500). CRC Press. DOI: 10.1201/9781003164395-61

Pereira, P., Bašić, F., Bogunovic, I., & Barcelo, D. (2022). Russian-Ukrainian war impacts the total environment. *The Science of the Total Environment*, 837, 155865. DOI: 10.1016/j.scitotenv.2022.155865 PMID: 35569661

Sánchez, F., & Hartlieb, P. (2020). Innovation in the mining industry: Technological trends and a case study of the challenges of disruptive innovation. *Mining, Metallurgy & Exploration*, 37(5), 1385–1399. DOI: 10.1007/s42461-020-00262-1

Sánchez, F., & Hartlieb, P. (2020). Innovation in the mining industry: Technological trends and a case study of the challenges of disruptive innovation. *Mining, Metallurgy & Exploration*, 37(5), 1385–1399. DOI: 10.1007/s42461-020-00262-1

Savas, S. (2022). Digital Transformation from Data Mining to Big Data and Its Effects on Productivity. *Current Studies in Digital Transformation and Productivity*, 54.

Shvedina, S. A. (2020). Digital transformation of mining enterprises contributes to the rational use of resources. []. IOP Publishing.]. *IOP Conference Series. Earth and Environmental Science*, 408(1), 012064. DOI: 10.1088/1755-1315/408/1/012064

Sishi, M., & Telukdarie, A. (2020). Implementation of Industry 4.0 technologies in the mining industry-a case study. *International Journal of Mining and Mineral Engineering*, 11(1), 1–22. DOI: 10.1504/IJMME.2020.105852

Stenin, D. (2021). On possible changes in the calculation of parameters of transport technology of open pit mining with the use of autonomous heavy platforms. In *E3S Web of Conferences* (Vol. 315, p. 03015). EDP Sciences. DOI: 10.1051/e3sconf/202131503015

Ulewicz, R., KRSTIĆ, B., & Ingaldi, M. (2022). Mining Industry 4.0—Opportunities and Barriers. *Acta Montanistica Slovaca*, 27(2).

Van Hau, N., Khanh Ly, C. T., Quynh Nga, N., Hong Duyen, N. T., & Huong Hue, T. T. (2022). Digital Transformation in Mining Sector in Vietnam. *Inżynieria Mineralna*, (2), 21–30.

Xu, C., Chen, X., & Dai, W. (2022). Effects of digital transformation on environmental governance of mining enterprises: Evidence from China. *International Journal of Environmental Research and Public Health*, 19(24), 16474. DOI: 10.3390/ijerph192416474 PMID: 36554353

Xu, Y., Wang, L., Xiong, Y., Wang, M., & Xie, X. (2023). Does digital transformation foster corporate social responsibility? Evidence from Chinese mining industry. *Journal of Environmental Management*, 344, 118646. DOI: 10.1016/j.jenvman.2023.118646 PMID: 37481916

Yamashita, A. S., & Fujii, H. (2022). Trend and priority change of climate change mitigation technology in the global mining sector. *Resources Policy*, 78, 102870. DOI: 10.1016/j.resourpol.2022.102870

Young, A., & Rogers, P. (2019). A review of digital transformation in mining. *Mining, Metallurgy & Exploration*, 36(4), 683–699. DOI: 10.1007/s42461-019-00103-w

Zhang, L., Mu, R., Zhan, Y., Yu, J., Liu, L., Yu, Y., & Zhang, J. (2022). Digital economy, energy efficiency, and carbon emissions: Evidence from provincial panel data in China. *The Science of the Total Environment*, 852, 158403. DOI: 10.1016/j.scitotenv.2022.158403 PMID: 36057314

Zhironkin, S., & Ezdina, N. (2023). Review of transition from mining 4.0 to mining 5.0 innovative technologies. *Applied Sciences (Basel, Switzerland)*, 13(8), 4917. DOI: 10.3390/app13084917

Zhironkin, S., & Ezdina, N. (2023). Review of transition from mining 4.0 to mining 5.0 innovative technologies. *Applied Sciences (Basel, Switzerland)*, 13(8), 4917. DOI: 10.3390/app13084917

Zhironkin, S., & Ezdina, N. (2023). Review of transition from mining 4.0 to mining 5.0 innovative technologies. *Applied Sciences (Basel, Switzerland)*, 13(8), 4917. DOI: 10.3390/app13084917

Zhironkin, S., Gasanov, M., & Suslova, Y. (2022). Orderliness in Mining 4.0. *Energies*, 15(21), 8153. DOI: 10.3390/en15218153

Zhironkina, O., & Zhironkin, S. (2023). Technological and intellectual transition to mining 4.0: A review. *Energies*, 16(3), 1427. DOI: 10.3390/en16031427

Zhironkina, O., & Zhironkin, S. (2023). Technological and intellectual transition to mining 4.0: A review. *Energies*, 16(3), 1427. DOI: 10.3390/en16031427

Zhou, Y. (2024). Natural resources and green economic growth: A pathway to innovation and digital transformation in the mining industry. *Resources Policy*, 90, 104667. DOI: 10.1016/j.resourpol.2024.104667

Zhukova, O. (2019). *Education in modern Russia: policy and discourse* (Master's thesis).

Chapter 6
Digitalization of Business Sector and Commercial Banks Interaction in the Russian Federation:
Results of the Study

Elena I. Generalnitskaia
Dar-Cosmetic Ltd., Russia

Natalya A. Loginova
Russian Customs Academy the St. Petersburg Branch Named After Vladimir Bobkov, Russia

ABSTRACT

Based on the analysis of the Bank of Russia statistics, as well as data collected by the National Research University Higher School of Economics, the authors of the article identified 12 trends describing the business and banking sector interaction in the context of digital environment development. Based on questioning business sector and commercial banks representatives, the identified trends were ranked according to the degree of importance for each of the groups of respondents, which helped to identify their development priorities. Moreover, it helped to indicate one of the most important problems of digitalization of commercial banks and business sector interaction. Thus, it has been pointed out that credit institutions, characterized by a high degree of digitalization at present time, can become a tool for the diffusion of innovations for the business sector, in the case of corresponding business sector adaptive capabilities. The authors proposed a set of measures to increase the digitalization of economic interaction between credit institutions and the business sector.

DOI: 10.4018/979-8-3693-3423-2.ch006

1. INTRODUCTION

The digitalization of the Russian economy has become a central pillar of the nation's economic policy, with the potential to significantly influence its global competitiveness in the coming years. As McKinsey's report indicates, the ongoing digital transformation could contribute between 4.1 and 8.9 trillion rubles to Russia's GDP by 2025, potentially accounting for 19-34% of total GDP growth (Potapova et al., 2022; Liu et al., 2023). This digital shift is not merely a technical upgrade but is expected to define Russia's strategic economic future, ushering in a new technological order that hinges on the success of this transformation. At the heart of this transformation is the interaction between two critical sectors: commercial banks and the business sector. The efficiency and effectiveness of this relationship will determine how smoothly the Russian economy transitions into the digital age (Kolmykova et al., 2022; Raza, et al., 2024).

The nature of digital transformation presents many benefits for the business side and the banking sector (Liu et al., 2023). Some of the most significant advantages are the decrease in the cost of banking services as well as the improvement in the availability of services because of the shift to online banking even if many branches have been shut down (Bukhonova & Yablonskaya, 2022). The increase in the commitment to personal approach to the clients due to the integration of technologies; the improvement in the transparency of the domestic economy through the effective collection and analysis of the data; as well as the growth of the share of these sectors in All of these changes accentuate the need for establishing vibrant, digital partnership between the commercial banks and the business society. It has to be mentioned that these partnerships are crucial for the successful transition to the digital economy in Russia (Yuryeva et al., 2020). Given these changes, the aim of this paper is to examine the most important phenomena related to the digitalization of the relations between the business and banking sectors (Kanishcheva, 2021). Thus, this research will, through the interviews conducted on the sample of successful entrepreneurs and bankers, measure the current levels of digitalization in these sectors, evaluating the barriers to further advance and providing he solution for the intensification of this digitalization process. This will give us an insight of both sides thus giving us a bigger picture of how both sectors can help support and strengthen and contribute to the growth of the digital economy.

1.1. The Growing Importance of Digitalization in Russia

Over the last couple of years, there has been a rising trend in the number of research studies, policy consideration, and management discourses on digitalization (Romanyuk et al., 2021). The technological advancement and the prevailing feature of globalization has made the issue of digitalization a critical feature of economic discourse all over the world. Thus, Russia, being one of the richest countries of the world with good perspectives for economic growth, cannot be an exception. Thus, digitalization is viewed as one of the important driving forces that would help the country to build up the sustainable long-term economic development trajectory (Khalimon et al., 2019).

The authorities have identified digitalization as necessary for improving the economy and the nation's competitiveness on the global arena as one of the key factors contributing to its advancement in Russia. Digital transformation is perceived neither simply as a process of updating archaic systems but as an impulse for developing new sectors, increasing efficiency, and promoting innovation. As stated by Loucks et al. (2016), the digital process may cause the change of the dominating industries since the existing and the customers' experience may be improved. This disruption therefore holds a lot of significance especially the banking industry where service delivery and consumption is rapidly shifting towards the use of digital technologies. The government of the Russian federation has included major investments in the digital area since it is seen as important for sustaining competitiveness in modern global economy. They include five key areas, namely: The introduction of 5G-Networks; the expansion of the Digital Ecosystem; pushing forward the development of AI and Blockchain Technologies. These actions are in line with the worldwide trend when many countries continue to actively work on improving the level of digitalization to maintain a competitive advantage (Swaminathan & Meffert, 2017).

1.2. Interaction Between Commercial Banks and the Business Sector

The relations of commercial banks with the business sector are another factor characterizing the state of the digital economy in Russia. This means that there is a vital role that the banking systems, or the banks in particular, have to offer to the business organizations so that they can be in a position to offer their financial services as they are needed in the marketplace (Yurak et al., 2023). In the course of their digitization, the interactions between the two sectors are getting more diverse and symbiotic. For commercial banks, the digitalization means the chance to automate, decrease expenses and enhance the customers' satisfaction. Blockchain and artificial intelligence, Big data and analytics are some of the technologies that are helping the banks to deliver personalized services, manage the risks effectively, and improve the level of transparency. Furthermore, it has been identified that the availability of digital banking services increasing at each year, this is a factor that is significant given that a number of physical branches of banks are being shut down (Verbivska et al., 2023). The ability to operate through digital platforms enables banks address their customers' needs while at the same time mitigating cost implications of reaching out to outlaying branches.

On the business side, digitalization brings value within the increased availability of financial services as well as capital for the companies. In other words, working with the banks through digital means enable business to obtain loans, oversee accounts, and effect payments. Moreover, digital tools let the companies use data analytics as the main way to enhance the financial decision-making process (Mottaeva et al., 2023). The capability to leverage digital banking services in value delivery is emerging as a source of competitiveness in firms specifically in the digital era. However, as much as the increased use of technology to support the banking and business transactions, there are some specific problems associated with it. Challenges that are eliciting concern in both sectors include; regulatory hurdles, cybersecurity vulnerabilities, and high capital investments required on the digital terrain. Thirdly, there is an issue of digital transformation which might be challenging particularly to small enterprises since they may lack adequate resources or skills to manage the digital tools (Molchan et al., 2023). Piecemeal ignorance of these challenges is highly transformative of ensuring the overall economy gains from the benefits of digitalization.

1.3. Structure of the Chapter

The remainder of this chapter is organized as follows. Section 2 provides a detailed review of the literature on digitalization, focusing on its impact on the banking and business sectors. Section 3 outlines the methodology used to conduct the study, including the survey design and data collection process. Section 4 presents the findings of the study, highlighting the key trends in digitalization and the challenges faced by both bankers and entrepreneurs. Finally, Section 5 discusses the discussion findings for the future of digitalization in Russia and offers recommendations for policymakers and industry leaders and section 6 discusses conclusion of this chapter.

2. LITERATURE REVIEW

Digital economy has become an important area of investigation only within the past few years with more emphasis on the financial and the business dimensions (Glebova et al., 2023). Academics have looked at how firms and industries are being transformed into digital ones, how markets are being transformed and new possibilities being opened for contributing to the growth of the economy. This section presents the literature review cultivating an understanding of the concept of digitalization particularly in business and banking segment especially in Russia (Kostomakhin et al., 2023). The review is divided into two subsections: the first one that discusses general tendencies of the digitalization in banking and businesses all over the world and the second one that considers the results of digitalization in Russia, and its impact on relations between commercial banks and businesses.

2.1 Global Trends in Digital Transformation: Banking and Business Sector Interactions

Currently, the world of banking and businesses has enormous transformational changes attributed to developments in the digital world. The trends of digitalization driving AI blockchain and big data analytics are significantly disrupting the industry's standardized and time-honored business models among others in banking. Empirical studies show that banks can enhance organizational performance and productivity objectives by implementing digital business strategies besides cutting on expenses and customizing on clients (Osei et al., 2023). In particular, the fourth wave involving AI and machine learning has helped the banking sector to improve risk assessment, automation, and improve or expand customer services making banks more sensitive and more dynamic (Rodrigues, et al., 2023). Another innovation of huge importance to the banking industry is the concept of block chain technology. It has been advocated for because it has the potential of transforming the current financial transaction methodologies to improved security, transparency and effectiveness (Khoso et al., 2022; Darazi et al., 2023). In the business perspective, the blockchain enables the enhanced security in transactions between the company and the bank as well as fighting with fraud in the digital exchanges (Shanti et al., 2023). One of the major benefits of blockchain involves the reduction of middlemen in financial dealings hence reducing costs and time spent. Therefore, banks are searching for the potential application of the blockchain in order to optimize the processes and consumers' satisfaction.

The use of digital platforms also has a positive impact in the interaction between the banks with the businesses through enhancing means and ways of delivering financial services. E-business channels are useful for banking products including loans, payment system and even financial planning tools for the business, especially SMEs through screen-based interfaces (Kryvovyazyuk, et al., 2009). Out of all these platforms, they have been very helpful in deepening the financial landscape since they eradicated the need for physical branches and enable business entities to perform so many activities without necessarily visiting a bank. It is therefore evident that while there are advantages in the processes of digital transformation or banking and business, there are also disadvantages too. Challenges that hamper the improvement of integration of digital technologies include digital divide, cybersecurity challenges, and regulatory issues are some of the challenges (Liu et al., 2023;

Khoso et al., 2024). Specifically, cybersecurity is still an issue that needs attention since more and more banks and businesses use digital technologies. The growing digitization of financial services creates new kinds of risks in terms of cyber threats that may reduce the level of trust in the new digital economy. Thus, the banks and other financial institutions need to develop and strengthen methods of protecting customer and company information in order to expand production and availability of such financial services.

2.2 Digitalization in Russia: Commercial Banks and Business Sector Interactions

Thus, in Russia, digitalization is gradually becoming one of the main strategic priorities for the further evolution of economic activity with emphasis on banking and businesses. Digitalization is among the Russian government's strategic goals coupled with the processes of modernizing the economy with an aim of improving competitiveness at the international level (Timkina et al., 2023; Liu et al., 2023). This emphasis on digitalization is seen in policies like the Federal Program 'Digital Economy of the Russian Federation' which seeks to facilitate the use of digital solutions everyday products and services including the banking and finance industries (Bershidsky, 2018; Darazi et al., 2023; Khoso et al., 2022). Russian commercial banks are gradually inclining towards digitalization as they understand that it can make more positive changes to client-end scenario and cut costs and time. Galazova (2023), pointed out that Russian banks had advanced in the use of technological gadgets like mobile banking and online payment systems and use of artificial intelligence in customer relations. Leading Russian banks such as Sberbank has not been left in this kind of situation by having the most advanced and modern investment in artificial intelligence as well as blockchain technologies for improving its digitized services (Semenyuta & Shapiro, 2023). Such investments have contributed to increased efficiency and effectiveness in customers' services provision while at the same time, decreasing bank's operating costs.

Still numerous obstacles can be mentioned and they all denote the fact that Russian banks are still far away from total integration of digital technologies into their business (Abdullaev & Bekmurodova, 2023). That is one of the reasons that we may speak about the regulation, which often does not keep up with the developments in modern technologies. Russian authorities have been rather restrained in their activity regarding digital finance especially with regards to cryptocurrencies and blockchain (Yangibaevich, 2023). Although the Central Bank of Russia is interested in introducing the application of blockchain technologies they have also noted certain risks potentially inherent in digital currencies with regard to financial stability and consumer protection (Melnyk, 2024). Such regulatory uncertainties thus present difficulties for the banks as well as businesses who wish to harness the full potential offered by digital technologies. Digitalization has also affected the business sphere in Russia for instance, the interaction between the companies and the banks. SMEs are also elongating its arm to get banking services through website since the banking institution has become a center of attraction in the modern business world. This increase in efficiency has proven to lead to cut on expenses as well as easy access to financing since most of them could not afford to access financial services due to geographical or financial barriers (Fayziev, 2024). For instance, players in the finch industry in Russia including firms that are offering business focused services such as electronic wallet and payment systems and peer to peer lending services and crowdfunding services to businesses (Shinkevich et al., 2023).

But like the other sectors especially in banking, there are some barriers that prevent business from fully adopting to digitization. The first is the issue of access to Information Technology infrastructure, this is evident in small businesses and those located in rural areas. Melnyk (2024) reported that challenges of digitalization in Russian SMEs remained high because organization have low levels of digital competence, inadequate funds for technology purchases and inadequate access to digital technologies. This has led to the emergence of what can be termed as the 'Digital Divide' hence affecting the advancement and adoption of digital banking services, thereby inhibiting the three cardinal objectives of financial inclusion. Other issue is related to security threat linked with the digitalization process. As more of the Russian businesses rely on online services for their banking operations and other financial activities they exposed to the risks of cybercrimes (Galazova, 2023). Implementing and maintaining strong cybersecurity is important to instill confidence and faith in people towards the digital financial services which will lead to an increased number of firms using the digital banking services. Businesses, banks, and regulators, thus, have to synergistically enhance cybersecurity from the effects of digital transformation.

3. METHODOLOGY

This study employed a straightforward approach to explore the interaction between commercial banks and the business sector in the context of digitalization in the Russian economy. Both secondary and primary data were collected to provide a comprehensive understanding of the key trends, challenges, and benefits of digitalization from the perspectives of entrepreneurs and bankers. The analysis was conducted using simple statistical techniques to ensure clarity and accuracy.

3.1 Data Collection

Data for this study were gathered from two main sources. First, secondary data were sourced from the Bank of Russia's official website, which provides extensive statistical information on the digitalization of the banking sector. This data spans from 2008 to 2020, offering insights into the long-term trends in digital banking services. Additionally, data were collected from the National Research University Higher School of Economics (HSE) through their "Digital Economy Indicators in the Russian Federation" report. This report tracks the digitalization of various sectors of the Russian economy, with a particular focus on the interaction between banks and businesses. The secondary data provided a foundation for understanding how digital tools and services have evolved in these sectors. Along with secondary data, primary data were collected through an online survey targeting two key groups: entrepreneurs and bankers. The survey was distributed to 150 entrepreneurs and 100 bankers between January and March 2023. This survey aimed to gather first-hand information about their experiences with digitalization, as well as their perceptions of how digital transformation has impacted their operations. The combination of secondary and primary data allowed for a comprehensive view of the ongoing digitalization process in the Russian banking and business sectors.

3.2 Survey Design

The survey used in this study was structured into three key sections. The first section focused on collecting demographic information from the respondents. This included basic details such as their age, the size of their business or bank, and their years of experience. This demographic data provided a context for the responses and allowed for an analysis of differences based on these characteristics. The second section of the survey asked respondents about their perceptions of digitalization. They were required to rate a series of statements regarding how digitalization had impacted their business or banking operations. Statements such as "Digitalization has reduced the cost of services" or "Digital tools have improved efficiency" were rated on a Likert scale from 1 (strongly disagree) to 5 (strongly agree). This section provided valuable insight into how digital transformation was viewed by both entrepreneurs and bankers in terms of benefits and challenges. In the final section, respondents were asked about the specific challenges they faced in adopting digital technologies and their expectations for future trends in digitalization. This section explored the respondents' thoughts on the future of digital banking and business interactions, including the role of emerging technologies such as artificial intelligence and blockchain. The survey responses helped highlight the main concerns and future directions for digitalization in Russia.

3.3 Analytical Methods

To analyze the data collected from both secondary sources and the survey, simple statistical methods were applied. Descriptive statistics were used to summarize the survey responses, providing an overview of the general attitudes of entrepreneurs and bankers towards digitalization. This included calculating the average (mean) responses for each question and determining the percentage of respondents who agreed or disagreed with the statements. A simple comparison was also conducted to explore any differences in the views of entrepreneurs and bankers. For example, the study compared the average scores given by both groups regarding the benefits of mobile banking or online payment platforms. This comparison helped identify any significant differences in how digitalization was experienced by businesses versus banks. In addition, a trend analysis was conducted using the secondary data from the Bank of Russia and HSE. This analysis tracked changes in digital banking services over time, particularly from 2008 to 2020. The analysis focused on understanding how the adoption of digital tools had progressed in both the banking and business sectors. To make the findings more accessible, graphical representations such as bar charts and line graphs were used to visually display the trends in digitalization and the survey responses.

3.4 Formulation of Key Trends

As a result of comparing the results of the analysis of the secondary data and the results of the survey, 12 trends concerning the digitalization of the identified interaction were revealed. These trends showed positive and negative aspects of digital transformation which this paper sought to define. As emerging trends, the study highlighted among others the enhanced numbers of the frequent online banking services, the gradual trend towards contraction of physical banking facilities, and enhanced exploitation of the mobile and digital payments platforms. However, the is also brought a few limitations that are restraining further moving towards digitalization including, cyber security threats, and legal constraints. These challenges make it crucial for the banks and the businesses to invest in the digital frontages and to actively coordinate with the legal bodies to make a safer and efficient digital economy.

3.5 Ethical Considerations

Issues of ethics were relevant to this particular study more so in the process of conducting the survey to gather primary data. It is also important to note that response to the survey was voluntary and all individual respondents were made to understand that their responses would not attract any form of identification. Participants' consent to participate in the study was sought before completing the survey, and no identification information was collected during the study. The survey data collected were employed for analysis in various researches, and all the results given were anonymized to ensure the respondents' confidentiality. Thus, following these ethical rules the rights and privacy of participants were honored throughout the course of the study. Their collection, basic calculations, and application of both secondary and primary data research provided a reliable insight into the digitalization processes concerning the interaction of banks and businesses in Russia.

4. RESULTS

The results of the analysis, drawn from both secondary data and survey responses, provide a clear picture of the current trends, benefits, and challenges of digitalization in the interaction between commercial banks and the business sector in Russia. This section is structured to present the key findings through tables and figures, offering both quantitative data and a visual representation of the results.

4.1 Key Trends in Digitalization

Based on the gather data from the Bank of Russia and HSE, this study identifies the following trends of digitalization of banking & business sectors across 2008–2020. One of the most obvious trends that can be identified is the sharp growth in the number of online and mobile banking services available to the client and its popularity. During this period, accessibility of the Internet service increased significantly; by 2020 more than 70% of Russian banks offered mobile banking services. Such a change has resulted in shutting down of many brick and mortar bank branches; as online services are now the primary means though which firms are able to obtain credit facilities. It has also introduced efficiency in the business practices through the reduction of transaction costs especially among the SMSe's through the adoption of the digital payment platforms.

Table 1. Key trends in digitalization (2008–2020)

Trend	2008	2012	2016	2020
% of banks offering online services	30%	45%	60%	75%
Mobile banking adoption by businesses	10%	25%	50%	70%
% of financial transactions via digital platforms	15%	35%	60%	80%

The data in Table 1 shows a clear upward trend in the availability and use of digital services in the banking sector. These services have become a crucial part of the interaction between banks and businesses, enabling easier access to financial tools and reducing reliance on traditional banking infrastructure.

4.2 Perceptions of Digitalization

The results of the survey can reveal the need and perception of digitalization as viewed by the entrepreneurs and bankers. Altogether, two groups show high readiness in digitalization emphasizing such values as effectiveness and cost reduction. This is true especially with entrepreneurs who mentioned that through the use of digital banking services financial management has become easier without necessarily having to transact physically within the bank. On the other hand, bankers focused on the issue of improving customer service through the process of digitalization particularly mobile banking.

Table 2. Survey responses – perceptions of digitalization (entrepreneurs vs. bankers)

Statement	Entrepreneurs (Mean Score)	Bankers (Mean Score)
Digitalization has reduced operational costs	4.2	3.8
Digital tools have improved efficiency	4.5	4.3
Mobile banking has enhanced customer experience	4	4.8
Cybersecurity is a major concern	4.6	4.7

Table 2 reiterated that the entrepreneurs and the bankers have agreed on the benefits of digitalization where the bankers appreciated more on the improvements on the customer service. Concerns, Mean, CV% About, about cybersecurity, cybersecurity, both groups, both groups high relatively relatively Similarly, both groups identified internal threats as another problem that has been well recognized by organizations. Figure 1. depicts the general development of the types of Internet banking services in Russia for the years 2008 – 2020. It emphasizes increased tendencies in the provision of both, online as well as mobile banking services among existed banks. This has been a trend noticing a sharp inclination towards the utilization of digital techniques in the banking system which underlines their significance in present – day financial activity.

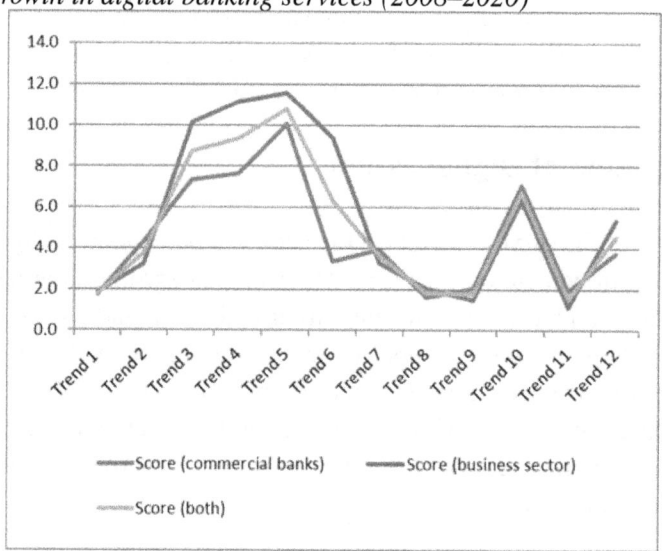

Figure 1. Growth in digital banking services (2008–2020)

4.3 Comparison Between Entrepreneurs and Bankers

Based on the survey, the findings show that different factors are considered by entrepreneurs and bankers and the way they view the impact of digitalization is unlike. Business owners were more concerned with the cost-savings of digital tools for the business with 75% of the respondents agreeing that digitalization had shaved down their operational costs. On the other hand, the bankers claimed that there has been an increase in the responsiveness on Mobile Banking with 80% of the respondents believing that the Mobile Banking has greatly improved the customer services' experience.

Table 3. Comparison of key perceptions (entrepreneurs vs. bankers)

Aspect	Entrepreneurs (%)	Bankers (%)
Cost-saving benefits	75%	60%
Improvement in customer service	70%	80%
Cybersecurity concerns	62%	70%

Table 3 shows that while both groups see benefits in digitalization, their focus differs, with entrepreneurs placing more emphasis on cost reduction and bankers on customer service improvements.

Figure 2 shows a relative ranking of what was valued most of least by entrepreneurs and bankers in matters concerning digitalization. This bar chart can exemplify the dissimilarities of their concerns with reference to the cost-saving opportunities, efficiency gains, customer service and impact of cyber-security issues. The two sets of perceptions are fascinating as they have revealed how differently the two important groups regard the aspects of digital tools.

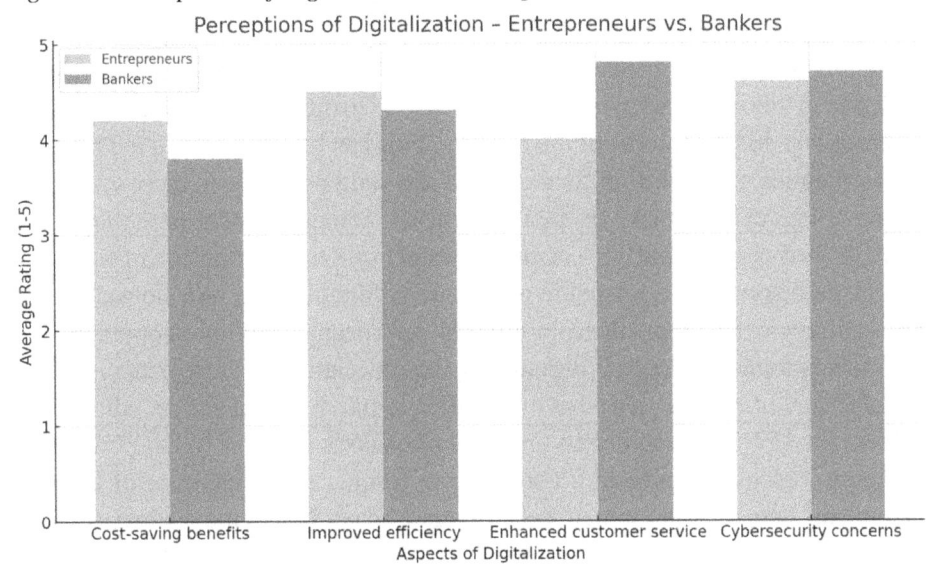

Figure 2. Perceptions of digitalization – entrepreneurs vs. bankers

4.4 Challenges in Digitalization

However, both the groups came up with the following challenges while advancing the benefits of digitalization. IT security was the only risk indicated by over two-thirds of all the respondents with 65% citing concerns particularly with regards to cyber security threats such as hacks and leaks. A challenge that was highlighted by the entrepreneurs in rural areas was the issue of reliable internet connection that prevents them from optimizing on the digital banking services. On the same note, lack of regulatory clarity especially regarding digital currencies and block chain was viewed by 45% of the bankers as posing further challenge in the adoption of these technologies.

Table 4. Challenges in digitalization

Challenge	Entrepreneurs (%)	Bankers (%)
Cybersecurity concerns	62%	70%
Regulatory uncertainties	45%	40%
Limited digital infrastructure	40%	30%
Digital literacy gaps	35%	25%

Table 4 summarizes down the problems in digitalization captured according to the view of the entrepreneurs and bankers. These two issues relate to problems that both entrepreneurs and bankers face in their operations; the most overwhelming challenge mentioned is cybersecurity, which was mentioned by 62 percent of the entrepreneurs and 70 percent of the bankers. This can be attributed to the high rate of uncertainties with a shift in financial services online and hence more vulnerable to threats like cyber-attacks or data breaches. Another problem is the legal risk where 40% of bankers and 45% of entrepreneurs said they are unsure of the future legal landscape on issues related to application of innovations like blockchain and cryptocurrencies. Further, there is a dearth of adequate digital support as 40% respondents found inadequate digital support especially the entrepreneur. Lastly, lack of digital literacy is viewed as a limitation to future digitalization, which is the problem for 35% of entrepreneurs and 25% of bankers.

Figure 3 also illustrates in a comparative manner the percentage of the total respondents who are entrepreneurs or bankers who opinion that cybersecurity is an issue that hampers digitalization efforts. This bar chart focuses on the awareness of these two groups with the security of digital environments, both realizing the growing threats of cyber threats and data leakage while implementing more digital tools.

Figure 3. Cybersecurity concerns in digitalization

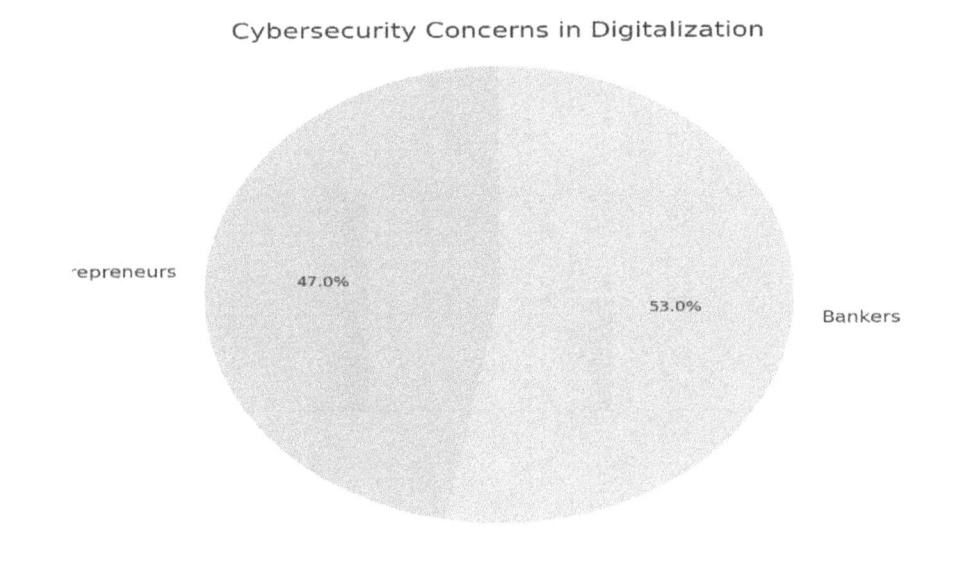

Cybersecurity Concerns in Digitalization

epreneurs 47.0%

53.0% Bankers

4.5 Adoption of Mobile and Digital Payment Platforms

The mobile and digital platforms were also identified in the survey and secondary data as areas that are beginning to experience increased use with regard to payments by businesses. This trend however is especially pronounced in SMEs that have adopted digital technology to enhance its operations and cut on costs. Figure 4. The second graph perfectly illustrates the trends of mobile banking and the digital payment platforms have been embraced by businesses in the recent past years. This bar chart shows that more people are using these tools within their business and financial activities as they become more automated in today's financial world. The figure shows how these technologies have assumed centrality in efforts to enhance efficiency while decreasing transaction costs in organizations of all sizes.

Figure 4. Increasing adoption of mobile banking and digital payment platforms by businesses (2008–2020)

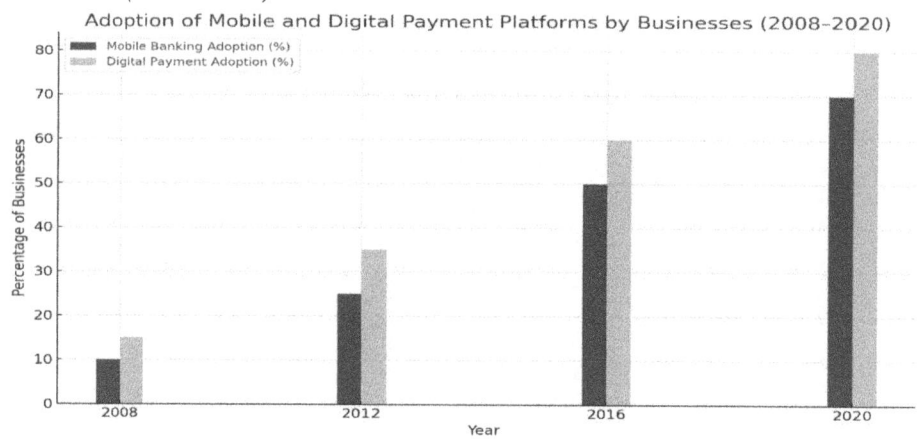

4. DISCUSSION

The results of this study give a clear sense of digitalization tendencies of the relations between commercial banks and the business in the Russian Federation. The findings further suggest a paradigm change in favor of digital intermediation with increased use of online banking, mobile banking and other digital payment methods in the past decade. This change mirrors the trend of digitalization of all businesses and financial services throughout the world where all the service providers are opting for new technology that will help them do their work efficiently and at reduced cost while offering improved service to customers. Conclusions based on this study can be made with the help of studying the changes that occurred in the banking system of Russia and in the business as a result of digitalization – mobile banking and digital platforms for making payments are common in the territory of Russia. Today, these technologies are considered as core instruments, especially for SMEs as these solutions provide evident cost efficiencies and work improvements. As suggested by the survey, it is especially the businesspeople who understand the benefits digital tools provide for access and speed. This conclusion can be made in view of the global tendencies that treat digitalization as a key driver to enhance the competitive advantage and the corresponding business outcomes.

In the case of the commercial banks, the integration of digitized services has enabled the banks to cut down on several operational costs and the need for the establishment of many branches; in addition, the provision of customized services to the clients. Mobile banking services available and the increased availability, as described by bankers as a factor that enhances customer satisfaction, also point to the shifts in the nature of customer relations of the firms. E–customers long for easy mobile solutions for their regular purchases, smaller financial transactions and other needs, therefore banks seek to improve possibilities and widen the number of modern devices for serving customers. This shift has been strongly evidenced by the increased percentage of the banks that have adopted the innovation in the mobile and online services over the last one decade. In light of these findings, it can be concluded that digital platforms are now part of the banks' corporate structure and will remain significant in defining the further development of the financial market.

However, the study also brings out some few difficulties that cause a gauge in the real attainment of the worth of digitalization in the organization. The second major issue highlighted by both the entrepreneurs and the bankers is the issue of security with special emphasis on cyber security. This rise in the use of different online avenues has put both companies and financial institutions in a vulnerable position to being attacked by hackers, have their databases infiltrated or be subjected to fraudulent activities. As the experimental data presented above demonstrate, the majority of respondents mentioned the cybersecurity problem as one of the most critical threats in relation to digital banking systems, which people are afraid of. This concern has support from earlier studies which recommend that organizations should enhance cybersecurity to ensure the safe custody of financial information and to establish more trust in E-Systems. The necessity of solid cybersecurity measures is rising with the ever-increasing use of new technology tools and platforms in both sectors.

Another issue which was touched upon in the study is the legal unpredictability of new generations of digital assets including blockchain and cryptocurrencies. Interviews held with both the bankers and entrepreneurs highlighted the issues of uncertainty with the rules and regulations surrounding these technologies as one of the reasons why they have not been implemented widely. For instance, according to the bankers, lack of legal regulation of digital currencies and blockchain technology is a major challenge limiting the financial sector's ability to incorporate these innovations into practice. This has been a recurring phenomenon in Russia where the formulation of regulations of digital finance have been slow to catch up with the rates of technology innovation. It is therefore important to meet these regulatory challenges if there is going to be room for innovation whilst at the same time preventing or mitigating on the risks associated with newer technologies.

5. CONCLUSION

The aim of this work is to present an overview of the trends in digitalization of the interaction between the Russian commercial banks and the business sector. The findings prove that the digital technologies, including mobile banking as well as the digital payment systems, have brought about significant improvements in reducing operational costs, increasing operation efficiencies and enhancing customer services to both the businesses and the financial institutions. The current shift towards the use of digital tools has been advantageous to the entrepreneurs and bankers by enhancing the provision of financial services and products particularly to the SMEs. The study also establishes that organizations in both sectors are progressively integrating the use of these e-business platforms into core functions. Nonetheless there are a number of barriers that need to be addressed though the gains realized through digitalization are apparent. Cybersecurity threats have evolved to a key consideration for any business and the banks in particular since digital technology presents the systems to these dangers. Furthermore, the regulatory risks about the new technologies including the blockchain and the cryptocurrencies have been other limitations to digitalization. Because of this ambiguity, efforts to incorporate such technologies into the offering of such financial institutions have remained subpar. Further, the digital divide and more so in the rural area hinders the effectiveness of businesses in leveraging on digital banking services because of poor internet connectivity.

To unlock digitalization in the Russian banking and business sectors it will be imperative to deal with these challenges. Enhancing cybersecurity, bringing out effective and friendly regulative policies for the digital money, and embracing digital facilities especially in excluded areas would be decisive for the inclusive and sustainable growth of the digital economy. The major implication that could be drawn is that while businesses will have to advance and become proactive to fully exploit the opportunities offered by new digital technologies in enhancing their services and function, so will the banks. This work adds to knowledge in the digitalization domain by offering information on the strengths and weaknesses of digitalization processes in the financial and business sectors of Russia.

REFERENCES

Abdullaev, A., & Bekmurodova, S. (2023). Transformation of the regulation of commercial banks in the conditions of the development of the digital economy of Uzbekistan. Scientific Collection. *InterConf*, (142), 93–102.

Bukhonova, S., & Yablonskaya, A. (2022). Study of the Level of Digitalization of the Banking Sector of Russia in the Context of the Pandemic and the Development of Transport Technologies. In *International School on Neural Networks, Initiated by IIASS and EMFCSC* (pp. 703–710). Springer International Publishing.

Darazi, M. A., Khoso, A. K., & Mahesar, K. A. (2023). INVESTIGATING THE EFFECTS OF ESL TEACHERS' FEEDBACK ON ESL UNDERGRADUATE STUDENTS' LEVEL OF MOTIVATION, ACADEMIC PERFORMANCE, AND SATISFACTION: MEDIATING ROLE OF STUDENTS' MOTIVATION. *Pakistan Journal of Educational Research*, 6(2).

Fayziev, S. (2024). Complaens Risks of Using It Technologies in Banks in the Period of Digitization. Asian Journal of Technology & Management Research (AJTMR) ISSN, 2249(0892).

Galazova, S. S. (2023). Digital banking ecosystems: Comparative analysis and competition regulation in Russia. [переводная версия]. *Journal of New Economy*, 24(4), 82–106. DOI: 10.29141/2658-5081-2023-24-4-5

Glebova, I., Berman, S., Khafizova, L., Biktimirova, A., & Alhasov, Z. (2023). Digital Divide of Regions: Possible Growth Points for Their Digital Maturity. *International Journal of Sustainable Development and Planning*, 18(5), 1457–1465. DOI: 10.18280/ijsdp.180516

Kanishcheva, N. A. (2021, February). Current state of commercial banks in a digital economy. In International Scientific and Practical Conference "Russia 2020-a new reality: economy and society"(ISPCR 2020) (pp. 169-172). Atlantis Press. DOI: 10.2991/aebmr.k.210222.033

Khalimon, E. A., Guseva, M. N., Kogotkova, I. Z., & Brikoshina, I. S. (2019). Digitalization of the Russian economy: first results. European Proceedings of Social and Behavioural Sciences.

Khoso, A. K., Darazi, M. A., Mahesar, K. A., Memon, M. A., & Nawaz, F. (2022). The impact of ESL teachers' emotional intelligence on ESL Students academic engagement, reading and writing proficiency: Mediating role of ESL students motivation. *Int. J. Early Childhood Spec. Educ*, 14, 3267–3280.

Khoso, A. K., Khurram, S., & Chachar, Z. A. (2024). Exploring the Effects of Embeddedness-Emanation Feminist Identity on Language Learning Anxiety: A Case Study of Female English as A Foreign Language (EFL) Learners in Higher Education Institutions of Karachi. *International Journal of Contemporary Issues in Social Sciences*, 3(1), 1277–1290.

Kolmykova, T. S., Sirotkina, N. V., Serebryakova, N. A., Sitnikova, E. V., & Tretyakova, I. N. (2022). Modern Tendencies of Digitalization of Banking Activities in the Russian Economy. In *Business 4.0 as a Subject of the Digital Economy* (pp. 469–474). Springer International Publishing. DOI: 10.1007/978-3-030-90324-4_75

Kostomakhin, M., Kostomakhin, N., & Tseiko, M. (2023). Impact of digitalization on the effectiveness of management in the field of agricultural development. In *E3S Web of Conferences* (Vol. 402, p. 13004). EDP Sciences. DOI: 10.1051/e3sconf/202340213004

Kryvovyazyuk, I., Britchenko, I., Smerichevskyi, S., Kovalska, L., Dorosh, V., & Kravchuk, P. (2023). Digital transformation and innovation in business: the impact of strategic alliances and their success factors.

Liu, H., Zhu, Q., Khoso, W. M., & Khoso, A. K. (2023). Spatial pattern and the development of green finance trends in China. *Renewable Energy*, 211, 370–378. DOI: 10.1016/j.renene.2023.05.014

Loucks, J., Macaulay, J., Noronha, A., & Wade, M. (2016). *Digital Vortex: How Today's Market Leaders Can Beat Disruptive Competitors at Their Own Game*. DBT Center Press.

Melnyk, V. (2024). Transforming the nature of trust between banks and young clients: From traditional to digital banking. *Qualitative Research in Financial Markets*, 16(4), 618–635. DOI: 10.1108/QRFM-08-2022-0129

Molchan, A. S., Osadchuk, L. M., Anichkina, O. A., Ponomarev, S. V., & Kuzmenko, N. I. (2023). The'Digitalisation trap'of Russian regions. International Journal of Technology. *Policy and Management*, 23(1), 20–41.

Mottaeva, A., Khussainova, Z., & Gordeyeva, Y. (2023). Impact of the digital economy on the development of economic systems. In *E3S Web of Conferences* (Vol. 381, p. 02011). EDP Sciences. DOI: 10.1051/e3sconf/202338102011

Osei, L. K., Cherkasova, Y., & Oware, K. M. (2023). Unlocking the full potential of digital transformation in banking: A bibliometric review and emerging trend. *Future Business Journal*, 9(1), 30. DOI: 10.1186/s43093-023-00207-2

Potapova, E. A., Iskoskov, M. O., & Mukhanova, N. V. (2022). The impact of digitalization on performance indicators of Russian commercial banks in 2021. *Journal of Risk and Financial Management*, 15(10), 452. DOI: 10.3390/jrfm15100452

Raza, M. H. (2024). The Effect of Remittances on Economic Expansion and Poverty Reduction: Evidence from Pakistan. *International Journal of Management Thinking*, 2(1), 75–92. DOI: 10.56868/ijmt.v2i1.47

Rodrigues, L. F., Oliveira, A., & Rodrigues, H. (2023). Technology management has a significant impact on digital transformation in the banking sector. *International Review of Economics & Finance*, 88, 1375–1388. DOI: 10.1016/j.iref.2023.07.040

Romanyuk, M., Sukharnikova, M., & Chekmareva, N. (2021, March). Trends of the digital economy development in Russia. []. IOP Publishing.]. *IOP Conference Series. Earth and Environmental Science*, 650(1), 012017. DOI: 10.1088/1755-1315/650/1/012017

Semenyuta, O. G., & Shapiro, I. E. (2023). Risks of Banking Transformation in Digitalization. In *Technological Trends in the AI Economy: International Review and Ways of Adaptation* (pp. 357–362). Springer Nature Singapore. DOI: 10.1007/978-981-19-7411-3_38

Shanti, R., Siregar, H., Zulbainarni, N., & Tony, . (2023). Role of digital transformation on digital business model banks. *Sustainability (Basel)*, 15(23), 16293. DOI: 10.3390/su152316293

Shinkevich, A. I., Kudryavtseva, S. S., & Samarina, V. P. (2023). Ecosystems as an Innovative Tool for the Development of the Financial Sector in the Digital Economy. *Journal of Risk and Financial Management*, 16(2), 72. DOI: 10.3390/jrfm16020072

Timkina, T. A., Savelyeva, N. K., & Kryukova, A. D. (2023). Innovative Activity of a Commercial Bank During the Period of Economic Transition. In *Anti-Crisis Approach to the Provision of the Environmental Sustainability of Economy* (pp. 367–373). Springer Nature Singapore. DOI: 10.1007/978-981-99-2198-0_40

Verbivska, L., Abramova, M., Gudz, M., Lyfar, V., & Khilukha, O. (2023). Digitalization of the Ukrainian economy during a state of war is a necessity of the time. *Amazonia Investiga*, 12(68), 184–194. DOI: 10.34069/AI/2023.68.08.17

Yangibaevich, A. A. (2023). Transformation of the regulation of commercial banks in the conditions of the development of the digital economy of Uzbekistan. *SAARJ Journal on Banking & Insurance Research*, 12(2), 11–18. DOI: 10.5958/2319-1422.2023.00005.X

Yurak, V. V., Polyanskaya, I. G., & Malyshev, A. N. (2023). The assessment of the level of digitalization and digital transformation of oil and gas industry of the Russian Federation. Gornye nauki i tekhnologii= Mining Science and Technology (Russia), 8(1), 87-110.

Yuryeva, O., Pudeyan, L., Medvedskaya, T., Zaporozceva, E., & Zemlyakova, N. (2020). The impact of the digital revolution on the Russian financial sector development and the results of economic transformation. In *E3S Web of Conferences* (Vol. 210, p. 02006). EDP Sciences. DOI: 10.1051/e3sconf/202021002006

Chapter 7
Factor Analysis of the Economic Security of Regions:
The Level of Economic Security of Regions

Yulia V. Granitsa
National Research Lobachevsky State University of Nizhny Novgorod, Russia

ABSTRACT

The level of economic security is assessed on the basis of resource, labor, financial, innovation, investment, budgetary and regional security, which together represent the socio-economic security or socio-economic potential of the regions. In the study, the regions of Russia are ranked by the level of economic security on the basis of the selected group of indicators for assessing the socio-economic potential of the regions. The tests carried out by Mann-Whitney, Fligner-Keelin, and split testing allowed us to conclude that the selected regional groups in terms of economic security are heterogeneous, that is, the distribution of indicators in groups and their mean values differ significantly. An adequate logit model has been formed, which makes it possible to predict the class of economic security of a region. The receiver operating characteristic has been constructed, and the quality of the model has been assessed by calculating the area under the curve. The key indicators of economic security are determined by the method of principal components.

DOI: 10.4018/979-8-3693-3423-2.ch007

1. INTRODUCTION

The modern model of Russia's economic growth demonstrates a directly high dependence on the external conditions for the most important sectors, including export-import operations. Such dependence worsens the economic condition of the country and makes it hard to manage or estimate the change due to variation in the external demand (Khan et al., 2023). At the same time, it results in discrepancies of the geographical distribution of economic performance, thus contributing to performance disparities across regions (Payne et al., 2023). In this context socio-economic potential has the key role as analyses the interaction of public policies with the process of creation and application of economic effects. Overall, such a broad assessment of the potential helps to get a more adequate picture of how regions operate as Verb economic subjects. The investigation of potential of the regional economy differs greatly according to the characteristics and the level of disaggregation of the assessment (Su et al., 2023). That is why some works are concentrated on studying certain facets of regional activity and may miss other factors that can also be critical for evaluation, but are not associated with this subject in one way or another (Fei et al., 2023). Many authors have pointed out that this selective focus may lead to the wrong perception of a region's capacity to act independently self-sustaining, particularly when the factors, including investment capacity, social stability and innovation capability, are excluded.

Funds provided to state programs by the government which are designed to reduce interregional disparities are important sources that enhance regional capacity (Chu et al., 2023). However, it is obvious that with the onset of the COVID-19 economic crisis, a certain number of restrictions have been introduced to the utilization of federal budget funds.

Therefore, regional economies have emerged as our more essential components of the social formation as autonomous categories for production, reproduction and capital formation. This research emphasizes that policy makers must gain knowledge on factors that support socio-economic security of regions in order to achieve economic stability and to avert any-sector shocks in the country (Li et al., 2023). In the previous research, authors used the transition possibilities of regional economies to maintain capital reproduction and tied it to the investment demands that can be seen as the means of regional economy's resistance in facing external challenges (LA et al., 2024; Liu et al., 2023). Sustainability of regions' socio-economic performance when and if faced with varied external shocks is crucial in ensuring regions sustain their socio-economic status, improve the standards of living of the community, and ultimately upgrade the human conveniences of the inhabitants. Altogether these factors guarantee the economic safety of the region (Khoso et al., 2022; Darazi et al., 2023).

Investment climate is even more an important element of regional economic security because it defines the financial situation of a region and its ability to attract investors (Jia et al., 2023). Investment potential to the financial capacity of the region which can be measured such as the indicators of enterprise profitability own-income sources, and the degree of inter-budgetary transfers. These factors afford deeper insights in the dynamics of regional stability and economic security although it is possible to gauge security of resources, manpower, financial, innovation and budgetary securities in the region (Cai et al., 2023). All these aspects are incorporated in this chapter, under socio-economic security which encompasses the capability of a regional system to resist external shocks as well as sustaining economic stability. The study is based on the literature review in order to test the research hypotheses and to develop a factor analysis of the concept of economic security that investigates other aspects of socio-economic security in regions; the findings of the study contribute to the development of new ideas about the evaluation of regions and supportive measures for sustainability.

1.1 Theoretical Foundations of Regional Economic Security

Scholars have devoted much attention towards the analysis of economic security at the regional level from different angles. Economic insecurity can be defined across several dimensions for many researchers including; Financial security, job security, and availability of resources among others. Thus, socio-economic security is defined Cao et al. (2019) as the Region's financial self-sufficiency both through domestic production potential and through external direct investment. In this respect economic security is a concept that goes beyond the notion of stability and aims at the position that allows coping with, and adapting to, external shocks to enhance economic growth. Some of the other scholars also stress on the need for the investment in Regional economic security (Liu et al., 2023).

According to Song et al. (2017), investment climate stands out as one of the most significant factors that define a region's socio-economic security since investment influences the economic capability to innovate, expand and create employment opportunities. The levels of investment to the stability of regions; they argue that the regions with good investment environment have stronger economic base and can easily cope with adverse economic conditions. In this regard, one can refer to the level of investment as the key characteristic of economic security in a given region. Furthermore, Kryshtanovych et al. (2023) indicate that economic security means not only the financial but social and political component as well. They find out that when these areas have high social capital which includes people's trust in institutions and their civic participation, then these areas are likely to rebound quickly from any downturns in the economy. With the help of this broader perspective of economic security as a socio-economic phenomenon, it is possible to identify factors that would affect stability and sustainability of regions.

1.2 Factors Influencing the Economic Security of Regions

There are different factors which determine the level of economic security of regions, and depending on specifics of some or other factors it can be higher or lower. There is the sub-criterion of financial capacity including the regional own-source revenues and sensitivity to the inter-budgetary transfers. According to Yu et al. (2023), it is necessary to note that the level of financial decentralization allows regions to adapt to economic risks and fluctuations. Autonomy in financial decision-making enable regional authorities to better managed financial resources and provide an appropriate economic stimulus to internal and external milieu.

The other aspect is the labor market that requires qualification in order to qualify for any job. Popelo et al., (2023) note that the factor of human capital, which includes flexibility and skill, is also pivotal to establish the economic security, because the flexible and skilled regions make the necessary changes in global and national economy. Human capital together with innovation and the use of technology is hence, an influential determinant of the economic development of regions. In the same way, Zamira (2024) stress that innovation potential can be seen as the vital need for regional development because those areas that provide funds for R&D shall demonstrate constant and steady economic growth. However, the quality of those forms of regional governance structures is another that defines economic security. The integrated policies characterized by clear and efficient governance are highly resistant and can create policies for sustainable development and managing and avoiding economic risks. However, places with bad governance and high levels of corruption face a lot of difficulties in maintaining economic stability, because main development indicators suffer from bad decisions and misallocations.

2. LITERATURE REVIEW

The idea of the economic security has attracted much attention in the recent past and especially in the facet of the region al development (Feng et al., 2023). That is 'as globalization, technological advancement and interdependent economy advance, regions are being viewed as key determinants of economic diversification and stability'. Several studies have attempted to define and measure economic security with focus on the characteristics, determinants or causes of economic insecurity in regions. This literature review aims at discussing the literature available in the current state regarding economic security of regions with special regard to socioeconomic characteristics and investment climate that determine region's stability. The review is structured into two sections: Firstly, socio-economic aspects of regional economic security; secondly, condition and role of investment climate and governance for the economic security.

2.1 Socio-Economic Dimensions of Regional Economic Security

The concept of economic security in the regional context is complex and has many layers, which include financial, employment, resource, and social perspectives (Pakhucha et al., 2023). The role that socio-economic factors play in determining regional economic security has also received attention in literature. Stability of funds is one of the most paramount components of the regional economic security and is defined as the extent at which a particular region generates adequate revenue to finance its expenditure (Sun et al., 2023). This is normally done either through local taxation, Intergovernmental transfers and or through revenue from economic activities (Liu et al., 2023). Economically resilient regions are likely to sustain any financial shock and warrant continuity of service delivery for the populace. Another significant factor is employment as it is one of the core sectors that influence the socio-economic equilibrium. Postulate that volatile regions create economic security because the labor force is adaptable to global economic volatility. In addition, human capital is not only an essential factor to stimulate economic growth but also able to minimize the potential of a social crisis or disturbance. Likewise, low unemployment accompanied by high workforce engagement rates tends to enhance regions' economic security (Ramskyi et al., 2023).

Therefore, social cohesiveness can also be considered as a parameter of regional economic security. According to Darazi et al. (2023) economic resilience is supported by the social capital. In their work they have established the fact that regions that are most trusted, covenanted, and cooperative are most likely to cope with economic shocks. Social integration encourages peoples' coherent obligation hence encouraging fruitful use of the resources and harmonization of economic policies in the whole economic fraternity. Also, it supports social stability that in result ensures the favorable climate for the counterparts and investments, as well as ensures the further economic development in the long term (Mishchuk, 2023). Availability of resources which include natural resources and capital is another important factor that defines regional economic security. Geographical territories that are endowed with natural resources including energy or minerals enjoy an added economic boon in the sense that their natural resources turn into cash generating machines for economic growth (Karanina & Karaulov, 2023). However, the use of natural resources has a flip side and this is more especially when it comes to the problem of fluctuating price and environmental problems. Such regions may suffer variations in their economic growth at certain periods in the volatile Global markets or when strict rules and regulations on the use of natural resources are implemented.

On the financial side, regions that have accesses to the credit and financial market are able to invest more on infrastructures, educations and health care, which are the building blocks for long term prosperity. Koval (2023) mentioned some views points, financial autonomy becomes crucial for creating stability within regions since the regions are able respond to shocks. Such regions may be vulnerable to fluctuation in economic security since they are highly dependent on inter-governmental transfers or external financier. Concluding, it is possible to state that socio-economic factors of regional gross are multiple and are interconnected. Economic vulnerability of the regions depends on aspects such as; financial stability, employment, social cohesiveness, and availability of resource (Mulska et al., 2023). In line with literature, research has found that countries that invest in these sectors are likely to have long term economic stability and cope with external volatile economic environment.

2.2 The Role of Investment Climate and Governance in Regional Economic Security

The investment climate as well as governance are considered to be the key determinants of economic security of the regions (Nesadurai, 2004). A good investment climate shapes capital, encourages the emergence of innovations and drives economic development, whereas governance properly distributes money and ensures correct execution of public policies (Nikitina et al., 2018). The connection between investment climate, governance and regional economic security has been examined, whereas researchers emphasized on the given factors for development. An investment climate of a particular region comprises of infrastructure, regulation regime, and political risk. The infrastructure as transportation, energy and communications are the factors that define the areas where investments will be directed and where the development of economy could be expected (Danylyshyn et al., 2019). Infrastructure not only sustains the economic activities but also creates better living standards of people and thus make the region more desirable for investment and other economic ventures. In the same context, there is need to address key legal risks such as regulatory legal risks by having stable political environment that supports and protects the rights of investors over their properties, the ability to enforce contractual terms that have been agreed to between the business and its counterparties and the absence of bureaucracies that hampers the growth of business (Poltorak et al., 2023). This is because areas that have bright and well-coordinated laws are the ones that will attract both; local and international investors.

Political stability is another of the most important factors of a good investment climate. Countries with stable institutions of governance that have low corruption levels and efficient policy institutions are likely to attain the economic security as they can institute policies that will foster growth of their economy in the long run. Reflecting on the same, from the perspective of supporting regional development, Belaïd et al. (2023) states that regions that boast of good governance practices of the use of funds have higher economic security in their regions. This means that the resources are well utilized, the services to the public are well provided and the legal framework is well observed. This paper rationally discusses the role of governance in the promotion of regional economic security with special reference to the Inter-Governmental Relations. According to the study by Eldor & Mamlakat, (2024), those regions that have a higher level of fiscal decentralization and those that have the authority to make decision regarding the financial resources generally have better chances of attaining the prospect of economic security. Fiscal decentralization enables administration to target policies at regional requirements, finance development of bore, expand human capital and manage externalities affecting the flow of funds. On the other hand, some regions may rely on inter-governmental transfers or central government decisions, and they may fail to achieve stability in their economy, especially in this current era of fiscal consolidation.

However, it is quality of governance that sheds further light on how the economic security is being advanced apart from fiscal decentralization. Hence it is concluded that governance institutions in a particular region have potential to influence its capacity to manage resources, execute policies and advance the process of development. Elnaiem et al. (2023) opined that where there is political will and accountability it results to higher investment, economic growth and social order. Governance helps to strengthen confidence in the state administration, minimize corruption and guarantee that both public policies and related implementation will lead to sustained development (Doğan et al., 2023). The social fabric which a region presents in form of its investment climate and governance structures can be seen to play significant roles in defining its economic security (Frick & Rodríguez-Pose, 2023; Aneslagon et al., 2024). Thus, the areas with good investment climate status and good institutions of governance are well positioned to mobilize capital, support growth, and sustain growth in the long run. On the other hand, the areas encountering problems with weak governance, corruption, and a bad climate for investment may not sustain the economic security since it will be hard for them to attract investors and hence more susceptible to fluctuations.

3. MATERIALS AND METHODS

This section describes the multi-stage methodology employed to assess and classify the economic security of Russia's regions. The approach involves clustering regions based on selected key indicators and then ranking them to determine their overall level of economic security.

3.1 Indicator Groups and Clustering

The first step of the analysis was conducted with the goal of clustering Russia's regions with the help of three sets of indicators that outlined three integrative groups of factors that were considered to be the key dimensions of regional economic security. The first group of indicators was related to investment security that included factors such as volume and stability of investment, investment friendly region and domestic as well as international investment. Sustainability of investments is important in determining the capacity of a given area to attract and sustain investment which is a key component in formation of a strong economy in the long run. Thus expecting that regions that attract a lot of investment will have high levels of economic security, since they have the means to support growth, generation of new ideas and putting up of structures. The second category included those that addressed resource readiness, budget solvency, innovativeness, social order and financial strength. These indicators paint an overall picture on the capacity of the region to sustain a socio-economic equilibrium and therefore, development. The socio-economic stability of each region was also captured by including a wider number of metrics by this group such as the capacity to support innovation, balance public finances and social cohesiveness. All of these factors are crucial for the sustainability of the region since areas that can forecast the use of resources and plan the appropriate financing are more capable of regulating the impact of economic crises.

The third group of indicators related to the assessment of regional economic performance which was defined as the balanced financial result divided by the gross regional product (GRP). This ratio offers information on financial solvency and efficiency of activity of the region and presents a quantitative characteristic on effectiveness of economic activity and its results in financial terms. High performers in this measure ought to have a better economic health than the less performing regions, and thus enhanced economic resilience. For each of these groups, regions were divided into three categories on the basis of their effective performances. Evidently, this clustering technique enabled the assessment of the regions' level of economic security. Security index equations: In assigning the best, average and worst regions the higher, average and lower scores respectively across the selected metrics the following high performing regions were sorted into the top cluster while regions with moderate and lower performances entered the related intermediate and lower clusters. As such, this approach offered some direction on how to evaluate and rank regions in terms of security in its multiple dimensions.

3.2 Scoring and Classification

After the regions were grouped according to the three groups of indicators, scoring was employed in order to determine their overall economic security score. The performance of each region on the three clusters was then measured on the basis of a point scheme. The best performing regions in each cluster, which had the corresponding highest security level, were assigned with the lowest value. However, moderate and low security regions were allocated higher scores with a view of showing that these areas are insecure as compared to the more secure regions. This scoring system enabled a summed up assessment of economic security of each region with performance on the indicators used. After converting the scores of each indicator in the three groups indicated above, the scores were added together to come up with an economic security score for all the regions. Based on the total score, regions were classified into two categories: enduring and fragile or secure and insecure can be used also as two different adjectives for the same noun. Those with scores of three or lower were deemed secure meaning that the area was well secured in all of the security facets examined; financial, investment and socio-economic. These regions were assumed to be relatively immune to the shocks that deform the external environment and internally challenge us.

On the other hand, the total score of more than three pointed the region as insecure. These regions were seen to have economic insecurity in one or more domains, which implied that these locations could prove unstable, and half-baked in terms of economic growth. This classification is thus helpful to evaluate where certain areas may benefit specific and special measure to improve economic resilience and functioning for the long-run. Thus, the use of this structured approach to segmentation, scoring, and classification allows presenting a comprehensive analysis of the regional economic security situation in Russia. In this approach, it becomes possible to define regional distinctions in security levels to establish the basis for further investigations and policies in the sphere of perceived disparities. The above method also enables a clear alignment with the regions so that the stakeholders can easily know the strong and the weak areas in case of economic shrinkage or insecurity.

4. RESULTS

The results of this study provide a detailed analysis of the economic security of Russia's regions based on the three primary groups of indicators: foreign investment guarantee, socio-economic stability and regional economic. output. Most of the clustering and scoring techniques applied in the analysis of the regions based on these indicators have led to effective categorization of regions in terms of security and insecurity. The purpose of this section is to discuss the results of the research and to reveal the disparities in the degree of economic security and to define regions with high and low levels of economic security. The results are presented in several sections. First, the performance of regions in each of the indicator groups is presented – how did regions perform concerning investment, socio-economic, and financial stabilities? After this, the overall scores are compared and then summated and this gives information which regions are most secure based on the scores given to the regions. Such characteristics indicate the level of economic security in the regions of Russia, allowing to determine which zone is ready to develop sustainably, what zone may need targeted stimuli and measures.

Table 1. Regional security factors

The aspect of regional security	Panel indicators
Labor security	Educational attainment (Ed), Human Development Index (HDI)
Resource security	Consumption per capita (Pt), share of regional or intermunicipal public roads that meet regulatory requirements (NR), share of goods shipped, works, manufacturing services in GRP (MI), ratio of shipped goods from extractive industries to GRP (EI), industrial production index (IPP), the ratio of investment in fixed assets to gross regional product (INV)
Social security	Share of urban population in total population (Urb), dependency ratio (DN), employment rate (UZ), Gini coefficient (Gini)
Innovative security	The share of innovative products in the total volume of goods shipped (In) and the share of products of high-tech and knowledge-intensive industries in the gross regional product (HT)
Financial security	Balanced financial result to GRP ratio (SFR), consumer price index (IPC), debt on loans per capita (Zd), share of small and medium-sized businesses in gross regional product (MSP)
Budgetary security	Regional budget expenditures per capita (Bd)
Investment security	Investment potential (P), investment risk (Risk)

The panel indicators that we have chosen – characteristics of the socioeconomic security of the region, on the basis of which we will draw a conclusion about the level of regional economic security, are presented in Table 1.

Table 2. Classification of regions based on investment security indicators

Region	Investment Security Rank	Cluster (0-2 Points)
Moscow	High	0
St. Petersburg	High	0
Kaluga	Moderate	1
Khabarovsk	Low	2
Tyumen	High	0
Magadan	Low	2

4.1 Investment Security

When evaluating investment security indicators one can conclude that the amount of foreign and domestic investments reflects the higher levels of economic security. Sustainable foreign investment inflows have been ascertained in these regions hence boosting long-term economic growth and development. For example, as shown in table 4 the highest cluster included the regions of Moscow and St. Petersburg as they recorded high performance in the attraction of both domestic and foreign investment. These regions have remained very robust in the sense that they always ensure that they have stable capital coming in and they provide sound operating environment for business. Conversely, higher levels of investment risk conditions were discovered in such regions as where investment flows were erratic or insignificant. For instance, some of the geographically isolated or remote or resource scarce zones could not sustain a steady flow of investment and have been classified in the lowest security recourse. This underinvestment limits their capacity to expand and, in particular, to buoy themselves up in the face of a downturn, identifying a sector where policymaker-led efforts to alter regional investment environments could be of use.

Figure 1. shows the Investment Security Index of selected regions over time, from 2010 to 2020. Each line represents a different region, highlighting the variation in their ability to attract and maintain investments. The graph helps visualize how regions like Moscow and St. Petersburg have consistently performed well, while others such as Khabarovsk and Magadan show more modest growth.

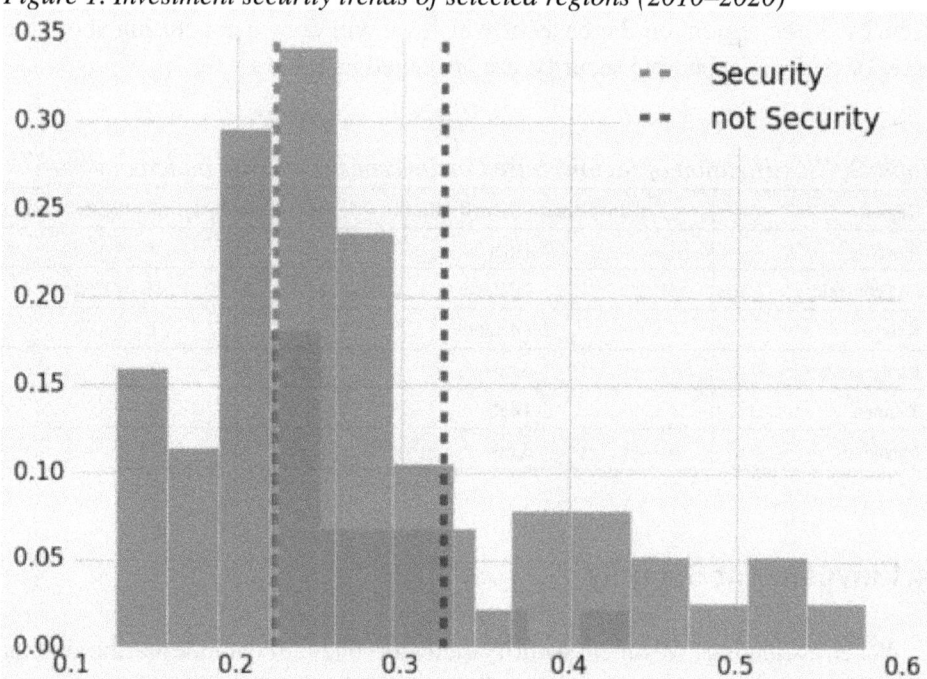

Figure 1. Investment security trends of selected regions (2010–2020)

4.2 Socio-Economic Stability

It is also worthy of note, that the second group of indicators, or socio-economic stability, presented noticeable differences in the rankings of the regions. Some of the ways that countries were sorted into the secure category included availability of infrastructure, stability in the labor market and good financial system. In these regions, they also proved their efficiency in managing public finances, low levels of social unrest, evidenced by low unemployment rates and high levels of public services. It is deemed necessary to note that some regions like Tatarstan or Kaluga occupy the place in the top cluster in terms of the balanced socio-economic environment. These regions have clearly shown that they have been performing well in employment generation, innovations and delivery of public services hence enhancing their socio-economic security. On the other hand, the ine3rf regions which was characterized by high une3r employment rates, social unrest or deficits in budget, was regarded as insecure. Some of the regions include several regions in the Far East and Siberia revealed critical issues, which hinder their socio-economic growth prospects such as resource constraints and shrinking population.

Table 2. Socio-economic stability indicators by region

Region	Unemployment Rate (%)	Budget Deficit/Surplus (%)	Social Stability (1-5 Scale)	Cluster (0-2 Points)
Moscow	2.1	5.2	4.8	0
Tatarstan	3	3	4.5	0
Kaluga	4.5	2.5	3.9	1
Khabarovsk	7.8	-4	2.8	2
Magadan	9.5	-5.5	2.5	2

Table 2 shows the SSI of selected regions based on unemployed rate, the budget deficit or surplus and social stability ranging from 1 to 5. There are three clusters or groupings of the regions depending on performance in terms of socio-economic returns. The districts such as Moscow and Tatarstan with the lowest indicators of unemployment and with budget which is more than excesses indicating high social stability belong to the highest cluster of security (0 points). Regions such as Khabarovsk and Magadan have poor socio-economic characteristics and even now have high unemployment and lack of financial resources with budget deficits and therefore have low scores in Social Stability and are in the 2nd security point. These reveal the level of socio-economic cyclic buffers where some regions evidenced high stability and others fragile and substantial vulnerability.

4.3 Regional Economic Performance

The performance indicators computed by dividing the balanced financial result to the gross regional product or GRP gave additional ideas about the financial adequacy and the economic efficiency of the regions in question. Some samples with a high GRP ratio were included in the highest security cluster, including the Moscow region and Tyumen. These regions demonstrated very high financial performance in comparison to their economic size indicating their capacity in generating and maintaining economic growth. On the other hand, companies with low GRP ratios had problems with getting the proper financing, the imperfect number reflecting underlying economic problems. These regions, relying heavily on the income derived from a rather limited set of industries or experiencing fiscal problems, were classified as the regions of lower economic performance and, therefore, were grouped into the less secured clusters.

Figure 1. Provides the world's economic performance of the selected regions, that was calculated by equation: balanced financial result gross regional product. Those ratios are higher for such regions as Moscow and St. Petersburg, which indicate higher financial efficiency to share of gross regional product, meaning that these regions possess increased financial performance in relation to their economic indicators. On the other hand, the figure is comparatively lower in the case of places such as Khabarovsk and Magadan – evidences of several areas struggling to maintain economic equilibrium in terms of financial profitability and productivity.

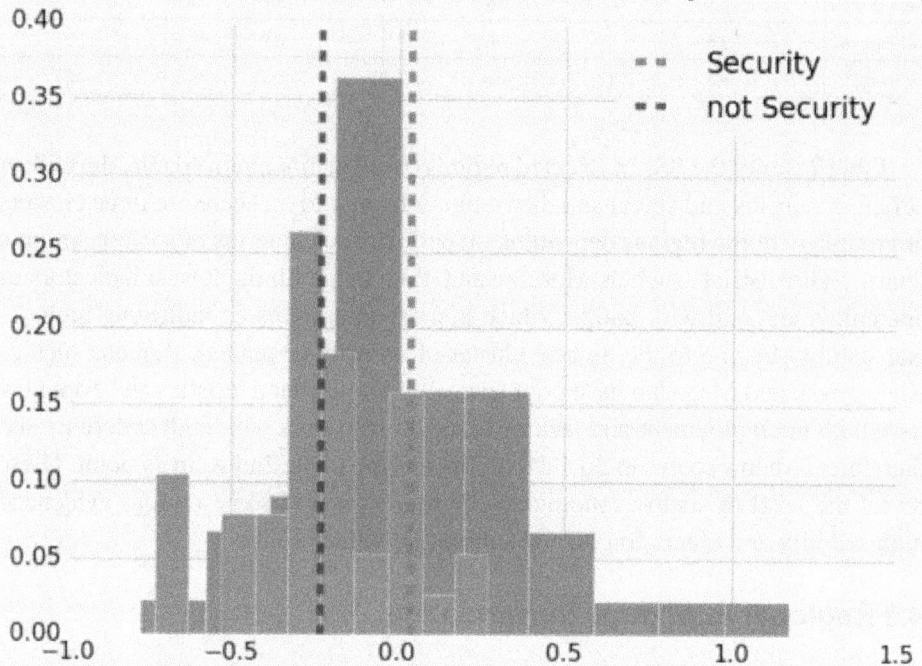

Figure 2. Economic performance of regions (GRP to balanced financial result ratio)

4.4 Overall Economic Security Classification

In general, the final classification of the assessing indicator groups highlighted significant differences in the level of economic security of Russia's regions when all three indicator groups were taken into account. Any regions with a total score of 3 points or below were considered secure from poor investment, socio-economic, and financial outlooks. These regions, which include Moscow, St. Petersburg, and Tatarstan, were able to recover and be prepared for a long-term and sustainable development. On the other hand, any area that had a total score of more than 3 points was categorized as insecure since it had challenges in one or more of the identified security pillars. Most of these areas especially in the Far East and certain areas of Siberia continue to grapple with problems relative to investment, financial issues, and socio-economic sustainability. These regions are likely to need specific economic policies suitable for their current state and needs with the aim of improving safety in the economy.

Table 3. Overall economic security scores and classification

Region	Investment Security Score	Socio-Economic Stability Score	Economic Performance Score	Total Score	Classification
Moscow	0	0	0	0	Secure
St. Petersburg	0	1	0	1	Secure
Kaluga	1	1	1	3	Secure
Khabarovsk	2	2	2	6	Insecure
Magadan	2	2	2	6	Insecure

Table 3 presents the composite Index of Economic Security and their respective classifications for investment security, socio-economic stability and economic performance of selected regions. Such areas as Moscow and St. Petersburg that have received very low scores in all the aspects are considered secure because they demonstrate good economic performance and stability. Same as the previous regions, intermediate scores in all the categories place Kaluga oblast in the secure, although there are some signs of insecurity detected. For example, how could the regions, such as Khabarovsk and Magadan with high scores on all the indices be categorized as insecure as they depict serious problems on investment, socio-economic and even economy performance. Such classification reflects the fact that the level of economic risk protection differs significantly in Russian regions.

5. DISCUSSION

The data obtained as the result of examination of the economic security of Russian regions reveal specific patterns of inequality and weakness in the country. The division of geographical areas into safe and unsafe zones is one of the major differences when it comes to investment security, socioeconomic development, and economic efficiency. This section considers the policy relevance of these findings, comparing the strengths and weaknesses of various regions and identifying potential targets for policy change. Another important empirical finding is that the grants targets regions that are clearly distinguishable in their level of development, from the highly developed centers of Moscow and St. Petersburg to the less safe Khabarovsk and Magadan. Both the security ratings and socio-economic stability and economic performance indices confirm high performance in the safe zones in terms of investment security, socio-economic stability, as well as economic performance. First of all, Moscow and St. Petersburg demonstrate high indicators of attraction of investments, low level of unemployment, and stable financial performance, as evidenced by high GRP coefficients to the balanced financial result.

Thus, insecure areas are the regions with high values of all the indicators and this is why Khabarovsk and Magadan with such results have difficulties in the country. These regions need to improve on the level of investment security that is lacking coupled with the socio-economic vulnerabilities and poor financial performance. The high levels of unemployment, budget deficit and low social stability reduce their capacity to sustain economic security. The low GRP ratios support the idea about the impossibility of these regions to build and maintain the economic growth rates. The presence of such weaknesses implies that more focused efforts with regard to investment promotion, social up lift and better financial management are required for this kind of region. Without these interventions, economic status between secured and insecure regions is expected to continue deteriorating thereby worsening the issue of rationality. The security of investments was identified as an essential component of the security of regions' economy. Thus, with regard to levels of investment security, OST, such as Moscow, Tyumen and St. Petersburg performed better in their global economic competence and socio-economic cohesiveness.

Another key factor that compiles the dimension of socio-economic stability in relationship to regional security. The results prove that wherever unemployment rate is low and budgets are surplus like in Moscow and Tatarstan social stability is high thus enhancing the economic stability of regions. On the same note, it is seen that socio-economic problems are much higher in places like Khabarovsk and Magadan that suffer from higher unemployment rates and budget deficits as well. The assumption that relates social stability to economic security implies that in the effort to make the insecure regions more secure, socio- economic conditions have to be boosted and this can be done by increasing employment opportunities as well as implementing improved fiscal policies. Education, health and social welfare consume a far greater share in a society hence their funding should be given due importance for the purposes of encouraging more investment and growth in a society's socio-economic base. Another aspect, which brings out the contrast between the secure areas and the insecure ones is the assessment of the economic performance by the means of economic capacities and the GRP to the balanced financial result ratio.

6. CONCLUSION

The paper offers a comprehensive evaluation of Russia's regional economic security, which contains much useful information comparing regional advantages and disadvantages with respect to various traits, such as investment security, socio-economic stability indices and economic effectiveness. The differences between secure regions that include Moscow and St. Petersburg though not very much challenged in terms of economic insecurity and the insecure regions that include Khabarovsk and Magadan seems quite critical. Annual savings and socio-economic stability are strong in secure destination with high investment inflows, sound financial management and efficient socio-economic environment creating continuity and moderate vulnerability to exogenous shocks.

On the other hand, insecure regions have a problematic investment environment, unemployment, budget deficits and low financial results and all these results suggest that these areas require policy adjustment. These regions need effective solutions for attraction of the investments, maintenance of the socio-economic stability, and the improvement of the financial performance. The authors believe that this theoretical and methodological approach indicates that the development of human capital, infrastructure, and the diversification of regional economies can improve the economic security of threatened areas. Therefore, it can be concluded that regional economic imbalances are critical; they have to be solved in order to build equally effective and efficient economy across Russian Federation. Thus, it's suggested to the policymakers to focus more on the treatment of insecure areas with concentrating on the capabilities of the safe areas. Closeness of the course of action, with the upgrading of investment climate, stability of social environment, and improvement of financial results will be critical to sustain long-term economic growth in all the regions.

REFERENCES

Aneslagon, D. M. C., & Lim, B., L. B., Tomongha, M. A. S., R. Legaspi, J. E., P. Limbaga, A. J., Casayas, J. S., & Talaboc, D. Q. (. (2024). Assessing the Nexus between Social Responsibility, Environmental Initiatives, and Profitability: A Sustainable Finance Perspective of the Universal Banks in the Philippines. *International Journal of Management Thinking*, 2(1), 38–51. DOI: 10.56868/ijmt.v2i1.40

Belaïd, F., Al-Sarihi, A., & Al-Mestneer, R. (2023). Balancing climate mitigation and energy security goals amid converging global energy crises: The role of green investments. *Renewable Energy*, 205, 534–542. DOI: 10.1016/j.renene.2023.01.083

Cai, D., Hu, J., Jiang, H., Ai, F., & Bai, T. (2023). Research on the relationship between defense technology innovation and high-quality economic development: Gray correlation analysis based on panel data. *MDE. Managerial and Decision Economics*, 44(7), 3867–3877. DOI: 10.1002/mde.3925

Cao, L., Gu, M., Jin, D., & Wang, C. (2023). Geopolitical risk and economic security: Exploring natural resources extraction from BRICS region. *Resources Policy*, 85, 103800. DOI: 10.1016/j.resourpol.2023.103800

Chu, L. K., Ghosh, S., Doğan, B., Nguyen, N. H., & Shahbaz, M. (2023). Energy security as new determinant of renewable energy: The role of economic complexity in top energy users. *Energy*, 263, 125799. DOI: 10.1016/j.energy.2022.125799

Danylyshyn, B., Onyshchenko, S., & Maslii, O. (2019). Socio-economic security: modern approach to ensuring the socio-economic development of the region. *Науковий журнал «Економіка і регіон»*, (4 (75)), 6-13.

Darazi, M. A., Khoso, A. K., & Mahesar, K. A. (2023). INVESTIGATING THE EFFECTS OF ESL TEACHERS' FEEDBACK ON ESL UNDERGRADUATE STUDENTS' LEVEL OF MOTIVATION, ACADEMIC PERFORMANCE, AND SATISFACTION: MEDIATING ROLE OF STUDENTS' MOTIVATION. *Pakistan Journal of Educational Research*, 6(2).

Doğan, B., Shahbaz, M., Bashir, M. F., Abbas, S., & Ghosh, S. (2023). Formulating energy security strategies for a sustainable environment: Evidence from the newly industrialized economies. *Renewable & Sustainable Energy Reviews*, 184, 113551. DOI: 10.1016/j.rser.2023.113551

Eldor, N., & Mamlakat, K. (2024). ISSUES OF INCREASING INVESTMENT ATTRACTIVENESS IN THE DEVELOPMENT OF THE COUNTRY'S ECONOMY. *EUROPEAN JOURNAL OF INNOVATION IN NONFORMAL EDUCATION*, 4(3), 474–478.

Elnaiem, A., Mohamed-Ahmed, O., Zumla, A., Mecaskey, J., Charron, N., Abakar, M. F., Raji, T., Bahalim, A., Manikam, L., Risk, O., Okereke, E., Squires, N., Nkengasong, J., Rüegg, S. R., Abdel Hamid, M. M., Osman, A. Y., Kapata, N., Alders, R., Heymann, D. L., & Dar, O. (2023). Global and regional governance of One Health and implications for global health security. *Lancet*, 401(10377), 688–704. DOI: 10.1016/S0140-6736(22)01597-5 PMID: 36682375

Fei, L., Shuang, M., & Xiaolin, L. (2023). Changing multi-scale spatiotemporal patterns in food security risk in China. *Journal of Cleaner Production*, 384, 135618. DOI: 10.1016/j.jclepro.2022.135618

Feng, Y., Lee, C. C., & Peng, D. (2023). Does regional integration improve economic resilience? Evidence from urban agglomerations in China. *Sustainable Cities and Society*, 88, 104273. DOI: 10.1016/j.scs.2022.104273

Frick, S. A., & Rodríguez-Pose, A. (2023). What draws investment to special economic zones? Lessons from developing countries. *Regional Studies*, 57(11), 2136–2147. DOI: 10.1080/00343404.2023.2185218

Jia, Q., Jiao, L., Lian, X., & Wang, W. (2023). Linking supply-demand balance of ecosystem services to identify ecological security patterns in urban agglomerations. *Sustainable Cities and Society*, 92, 104497. DOI: 10.1016/j.scs.2023.104497

Karanina, E. V., & Karaulov, V. M. (2023). Differentiated approach to the diagnostics of economic security and resilience of Russian regions (case of the Volga federal district). *R-Economy. 2023. Vol. 9. Iss. 1*, 9(1), 19-37.

Khan, K., Khurshid, A., & Cifuentes-Faura, J. (2023). Investigating the relationship between geopolitical risks and economic security: Empirical evidence from central and Eastern European countries. *Resources Policy*, 85, 103872. DOI: 10.1016/j.resourpol.2023.103872

Khoso, A. K., Darazi, M. A., Mahesar, K. A., Memon, M. A., & Nawaz, F. (2022). The impact of ESL teachers' emotional intelligence on ESL Students academic engagement, reading and writing proficiency: Mediating role of ESL students motivation. *Int. J. Early Childhood Spec. Educ*, 14, 3267–3280.

Khoso, A. K., Khurram, S., & Chachar, Z. A. (2024). Exploring the Effects of Embeddedness-Emanation Feminist Identity on Language Learning Anxiety: A Case Study of Female English as A Foreign Language (EFL) Learners in Higher Education Institutions of Karachi. *International Journal of Contemporary Issues in Social Sciences*, 3(1), 1277–1290.

Koval, Y. (2023). Directions management of economic security of the state in the context of globalization processes. *Public administration and law review*, (3), 39-48.

Kryshtanovych, M., Panfilova, T., Khomenko, A., Dziubenko, O., & Lukashuk, L. (2023). Optimization of state regulation in the field of safety and security of business: A local approach. *Business: Theory and Practice*, 24(2), 613–621. DOI: 10.3846/btp.2023.19563

LA, M., ZA, S., RS, A., IT, M., & Yu, A. N. (2024). Issues of improving the basis of tax administration. HOLDERS OF REASON, 2(6), 184-189.

Li, L., Huang, X., Wu, D., & Yang, H. (2023). Construction of ecological security pattern adapting to future land use change in Pearl River Delta, China. *Applied Geography (Sevenoaks, England)*, 154, 102946. DOI: 10.1016/j.apgeog.2023.102946

Liu, H., Zhu, Q., Khoso, W. M., & Khoso, A. K. (2023). Spatial pattern and the development of green finance trends in China. *Renewable Energy*, 211, 370–378. DOI: 10.1016/j.renene.2023.05.014

Mishchuk, I. (2023). Conceptual model of economic security formation and the place of the security process in this model. *Economics. Finance and Management Review*, (1), 40–49. DOI: 10.36690/2674-5208-2023-1-40

Mulska, O. P., Vasyltsiv, T. G., Kunytska-Iliash, M. V., & Baranyak, I. Y. (2023). Innovative Empirics of Migration-Economic Security Causal Nexus. *Science and innovation, 19*(3), 48-64.

Nesadurai, H. E. (2004). Introduction: Economic security, globalization and governance. *The Pacific Review*, 17(4), 459–484. DOI: 10.1080/0951274042000326023

Nikitina, M. G., Pobirchenko, V. V., Shutaieva, E. A., & Karlova, A. I. (2018). The investment component in a nation's economic security: the case of the Russian Federation. *Entrepreneurship and sustainability issues, 6*(2), 958.

Pakhucha, E., Sievidova, I. O., Romaniuk, I., Bilousko, T., Tkachenko, S. O., Diadin, A. S., & Babko, N. (2023). Investigating the impact of structural changes: the socio-economic security framework.

Payne, J. E., Truong, H. H. D., Chu, L. K., Doğan, B., & Ghosh, S. (2023). The effect of economic complexity and energy security on measures of energy efficiency: Evidence from panel quantile analysis. *Energy Policy*, 177, 113547. DOI: 10.1016/j.enpol.2023.113547

Poltorak, A., Volosyuk, Y., Tyshchenko, S., Khrystenko, O., & Rybachuk, V. (2023). DEVELOPMENT OF DIRECTIONS FOR IMPROVING THE MONITORING OF THE STATE ECONOMIC SECURITY UNDER CONDITIONS OF GLOBAL INSTABILITY. *Eastern-European Journal of Enterprise Technologies*, 122(13).

Popelo, O., Shaposhnykov, K., Popelo, O., Hrubliak, O., Malysh, V., & Lysenko, Z. (2023). The Influence of Digitalization on the Innovative Strategy of the Industrial Enterprises Development in the Context of Ensuring Economic Security. *International Journal of Safety and Security Engineering*, 13(1), 39–49. DOI: 10.18280/ijsse.130105

Ramskyi, A., Gontar, Z., Kazak, O., Podzihun, S., & Naumchuk, K. (2023). Formation of the security environment through manimization of the negative impact of threats in the socio-economic system.

Song, M., Xie, Q., Shahbaz, M., & Yao, X. (2023). Economic growth and security from the perspective of natural resource assets. *Resources Policy*, 80, 103153. DOI: 10.1016/j.resourpol.2022.103153

Su, F., Liu, Y., Chen, S. J., & Fahad, S. (2023). Towards the impact of economic policy uncertainty on food security: Introducing a comprehensive heterogeneous framework for assessment. *Journal of Cleaner Production*, 386, 135792. DOI: 10.1016/j.jclepro.2022.135792

Sun, Y., Dong, Y., Chen, X., & Song, M. (2023). Dynamic evaluation of ecological and economic security: Analysis of China. *Journal of Cleaner Production*, 387, 135922. DOI: 10.1016/j.jclepro.2023.135922

Yu, Z., Li, Y., & Dai, L. (2023). Digital finance and regional economic resilience: Theoretical framework and empirical test. *Finance Research Letters*, 55, 103920. DOI: 10.1016/j.frl.2023.103920

Zamira, J. (2024). ENSURING ECONOMIC SECURITY IN THE BANKING SECTOR. *Gospodarka i Innowacje.*, 47, 343–348.

Chapter 8
Human-Centered Technologies in the Development of Digital Energy:
Integrated Power Management Systems

Anatolii Gladkikh
Ulyanovsk State Technical University, Russia

Rafael Sayfutdinov
Ulyanovsk Civil Aviation Institute, Russia

Artem Nichunaev
Ulyanovsk State Technical University, Russia

Tatyana Ulasyuk
Ulyanovsk Civil Aviation Institute, Russia

ABSTRACT

The fundamental factor in ensuring the development of digital energy at the present stage is the quality of life of the population, where human-oriented technologies are fundamental in the information society. Integrated Power Management Systems (IEMS) are one of the most effective areas of digital technologies and processes in the energy sector. The most important task of IEMS is timely identification, reduction and elimination of hazards and risks in the system of energy generating enterprises. In the conditions of digital energy as an integral part of the digital economy, data

DOI: 10.4018/979-8-3693-3423-2.ch008

Copyright © 2025, IGI Global. Copying or distributing in print or electronic forms without written permission of IGI Global is prohibited.

on the state of managed subsystems become necessary for the IEMS resources management, where the role of human-oriented technologies is significantly increasing. These directions essentially determine the main content of the concept of intelligent energy systems, which are based on information and control technologies using a distributed data processing system. The paper proposes a method for protecting data from errors, which is based on the method of cluster partitioning of the space of code combinations.

NOVELTY STATEMENT

The novelty of this book chapter lies in its exploration of the integration of human-centered technologies within the evolving landscape of digital energy, specifically focusing on the development of advanced integrated power management systems. Unlike conventional approaches that primarily emphasize technical efficiency, this chapter highlights the approach of aligning digital energy innovations with human needs, user experience, and environmental sustainability. By delving into how these technologies can enhance system reliability, optimize energy efficiency, and foster greater user engagement, this chapter offers a fresh perspective on the future of digital energy management, positioning it at the intersection of technological advancement and human-centric design.

1. INTRODUCTION

Digital energy is a fundamental component of the digital economy. The emergence of new opportunities in information technology, the implementation of modern digital services at energy facilities, and the analysis of large volumes of data necessitate a shift in digital energy towards human-centered technologies. This reorientation aims to enhance efficiency, reliability, and environmental safety (Fang et al., 2011; Fischer et al., 2013). Leading electric power companies around the world are implementing digital transformation programs. The main goal of the digitalization of the energy complex is the introduction of intelligent control systems that ensure the receipt of economic effects, which in turn entails an improvement in people's quality of life. In this regard, such systems can be attributed to be human-oriented and aimed at improving the life of modern society (Li & Shi, 2013; Sridhar et al., 2011Sanchez et al., 2013).

The introduction of an intelligent risk-oriented model of industry management is highlighted as the main task for the near future. Within the framework of digitalization, tools should be created that facilitate the operation of energy facilities and increase the stability of their energy supply, one of which is noise-resistant coding methods. The use of noise-resistant coding in modern communication systems remains the only means of increasing the energy efficiency of such systems (Mitchell et al., 2015; Genga et al., 2021; Tang et al., 2012; Huang et al., 2007). This parameter tends to increase in conditions when the receiver of the communication system is able to correct errors of large multiplicity (Dong et al., 2019; Koike-Akino et al., 2019; Doan et al., 2019; Naeem et al., 2021; Liu et al., 2019 a).

At the same time, the existing experience of using various methods of decoding the received data to achieve a certain goal in the format of algebraic or iterative procedures does not give a noticeable effect. It leads to considerable time-draining costs and an exponential increase in the complexity of implementing the decoding processor. The reason for this situation is the passive position of the receiver. While processing each code vector, it remains a fixator of the image that has arisen in the communication channel and in general by composing a system of linear equations and its subsequent solution, ultimately trying to identify the error vector. (Nachmani et al., 2018; Doan et al., 2019; Xu et al., 2020). Some exceptions are permutation decoding systems. They proceed by selecting and using reliable symbols from the number received during reception, simulate the operation of their transmitter and compare the (almost error-free) result of such encoding with the resulting combination. With the growing influence of destructive factors, such methods are becoming ineffective. Consequently, a natural question follows that, whether there are modern solutions in neural network technologies capable of improving the characteristics of code vector recognition systems in order to obtain acceptable machine time costs and achieve an increase in the energy characteristics of communication systems. There are increased requirements for human-oriented technologies in the energy sector and, for obvious reasons expect from most of the digital information of this kind to ensure its reliability (Liu et al., 2019 b). The specificity of such data lies in protection protocols based on noise-resistant coding. This chapter aims to identify those opportunities for improving digital data processing schemes by modernizing methods for decoding redundant codes.

1.1 The Evolution of Digital Energy in the Context of the Digital Economy

Digitalization constitutes the modification of many industries, and the energy sector is not an exception to this effect. Digital energy refers to the application of information technologies in the production, distribution, and use of energy, and has been at the heart of this change (Bennatan et al., 2018). From digital services, and the large-scale data analytics that are now becoming the norm, it can be understood that there is a requirement to refocus these distributed energy systems towards human welfare. This marks not just a change of a piece of technology but a complete shift in the nature of energy systems with an emphasis on improved efficiency, reliability and safety from an environmental perspective (Liu and Poulin, 2019; Fischer et al., 2013). The drive behind this change is the consistent realization that the current conventional power structures remain insufficient to serve modern society. The rise in complexity of energy systems together with sustainable development initiatives have made electric power organizations globally pervasive to pursue digital initiatives. These measures are intended to install intelligent controlling systems to enhance the company's performance as well as yield concrete economic benefits to enhance the quality of life for individuals as well as the population (Li & Shi, 2013).

When it comes to digital energy systems, it is crucial to speak about human-oriented technologies as the advancement of digital energy systems is becoming a global phenomenon. All these technologies aim at envisaging user centricity to enhance the convenience and effectiveness of energy systems for the consumer. This approach is consistent with the aims of the digital economy, in which the consumer end is the primary focus for technological development. Thus, in energy, it involves creating systems that are not only reliable and efficient, but that can also be modified in correspondence with the user's needs or desires. Through incorporating human-centered technologies into the Digital Energy Systems, the firms can guarantee that the systems will positively impact the future of energy sustainability and resilience (Sridhar et al., 2011). Furthermore, the initiatives taken with due focus on the application of technologies for people is also becoming a trend in the digital economy. It highlights a paradigm shift from developing technologies as solutions to implementing technologies while keeping in view the social and natural environment and the consequences it has to put up with. In the energy sector, this transformation has manifested and increased digitalization of tools and services that facilitate operational interface between suppliers and consumers of energy for improved individualized energy management. This not only optimizes efficiency of energy systems, but also enhances consumer participation in the management of energy usage. This leads to enhanced sustainable development and environmental conservation (Fang et al., 2011; Fischer et al., 2013).

1.2 The Role of Intelligent Control Systems in Enhancing Energy Efficiency and Reliability

Automated control systems are an inherent part of intelligent digitalization processes occurring in the energy industry, and security of both efficiency and reliability are attributed to this function. It is a system that employs algorithms, real-time big data, and decision control functions that seek to enhance the efficiency of energy networks. The objective is to get a power system that adapts easily to conditions in the market and sustains other environmental factors. This is even more relevant when considering contemporary power systems, as these are becoming more complex and are incorporating ever larger shares of renewable resources (Liu et al., 2023). Effective usage of intelligent control systems is a step ahead from prior methods of energy management that were based on conventional styles and relied heavily on a large number of interventions. On the other hand, intelligent systems are capable of handling energy flow requirements, identifying and preventing disruptions with efficiency and play a key role in real-time resource management. Such an approach is critical in fundamentally ensuring the robustness of energy networks as these become even more integrated and complex, relying ever-more prominently on digital technologies (Wang & Wang, 2012).

Intelligent control systems offer the primary opportunity to introduce risk-oriented management models in the energy sector. These models focus heavily on the prevention of possible threats to the supply of energy along the network of pipes and wires such as equipment breakdowns, cyber terrorism, or other natural disasters. This real-time integration of data will enable intelligent control systems to give the operators of energy systems the kind of data that will help them avoid disruption and guarantee constant energy delivery (Wang et al., 2021). Apart from reliability, benefits from the use of intelligent control systems include energy efficiency. Since they help to make the best of the available resources and avoid any wastage, these systems can lower the energy needs for industrial production, consconstruction, other structures up to the municipal levels. This is particularly important in the context of global initiatives for the reduction of greenhouse emissions and combating climate change. Advanced energy management systems help the providers in balancing demand and supply, integration of end-use renewable capacity and minimizing carbon footprints in generation and utilization (Blanco et al., 2018; Farzaneh et al., 2021). In context of the energy domain, the application of intelligent control systems is consistent with the subjects of human-oriented technologies. When developed with the purpose of being optimally suitable to serve users, energy systems improve the general welfare of people and societies. For instance, intelligent systems can apply enhanced details of energy efficiency for consumers, and sustain varieties of opportunities to control energy consumption in real-time. It also assists in saving energy prices and informs consumers to make more sustainable judgements which will assist in achieving the goal towards a more resilient and sustainable power system (Rosset et al., 2017; Sridhar et al., 2011).

1.3 The Challenges and Opportunities of Noise-Resistant Coding in Digital Energy Systems

With the development of the new forms of digital energy systems, there is one of the crucial issues that appear critical i.e., the problem of data communication reliability in the new systems. Working with large volumes of data in modern communication systems is an important factor that has made noise-resistant coding methods an essential solution for this problem. These coding methods are crucial for keeping the overall energy utilization at an optimum or above levels and especially so in surroundings where data purity is a force of direction, guiding energy facilities (Boiko & Eromenko, 2014). The purpose of noise-resistant coding techniques is to provide a means of correcting errors that may have taken place at the time of transmission so as to deliver the correct message. This is particularly important in the situations when the communication systems are operating in severe environments or when they potentially might be affected by cyber threats (Grigoriev et al., 2020).

Nevertheless, there are several difficulties inherent in noise-resistant coding, regardless of their apparent benefits. Among the main problems, one can identify the fact that the decoding of the received data is a time-consuming process in terms of computational resources. Standard decoding techniques like the algebraic or iterative forms are quite intensive and time-consuming, demanding a lot of power which leads to time wastage and inevitable energy consumption. These inefficiencies can be detrimental, especially in the high-demand situations which require fast and precise data exchange (Kapranova et al., 2019). In response to these challenges, there are new solutions that have arisen from neural network technologies in recent years. Through using machine learning algorithms, it is possible to improve decoding methods in order to decrease time and effort needed for the data analysis. It not only optimizes the reliability or stability of digital energy systems but also enables the system to work in a more adaptive manner for various conditions and ensure the reliability of the communication in more complex scenario (Liu et al., 2019 a).

The incorporation of impartial and sensitive decoding for neural networks (as part of digital energy systems) is regarded as a massive prospect for enhancing such systems. In order to decrease computational demand that comes with the neural networks, the use of the decoding methods enhances the energy aspects of the communication networks in sustainability and cost. Furthermore, these advancements also corroborate with the humanistic smart technologies by ensuring that the digital energy systems remain robust and are in a position to fulfil the needs of its consumers (Genga et al., 2021). While the implementation of noise-resistant coding in digital energy systems presents certain challenges, the opportunities offered by new technologies such as neural networks provide a pathway to overcome these obstacles. By continuing with innovation in this area, it is possible to enhance the reliability, efficiency, and sustainability of digital energy systems, ensuring that they can meet the demands of the digital economy while also supporting the broader goals of environmental protection and social well-being (Zhang & Meng, 2009; Doan et al., 2019).

2. LITERATURE REVIEW

2.1 The Evolution and Impact of Digital Energy Systems

Smart energy systems have grown to be a crucial factor in the face of modernized industry of the global energy, mainly facilitated by the growth of information technology in the controlling and regulating aspect of energy production, distribution, and consumption (Bañales, 2020; Kolloch & Dellermann, 2018). Digital energy refers to the use of digital commodities including big data analytics, artificial intelligence, internet of things and other related technologies in managing and advancing the energy system efficiency, reliability, and sustainability (Ning & Xiong, 2024). This evolution is not unique and is a part of a global transition to smart grid and intelligent energy systems which are designed to increase system reliability and efficiency as well as address the new requirements related to energy sustainability (Kolloch & Dellermann, 2018).

There are several objective factors that have caused the current landscape of digital energy systems. However, the most significant factor is the complexity of the existing conventional energy infrastructure. The traditional centralized energy systems still mainly employ fossil energy resources and are featured with increased energy losses, extensive emission of pollution and restricted adaptability in response to the change of energy demands (Fang et al., 2011). While conventional energy systems are fixed and hard to manage, digital energy systems on the other hand use IT data to manage energy flows in real-time hence being able to minimize wastage and improve the efficiency of energy systems. Advanced digital technologies in connection with energy systems also enable new business models and value propositions. For example, DERs including solar power as well as battery storage systems has made consumers to be active players in the energy market rather than being mere recipients (Nazari & Musilek, 2023). This has however been reinforced by smart meters and demand response programs that enable both industrial and residential consumers to modify use in times when power is expensive or scarce in order to ensure grid stability and minimal peak demand (Alvarez-Herault et al., 2023).

Alongside substantial challenges, digital energy systems also have undeniable benefits. Among all the disruptions, one of them is cybersecurity, as the energy systems are prone to cyber threats and data hacks due to the incorporation of digital techniques in the sector (O'Dwyer et al., 2020). It is important for entities to safeguard and strengthen digital energy infrastructure because it affects energy reliability owing to the importance of data protection. Also, the implementation of digital energy systems has meant a significant level of investments on the infrastructure front and in the creation of new frameworks of regulation of the sector which allows for the development of such projects based on the protection of the public interest (Nazari & Musilek, 2023). Altogether, the transformation of energy systems into digital requires a closer look at the progress it delivers in the context of energy advancement. With the transformation, it has numerous chances to improve the industry's performance, stability, and sustainability (Ning & Xiong, 2024; Khoso et al., 2024). Nevertheless, it is imperative to weigh these against the disadvantages that organizations experience in relation to cybersecurity, compliance, and the requirement of investing a great deal of capital on infrastructure.

2.2 Human-Centered Technologies in Digital Energy Systems

It has recently become widely discussed how technologies should be designed human-centric and especially in digital energy systems. Humanistic technologies are an approach in designing technology that pays more attention to the requirements and experience of its user. It is developed to improve the quality of users' lives (da Silva, 2020). In the energy sector, this has been integrated in the creation of smart grids, advanced energy management systems along with other such applications and solutions which facilitate active involvement of the consumers. It enables them to become a much more engaging part of the energy usage process (Ardebili et al., 2021). By removing the restrictions imposed by the orthodox top-down approach, human-centered technologies make a way to control and regulate the energy consumption in digital energy systems in ways the people prefer. For instance, smart meters provide consumers with information about the actual use of energy, making it possible for them to determine whether consumption is inefficient and take measures to adjust it (Huang et al., 2022). This type of savings not only assists consumers in saving money, but also drives the initiative to decrease the overall energy demand and hence mitigate the adverse effects of energy consumption on the environment (Kwilinski et al., 2023).

Human-centered technologies also help in improving the robustness and dependability or digital energy systems. These technologies make use of feedback as well as the users' preferences within frameworks. They formulate the design and management of energy networks in an effort to establish more revolutionary and efficient energy networks that are suitable for responding to individual as well as collective fluctuations of demand and availability (Kloppenburg & Boekelo, 2019). For example, demand response programs involving real-time utilization data and consumer feedback to alter energy use in the peak hours have been proved to contribute towards increased reliability of the power supply and decreased likelihood of blackouts (Timchenko et al., 2019). However, the implementation of HC technologies in digital energy systems is particularly relevant to the era's value of sustainability and social responsibility. By making energy systems more user-friendly and easier to access, it becomes easier to address challenges of affordability whilst at the same time helping to make technology gap within segments of society narrower (Shahbaz et al., 2022). It addresses energy justice by retaining its focus on ending the lack of equitable access to affordable and reasonably reliable energy services (Chen et al., 2015).

Although the adoption of human-centered technology in digital energy systems can be effective, it comes with some losses as well. Some of them include the challenge of aligning technical features of energy networks with the stakeholders' demands and expectations. This calls for a systems approach that entails information from the fields of engineering, social sciences, and design when developing complex systems that are not only sustainable but also easy to work upon. Furthermore, energy providers, technology developers, policymakers, and consumers should work together to promote human-centered technologies in digital energy systems (Timchenko et al., 2019). Human-centered technologies have the potential to transform digital energy systems by enhancing user engagement, improving system resilience, and promoting social equity. However, realizing this potential will require a concerted effort to address the technical, social, and regulatory challenges associated with their implementation (Huang et al., 2022).

2.3 Advancement in Noise-Resistant Coding for Digital Energy Systems

Noise-resistant coding has become fundamental in the progressive design of digital energy systems, mainly regarding the reliability of data transmission. With regards to digital energy, there is need to use coding schemes that are immune to noise since data could be corrupted as a result of transmission through complex and usually noisy communication networks (Boiko & Eromenko, 2014). These methods are useful for the monitoring and control of digital energy systems as they progress and become dependent on incoming data and automated systems. A major characteristic of noisy coding is that the scheme should be capable of detecting and correcting all errors that may be present in the course of data transmission. This capability is of high importance in energy systems where data is vital for the functioning of essential infrastructure like smart grids and dispersed generation resources (Tyncherov et al., 2021). Using these advanced error correcting methods, the digital energy system can help to reduce the loss of data integrity even when the transmission is interfered. It will improve the communication network availability and reduce down time of energy systems (Nasiopoulos & Ward, 1994).

Some recent developments of noise-resistant coding have emerged. They concentrated on discovering the methods that are more effective and suitable for realization within high-speed as well as high-capacity communication networks.

For example, researchers have come up with new coding schemes to use machine learning software that makes floating decoder errors to minimize the computation and power consumption of decoding algorithms (Doan et al., 2019). These are quite useful in digital energy systems where timeliness of data processing and low latency of communication is critical (Grigoriev et al., 2020). Along with such improvements as efficiency of noise-resistant coding, there is also shifting focus on the ability to make it more flexible and robust. As it is observed, the use of adaptive coding that can tune itself to the network can enhance the ability of digital energy systems due to the unstable nature of communication channels to noise and interference (Liu et al., 2023). They help digital energy systems acquire and send three different types of data with low error rates and high data integrity to support the total reliability and stability of energy systems (Khoso et al., 2022).

Nonetheless, till today numerous issues are associated with the use and integration of noise-resistant coding in digital energy systems (Kapranova et al., 2019). One of the burning issues is the conflict of interests between code density and computational resources consumption. Higher coding schemes in development can offer better error correction. However, it comes with a danger of increased amount of power used for processing and thus present a potential threat to the energy efficiency factor. Overcoming this challenge entails continuous research with a focus to finding coding techniques that will be optimal in terms of performance and efficiency. As noise resilient coding is used in electrical energy transmission, they become a significant technology for the creation of precise and efficient energy networks (Zhang & Meng, 2009). In the current years, specific advancements are implemented and aimed at upgrading coding techniques in terms of time, space, and flexibility, hence increasing efficiency as well as reliability of the energy networks.

2.4 Cluster Formation Algorithm

The analysis of non-binary codes is usually considered in extended binary Galois fields. A field is a commutative ring ($a + b = b + a$) with a unit element with respect to multiplication (unit multiplicative element of the ring), in which each nonzero element has a multiplicative inverse element (inverse by multiplication). The subsequent reasoning is carried out using the elements of the field $GF(2^3)$, over which we construct a RS code with parameters $k = 3$ and $n = 7$. Hamming metric for such code is $d_{min} = n - k + 1 = 5$. The code (7, 3, 5), using classical decoding methods, is able to correct two errors. Considering inverse elements, it is possible to represent half of code vectors in Cartesian coordinates, then their inverse elements will be reflected as a second-order central symmetry.

A vector of a non-binary RS code is usually represented through a primitive field element α. Then $\alpha^0 = 001_2; \alpha^1 = 010_2; \alpha^2 = 100_2; \alpha^3 = 011_2; \alpha^4 = 110_2; \alpha^5 = 111_2; \alpha^6 = 101_2$. To represent a vector in Cartesian coordinates, it must be represented in binary form, split into two parts and declare one half of the data as the x coordinate and the rest of the vector as the y coordinate. For example, one of the vectors of the RS code, expressed through a primitive element, has the form: $\alpha^0 0\ 0\ \alpha^4 \alpha^0 \alpha^4 \alpha^5$. We represent this vector in binary form, then we have: 001 000 000 110 001 110 111$_2$. Vector contains 21 bits. To equal value order of x and y, we will check the entire vector for parity, then each coordinate will have 11 bits. Generally, the latter procedure is not decisive, and it can be omitted. Then in Cartesian coordinates we get $x = 259_{10}$ and $y = 239_{10}$. Now this point can be plotted on the plane with its own coordinates. Let us take into account that the information bits for this vector were the values 0 0 α^0, and their reverse sequence in the redundant vector is determined by the structure of the coding register, since the data from the information source enters such a register from right to left and the symbol α^0 is the first to enter the communication channel.

Let's go on a mission to collect all combinations of the code that have zero values in the first two positions on the left. These combinations are shown below in Table 1. Figure 1 shows the values of their Cartesian coordinates and their general configuration on the plane.

Table 1. Code combinations

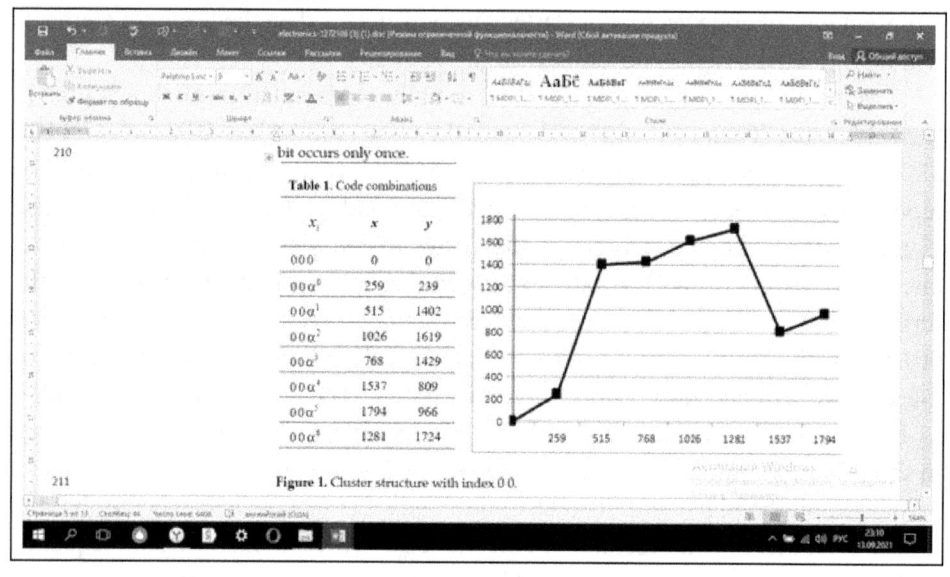

x_i	x	y
0 0 0	0	0
0 0 α^0	259	239
0 0 α^1	515	1402
0 0 α^2	1026	1619
0 0 α^3	768	1429
0 0 α^4	1537	809
0 0 α^5	1794	966
0 0 α^6	1281	1724

Figure 1. Cluster structure with index 0 0

$$\begin{matrix} 0 & 0 & 0 & 0 & 0 & 0 & 0 \\ \alpha^0 & 0 & 0 & \alpha^4 & \alpha^0 & \alpha^4 & \alpha^5 \\ \alpha^1 & 0 & 0 & \alpha^5 & \alpha^1 & \alpha^5 & \alpha^6 \end{matrix} \qquad \begin{matrix} \alpha^3 & 0 & 0 & \alpha^0 & \alpha^3 & \alpha^0 & \alpha^1 \\ \alpha^4 & 0 & 0 & \alpha^1 & \alpha^4 & \alpha^1 & \alpha^2 \\ \alpha^5 & 0 & 0 & \alpha^2 & \alpha^5 & \alpha^2 & \alpha^3 \end{matrix}$$

The analysis of the given cluster structures allows us to assert that in each check bit, there is a cyclic shift of the element sequence of the field GF(2^3), while each element in its bit occurs only once. It was shown that the clustering of code vectors has a number of features. The main features are:

– The unity of the cluster numbers for all its combinations (in our case, this value 0 0);
– The indispensable sequence of the given expansion degree after the Galois field cluster number (in our example, the number 0 0 is followed by all the field elements from the value 0 to the value α^6);
– To perform the clustering procedure, any positions of the code combinations space can be selected (not only the first two, as in the example, but also others, providing all corresponding field of the cluster number);
– In the columns of check digits, each element of the field is repeated only once;

– For any cluster, there is a prototype consisting of inverse elements.

From a theoretical perspective, this means that the auto-encoder does not need to «know» all the redundant code combinations. It generates them by the cluster number. The hidden layer, having received the cluster number in parallel mode, forms the corresponding elements of the Galois field and sends them to the output layer, where the general pattern of the cluster is formed. The vector that is accepted as $f(x_i^e, \theta)$ will be outside the cluster trajectory. It will indicate the presence of errors that acted in the user's biometric vector. Let's find a cluster with inverse elements for cluster number 0 0.

$$\begin{matrix} 0 & \alpha^5 & \alpha^5 & \alpha^3 & 0 & \alpha^3 & \alpha^2\alpha^3 & \alpha^5 & \alpha^5 & \alpha^1 & \alpha^2 & \alpha^1 & \alpha^4 \\ \alpha^0 & \alpha^5 & \alpha^5 & \alpha^6 & \alpha^0 & \alpha^6 & \alpha^3\alpha^4 & \alpha^5 & \alpha^5 & \alpha^0 & \alpha^4 & \alpha^0 & 0 \\ \alpha^1 & \alpha^5 & \alpha^5 & \alpha^2 & \alpha^1 & \alpha^2 & \alpha^0\alpha^5 & \alpha^5 & \alpha^5 & \alpha^5 & \alpha^5 & \alpha^5 & \alpha^5 \\ \alpha^2 & \alpha^5 & \alpha^5 & \alpha^4 & \alpha^2 & \alpha^4 & \alpha^6\alpha^6 & \alpha^5 & \alpha^5 & 0 & \alpha^6 & 0 & \alpha^1 \end{matrix}$$

The second example shows the validity of the assumption that each element of the field is repeated in the column only once. This feature can be used in a neural network decoding system to identify an information vector. This implies the statement: if one of the check columns of the cluster contains at least two identical elements $\alpha^i \in GF(2^N)$, it is an indicator that one vector contains an erroneous solution. To determine the erroneous element, an additional comparison of the received vector with an ideal cluster will be required. In the best case, after assuming that the number of the cluster is received correctly, the receiver should form all combinations of the cluster. Then compare the topology of the point of the received vector with the ideal trajectory of the processed cluster. For long code lengths, this procedure can reduce the efficiency of the receiver processor. Reduction of decoder complexity can be achieved by selecting not all combinations of the cluster; therefore, not the entire trajectory, but only a part of it about a point processed by the receiver can be selected. The search results for the full trajectory of the cluster from the second example are shown in Figure 2. Table 2 shows the corresponding code combinations.

Figure 2. The structure of the cluster with index 5 5

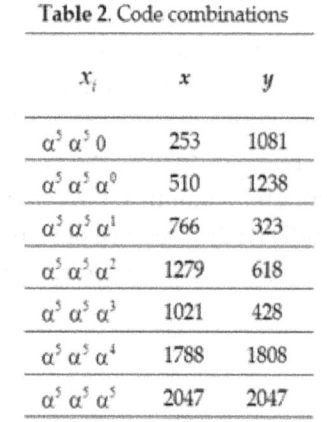

Table 2. Code combinations

x_i	x	y
$\alpha^5 \alpha^5 0$	253	1081
$\alpha^5 \alpha^5 \alpha^0$	510	1238
$\alpha^5 \alpha^5 \alpha^1$	766	323
$\alpha^5 \alpha^5 \alpha^2$	1279	618
$\alpha^5 \alpha^5 \alpha^3$	1021	428
$\alpha^5 \alpha^5 \alpha^4$	1788	1808
$\alpha^5 \alpha^5 \alpha^5$	2047	2047
$\alpha^5 \alpha^5 \alpha^6$	1532	645

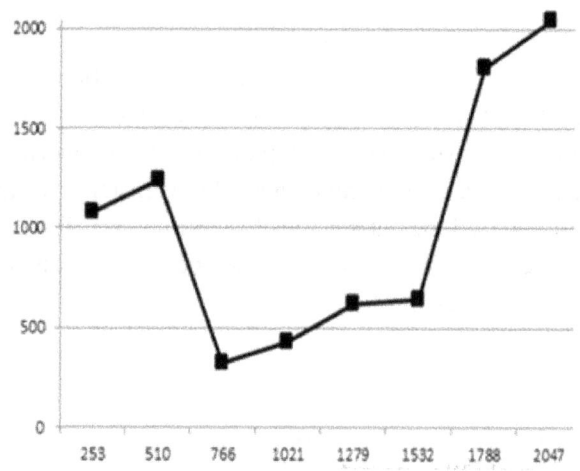

Opposite elements of the field are formed due to the inversion of their binary images. So, for an element 000 → 0, the opposite element in the field GF(2^3) is element 111 → α^5. Let's write it as 0 ↔ α^5. It is easy to check that α^0 ↔ α^4, α^1 ↔ α^6 and α^2 ↔ α^3. Thus, cluster 00 is the opposite of cluster $\alpha^5\alpha^5$ and 00 ↔ $\alpha^5\alpha^5$. Following this rule, it is possible to calculate the entire subset of clusters that can be formed after partitioning the set of redundant code vectors in the communication system. In this case, rotating the graph by 180° in Figure 2 leads to the graph in Figure 1. In this case, coordinates $x_{max} = y_{max} = 2047$ change to values $x_{min} = y_{min} = 0$. Naturally, the hidden layer of the neural network does not need to remember this order of elements, since each cluster is generated anew for a new received vector. In principle, with the large memory capacity of the receiver processor, the configuration of all clusters can be fixed in a special memory card after the completion of the learning procedure. If the received vector is not distorted by noise, then the corresponding point will be on the trajectory belonging to the cluster. Let us consider the configuration of points in the cluster trajectory system, provided that errors occur in the biometric vector. In the general case, the noise affects an arbitrary part of the vector. Let us consider three parts of the accepted vector. First, symbols related to the cluster number (at this stage of the analysis, we will assume that such symbols are received without distortion). Second, the distortion can be localized only in the x-coordinate, as shown in Figure 3a. Third, the interference affected only the y coordinate, as shown in Figure 3b. In any of these cases, finding the true vector does not cause any special difficulties. The simultaneous distortion of x and y coordinates is shown in Figure 4a.

Figure 3. The recovery of the vector which was received with errors: distortion in the area of the x coordinate (a), distortion in the area of the y coordinate) (b)

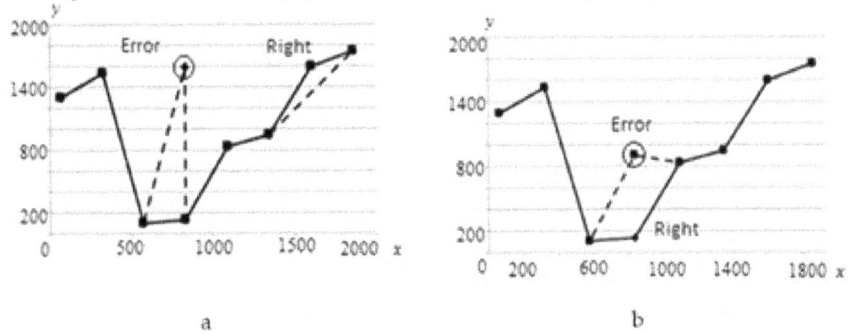

Figure 4. Examples of trajectories: simultaneous distortion of coordinates (a), trajectories of different clusters (b)

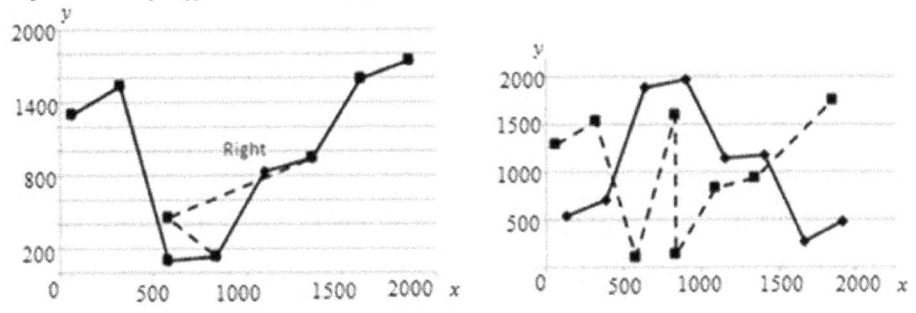

It becomes clear that a point outside the cluster trajectory is distorted, and since the position of such a point in the cluster is known, its correction is beyond doubt. In the same Figure, position 5b shows the trajectories of two different clusters. These trajectories are quite distinguishable, and the neural network always knows the position of the desired point in the cluster based on the shape of the primitive element of the Galois field.

2.5 Principle of Permutation Decoding (PD)

The computational process in the implementation of the classical PD algorithm is carried out in steps.

Step 1. Fix hard decisions of the code vector $V_{received}$ received from the channel with errors, accompanying each of them with the value of the soft decisions of the characters λ_i.

Step 2. Rank values of soft decisions and their corresponding bits in descending order so that the most reliable values of the character λ_i are in place of the highest digits (on the left). Here we take into account the left position of the unit matrix E in the generating matrices of systematic codes.

Step 3. Based on step 2, form a bijection $f: V_{received} \rightarrow V_{rearranged}$ and the corresponding permutation (commutative) matrix K, where $V_{rearranged}$ is a rearranged vector.

Step 4. Based on the results of the step 2 select the left k most reliable digits in vector $V_{rearranged}$ and remember them as a new information vector $V'_{инф}$.

Step 5. Multiply the column numbering of the generating matrix of the source code G by the matrix K to rearrange the columns of the matrix G according to step 2 and form a new rearranged matrix of the code $G_{rearranged}$.

Step 6. Select the first k columns in the matrix $G_{rearranged}$, get the square matrix $Q_{k \times k}$ and calculate the determinant of this matrix Δ. If $\Delta \neq 0$, go to step 7. If $\Delta = 0$, abandon decoding, go to step 2 and perform new permutations by swapping the column with the number k for the column with the symbol $k + 1$. In this case, the matrix K is adequately transformed. This step leads to additional time delays, so it is advisable to present such a combination in the form of erasure.

Step 7. For matrix $Q_{k \times k}$ calculate the matrix of minors M_Q (this step is not performed after the cognitive procedure in the new algorithm).

Step 8. Find the inverse matrix $Q^{-1}_{k \times k}$, by dividing the elements of matrix $Q^T_{k \times k}$ in value Δ (this step is not performed after cognitive procedure in the new algorithm).

Step 9. By the values of matrix $Q^{-1}_{k \times k}$ convert matrix $G_{rearranged}$ to the systematic form $G^{systematic}_{rearranged}$ (this step is not performed after cognitive procedure in the new algorithm).

Step 10. Multiply the vector of length k from step 4 V'_{inf} by the matrix $G^{systematic}_{rearranged}$ and calculate the vector of the equivalent code V_{equiv}.

Step 11. Multiply the vector V_{equiv} by K^T by performing the inverse bijective mapping $f: V_{received} \rightarrow V_{rearranged}$, and get the rearranged vector $V^{rearranged}_{equiv}$.

Step 12. We add the vectors $V_{received} \oplus V^{rearranged}_{equiv} = V_{error}$ bitwise to get the error vector that acted in the communication channel at the moments of fixing the hard solutions of vector $V_{received}$.

The classical algorithm's analysis shows that the decoder performance is significantly reduced from the sixth to the ninth step in the matrix computing system. The algorithm's main disadvantage is the need to perform the presented sequence of steps even if individual permutations are repeated during data processing.

A technical solution suggests remembering those permutations in the generating matrix G of the main code that do not lead to the degeneracy of the matrix $Q_{k \times k}$, and to keep in the decoder's memory the structure of the transformed matrix $G^{sys}_{rearranged}$, that corresponds to a specific permutation of the matrix $Q_{k \times k}$. Moreover, this solution allows you to pre – «train» the decoder to recognize repeated permutations and, by expanding the decoder's memory, implement its cognitive functions, creating a cognitive map of such permutations of the columns of the matrix G, which provides a positive and negative decoding result. In this process, three modes can be distinguished: the mode of operational data exchange with simultaneous filling of the decoder's cognitive map, the training mode, and the preliminary mode of filling the decoder's cognitive map from the system of external computing devices. Assuming $n \approx 2k$ By splitting the vector into parts x and y, it could be argued that the error correction will only be performed in the y region, as shown in Figure 5b. This significantly simplifies the search for the error vector.

The successful implementation of the permutation decoding method mainly depends on the number of hard solutions (binary or non-binary characters), accompanied by high rates of soft solutions. It is advisable that for the accepted code vector, the number of reliable characters corresponds to the ratio $k \geq (n + 1)/2$. Studies have shown that comparing histograms for their coincidence, proximity, and correlation dependence on the compared indicators are universal methods for solving the task. The mathematical apparatus of such a comparison is considered to be sufficiently developed. The evaluation of a certain reference histogram $H_э$ with an arbitrary histogram H_d can be carried out as follows. It can be carried out on the basis of correlation relationships assessment, using the criterion χ^2 (Chi-square), using the intersections of histograms method, using the Bhattacharya method. However, the most rational approach requires a digital format of blocks, which is the most suitable way for a system of non-binary codes. A set of biometric data which are divided into blocks and stored in a cloud technology system is taken as a reference histogram. The current data set is compared with the reference values. In the case of data distortion in the high-order region, the difference between $H_э$ and H_d will be significant, otherwise, it is minimal. This approach provides rational sorting of non-binary characters within the code vector and a quick implementation of the key recovery procedure.

3. SUMMARY

Human-centered and digital energy systems have thus emerged as the focus of this chapter in order to bring out the importance of these technologies in the energy industry. The digital economy is going forward, and hence the idea of digital energy is a factor that combines information technologies in the processes of producing, distributing and consuming energy. This integration is aiding in the upgrade of most of the conventional energy systems to more efficient, reliable and sustainable networks to meet current societal needs. The chapter also reflects the presentation of the background of digital energy in relation to the general perspectives of the digital economy. It was highlighted that the deployment of digital technologies within the energy systems is not in fact the simple substitution of a new technology for an old one, but rather a shift to the development of smart energy networks.

One of more apparent threads running through the chapter is that of human-centric technologies in digital energy systems. These technologies aim at enhancing the capacity to enable users and maximize growth in a way that the benefits derived from technology advancement commiserates with the consumer. The chapter looked at the ways in which devices like the smart meters and demand response programs provide the consumer with an opportunity to actively participate in decisions relating to energy use. This has not only given a chance to cut the cost of energy but also maintain the steadiness of the grid and other objectives of sustainability. Furthermore, these technologies foster the reliability and flexibility of energy systems, given the dynamism of the power demand and the supply. This discussion also pointed at practicalities of human sensitive technologies in digital energy systems setting.

Another major area of emphasis in this chapter was the progress made in noise-resistant coding for digital energy systems. Various noise resistant codes are required to effectively enhance the quality of data communication over different large signal networks. These methods safeguard data from the occurrence of errors in transit, something that is vital in ensuring that energy digital systems are effective and composed. This chapter described the most recent developments in this field, such as the application of machine learning algorithms that improve the effectiveness and expansiveness of error-correcting codes. These advancements are most useful and especially applied to high-speed and high-capacity communication systems in which real-time signal processing and fast response of the communication channels are important. However, the chapter did not deny that there were still some issues to be solved in noise-resistant coding, such as the conflict between coding rate and computational load or energy consumption.

4. CONCLUSION

In this chapter, we have explored the complex environment of digital systems in energy and people-oriented technologies, proving the scale of the shift in the contemporary energy industry. The analysis carried out when positioned in the intersection of technology and social demand indicates that the future of energy is substantially related to the implementation of discourses of digitalization as well as users' orientated design. The emerging systems of energy in the digital age are revolutionary in the sense that they are moving away from the fixed and sometimes outdated structures of energy systems to dynamic networks. These systems which are based on real time data and analytics, artificial intelligence, and communication technologies will provide exclusive opportunities to increase energy efficiency, reliability and sustainability. Such technologies when incorporated in energy management go further in enhancing operational performance while at the same time assisting in the management of some of the most stringent environmental issues of the current society.

Closely bound to this process is the implementation of human-oriented technologies, which implies technologies designed for users. The chapter has equally postulated the need to anchor technology and innovation into human values in a way that technology solutions such as digital energy systems do not just add value but do so in a way that is fair or equal and which meets the pluralistic need of the society. These technologies give consumers a greater measure of control over their energy usage, thus mediating the gap between available technology and consumers' willingness to engage with it, and thereby creating a wide range of socially-significant applications. These innovations in noise-resistant coding also highlight just how important robust data integrity and communication reliability is within these digital energy systems. Energy networks are getting more and more dependent on the exchange of data, thus data protection and correction in real-time is essential. Such exploration of state-of-the art methods, specifically machine learning and adaptive error-correction methods raises the bar of continuous development needed in energy communication systems.

REFERENCES

Alvarez-Herault, M. C., Gouin, V., Chardin-Segui, T., Malot, A., Coignard, J., Raison, B., & Coulet, J. (2023). *Distribution system planning: Evolution of methodologies and digital tools for energy transition* (1st ed.). Wiley-ISTE., DOI: 10.1002/9781394209477

Ardebili, A. A., Longo, A., & Ficarella, A. (2021, October). Digital twin (DT) in smart energy systems-systematic literature review of dt as a growing solution for energy internet of the things (EIoT). In *E3S Web of Conferences, 76th Italian National Congress ATI (ATI 2021),* Vol. 312(2), (pp. 1-18). https://doi.org/DOI: 10.1051/e3sconf/202131209002

Bañales, S. (2020). The enabling impact of digital technologies on distributed energy resources integration. *Journal of Renewable and Sustainable Energy*, 12(4), 045301. DOI: 10.1063/5.0009282

Bennatan, A., Choukroun, Y., & Kisilev, P. (2018, June). Deep learning for decoding of linear codes-a syndrome-based approach. In *IEEE International Symposium on Information Theory (ISIT)* (pp. 1595-1599). IEEE. https://doi.org/DOI: 10.1109/ISIT.2018.8437530

Blanco, J., García, A., & Morenas, J. D. L. (2018). Design and implementation of a wireless sensor and actuator network to support the intelligent control of efficient energy usage. *Sensors (Basel)*, 18(6), 1892. DOI: 10.3390/s18061892 PMID: 29890737

Boiko, J. M., & Eromenko, A. I. (2014). Improvements encoding energy benefit in protected telecommunication data transmission channels. *Communications*, 2(1), 7–14. DOI: 10.11648/j.com.20140201.12

Chen, D., Heyer, S., Ibbotson, S., Salonitis, K., Steingrímsson, J. G., & Thiede, S. (2015). Direct digital manufacturing: Definition, evolution, and sustainability implications. *Journal of Cleaner Production*, 107, 615–625. DOI: 10.1016/j.jclepro.2015.05.009

da Silva, M. M. (2020). *Power and gas asset management regulation, planning and operation of digital energy systems* [LNEN Series]. Springer., DOI: 10.1007/978-3-030-36200-3

Doan, N., Hashemi, S. A., Ercan, F., Tonnellier, T., & Gross, W. J. (2019, October). Neural dynamic successive cancellation flip decoding of polar codes. In *2019 IEEE International Workshop on Signal Processing Systems (SiPS)* (pp. 272-277). IEEE. https://doi.org/DOI: 10.1109/SiPS47522.2019.9020513

Dondossola, G., Garrone, F., & Szanto, J. (2011, July). *Cyber risk assessment of power control systems—A metrics weighed by attack experiments. In 2011 IEEE Power and Energy Society General Meeting.* IEEE., DOI: 10.1109/PES.2011.6039589

Dong, P., Zhang, H., Li, G. Y., Gaspar, I. S., & NaderiAlizadeh, N. (2019). Deep CNN-based channel estimation for mmwave massive MIMO systems. *IEEE Journal of Selected Topics in Signal Processing*, 13(5), 989–1000. DOI: 10.1109/JSTSP.2019.2925975

Fang, X., Misra, S., Xue, G., & Yang, D. (2011). Smart grid—The new and improved power grid: A survey. *IEEE Communications Surveys and Tutorials*, 14(4), 944–980. DOI: 10.1109/SURV.2011.101911.00087

Farzaneh, H., Malehmirchegini, L., Bejan, A., Afolabi, T., Mulumba, A., & Daka, P. P. (2021). Artificial intelligence evolution in smart buildings for energy efficiency. *Applied Sciences (Basel, Switzerland)*, 11(2), 763. DOI: 10.3390/app11020763

Fischer, J. R., González, S. A., Herran, M. A., Judewicz, M. G., & Carrica, D. O. (2013). Calculation-delay tolerant predictive current controller for three-phase inverters. *IEEE Transactions on Industrial Informatics*, 10(1), 233–242. DOI: 10.1109/TII.2013.2276104

Genga, Y., Oyerinde, O. O., & Versfeld, J. (2021). Iterative soft-input soft-output bit-level Reed-Solomon decoder based on information set decoding. *SAIEE Africa Research Journal*, 112(2), 52–65. DOI: 10.23919/SAIEE.2021.9432893

Gladkikh, A. A., Volkov, A. K., & Ulasyuk, T. G. (2021). Development of biometric systems for passenger identification based on noise-resistant coding means. *Civil Aviation High Technologies*, 24(2), 93–104. DOI: 10.26467/2079-0619-2021-24-2-93-104

Gladkikh, A. A., Volkov, A. K., Volkov, A. K., Andriyanov, N. A., & Shakhtanov, S. V. (2019, November). Development of network training complexes using fuzzy models and noise-resistant coding. In *International Conference on Aviamechanical Engineering and Transport (AviaENT 2019)* (pp. 373-379). Atlantis Press. https://doi.org/DOI: 10.2991/aviaent-19.2019.69

Grigoriev, E. K., Nenashev, V. A., Sergeev, A. M., & Nenashev, S. A. (2020, September). Research and analysis of methods for generating and processing new code structures for the problems of detection, synchronization and noise-resistant coding. In *Image and Signal Processing for Remote Sensing XXVI* (Vol. 11533, pp. 319–328). SPIE., DOI: 10.1117/12.2574238

Huang, J., Koroteev, D. D., Kharun, M., & Maksimenko, R. V. (2022, August). Impact analysis of digital technology on smart city and renewable energy management. In *AIP Conference Proceedings,* Vol. 2559(1). AIP Publishing. https://doi.org/DOI: 10.1063/5.0099010

Huang, Q., Wu, J., Zhao, C., & You, X. (2007, June). Waterfilling-like multiplicity assignment algorithm for algebraic soft-decision decoding of Reed-Solomon codes. In *2007 IEEE International Conference on Communications* (pp. 6210-6213). IEEE. https://doi.org/DOI: 10.1109/ICC.2007.1028

Kapranova, E. A., Nenashev, V. A., Sergeev, A. M., Burylev, D. A., & Nenashev, S. A. (2019, November). Distributed matrix methods of compression, masking and noise-resistant image encoding in a high-speed network of information exchange, information processing and aggregation. In *SPIE Future Sensing Technologies* (Vol. 11197, pp. 104–110). SPIE., DOI: 10.1117/12.2542677

Kazadaev, L. M., Sattarova, A. I., & Ryumshin, K. Y. (2024, July). Investigation of digital communication lines noise immunity in hydroacoustic channel using various types of product turbo code. In *2024 Systems of Signal Synchronization, Generating and Processing in Telecommunications (SYNCHROINFO)* (pp. 1-5). IEEE. https://doi.org/DOI: 10.1109/SYNCHROINFO61835.2024.10617661

Khoso, A. K., Darazi, M. A., Mahesar, K. A., Memon, M. A., & Nawaz, F. (2022). The impact of ESL teachers' emotional intelligence on ESL students academic engagement, reading and writing proficiency: Mediating role of ESL students motivation. *International Journal of Early Childhood Special Education (INT-JECSE), 14*, 3267-3280. https://doi.org/DOI: 10.9756/INT-JECSE/V14I1.393

Khoso, A. K., Khurram, S., & Chachar, Z. A. (2024). Exploring the effects of embeddedness-emanation feminist identity on language learning anxiety: A case study of female english as a foreign language (EFL) learners in higher education institutions of Karachi. *International Journal of Contemporary Issues in Social Sciences*, 3(1), 1277–1290.

Kloppenburg, S., & Boekelo, M. (2019). Digital platforms and the future of energy provisioning: Promises and perils for the next phase of the energy transition. *Energy Research & Social Science*, 49, 68–73. DOI: 10.1016/j.erss.2018.10.016

Koike-Akino, T., Wang, Y., Millar, D. S., Kojima, K., & Parsons, K. (2019, September). Neural turbo equalization to mitigate fiber nonlinearity. In *45th European Conference on Optical Communication (ECOC 2019)* (pp. 1-4). IET. https://doi.org/DOI: 10.1049/cp.2019.0803

Kolloch, M., & Dellermann, D. (2018). Digital innovation in the energy industry: The impact of controversies on the evolution of innovation ecosystems. *Technological Forecasting and Social Change*, 136, 254–264. DOI: 10.1016/j.techfore.2017.03.033

Kwilinski, A., Lyulyov, O., & Pimonenko, T. (2023). The impact of digital business on energy efficiency in EU countries. *Information (Basel)*, 14(9), 480. DOI: 10.3390/info14090480

Li, H., & Shi, Y. (2013). Network-based predictive control for constrained nonlinear systems with two-channel packet dropouts. *IEEE Transactions on Industrial Electronics*, 61(3), 1574–1582. DOI: 10.1109/TIE.2013.2261039

Liu, B., Li, S., Xie, Y., & Yuan, J. (2019 a, August). Deep learning assisted sum-product detection algorithm for faster-than-Nyquist signaling. In *2019 IEEE Information Theory Workshop (ITW)* (pp. 1-5). IEEE. https://doi.org/DOI: 10.1109/ITW44776.2019.8989271

Liu, H., Zhu, Q., Khoso, W. M., & Khoso, A. K. (2023). Spatial pattern and the development of green finance trends in China. *Renewable Energy*, 211, 370–378. DOI: 10.1016/j.renene.2023.05.014

Liu, Y. H., & Poulin, D. (2019b). Neural belief-propagation decoders for quantum error-correcting codes. *Physical Review Letters*, 122(20), 200501. DOI: 10.1103/PhysRevLett.122.200501 PMID: 31172756

Mitchell, D. G., Lentmaier, M., & Costello, D. J. (2015). Spatially coupled LDPC codes constructed from protographs. *IEEE Transactions on Information Theory*, 61(9), 4866–4889. DOI: 10.1109/TIT.2015.2453267

Nachmani, E., Bachar, Y., Marciano, E., Burshtein, D., & Be'ery, Y. (2018). Near maximum likelihood decoding with deep learning. *ArXiv, abs/1801.02726*. https://doi.org//arXiv.1801.02726DOI: 10.48550

Naeem, M., De Pietro, G., & Coronato, A. (2021). Application of reinforcement learning and deep learning in multiple-input and multiple-output (MIMO) systems. *Sensors (Basel)*, 22(1), 309. DOI: 10.3390/s22010309 PMID: 35009848

Nasiopoulos, P., & Ward, R. K. (1994). A noise resistant synchronization scheme for HDTV images. *IEEE Transactions on Broadcasting*, 40(4), 228–237. DOI: 10.1109/11.362935

Nazari, Z., & Musilek, P. (2023). Impact of digital transformation on the energy sector: A review. *Algorithms*, 16(4), 211. DOI: 10.3390/a16040211

Ning, J., & Xiong, L. (2024). Analysis of the dynamic evolution process of the digital transformation of renewable energy enterprises based on the cooperative and evolutionary game model. *Energy*, 288, 129758. DOI: 10.1016/j.energy.2023.129758

O'Dwyer, E., Pan, I., Charlesworth, R., Butler, S., & Shah, N. (2020). Integration of an energy management tool and digital twin for coordination and control of multi-vector smart energy systems. *Sustainable Cities and Society*, 62, 102412. DOI: 10.1016/j.scs.2020.102412

Rosset, V., Paulo, M. A., Cespedes, J. G., & Nascimento, M. C. (2017). Enhancing the reliability on data delivery and energy efficiency by combining swarm intelligence and community detection in large-scale WSNs. *Expert Systems with Applications*, 78, 89–102. DOI: 10.1016/j.eswa.2017.02.008

Sanchez, J., Caire, R., & Hadjsaid, N. (2013, June). *ICT and power distribution modeling using complex networks. In 2013 IEEE Grenoble Conference.* IEEE., DOI: 10.1109/PTC.2013.6652388

Shahbaz, M., Wang, J., Dong, K., & Zhao, J. (2022). The impact of digital economy on energy transition across the globe: The mediating role of government governance. *Renewable & Sustainable Energy Reviews*, 166, 112620. DOI: 10.1016/j.rser.2022.112620

Shmatok, A. N., Zolotenkova, M. K., & Egorov, V. V. (2024, May). Usage of noise-resistant coding for stable data transmission over household electrical networks. In *2024 IEEE Ural-Siberian Conference on Biomedical Engineering, Radioelectronics and Information Technology (USBEREIT)* (pp. 212-216). IEEE. https://doi.org/DOI: 10.1109/USBEREIT61901.2024.10583992

Sridhar, S., Hahn, A., & Govindarasu, M. (2011). Cyber–physical system security for the electric power grid. *Proceedings of the IEEE*, 100(1), 210–224. DOI: 10.1109/JPROC.2011.2165269

Tang, S., Chen, L., & Ma, X. (2012). Progressive list-enlarged algebraic soft decoding of Reed-Solomon codes. *IEEE Communications Letters*, 16(6), 901–904. DOI: 10.1109/LCOMM.2012.042512.112511

Timchenko, O., Nebrat, V. V., Lir, V., Bykonia, O., & Dubas, Y. (2019). Organizational and economic determinants of digital energy development in Ukraine. *Economy and forecasting*, Vol. 2019(3), 59-75. https://doi.org/DOI: 10.15407/econforecast2019.03.059

Tyncherov, K. T., Mukhametshin, V. S., & Rakhimov, N. (2021, February). Theoretical basis for constructing special codes for a noise-resistant downhole telemetry system. In *Journal of Physics: Conference Series,* Vol. 1753(1), 012081. IOP Publishing. DOI: 10.1088/1742-6596/1753/1/012081

Wang, K. J., Dagne, T. B., Lin, C. J., Woldegiorgis, B. H., & Nguyen, H. P. (2021). Intelligent control for energy conservation of air conditioning system in manufacturing systems. *Energy Reports*, 7, 2125–2137. DOI: 10.1016/j.egyr.2021.04.010

Wang, Z., & Wang, L. (2012, August). Occupancy pattern based intelligent control for improving energy efficiency in buildings. In *2012 IEEE International Conference on Automation Science and Engineering (CASE)* (pp. 804-809). IEEE. DOI: 10.1109/CoASE.2012.6386336

Xu, W., Tan, X., Be'ery, Y., Ueng, Y. L., Huang, Y., You, X., & Zhang, C. (2020). Deep learning-aided belief propagation decoder for polar codes. *IEEE Journal on Emerging and Selected Topics in Circuits and Systems*, 10(2), 189–203. DOI: 10.1109/JETCAS.2020.2995962

Zhang, J., & Meng, J. (2009). Noise resistant OFDM for power-line communication systems. *IEEE Transactions on Power Delivery*, 25(2), 693–701. DOI: 10.1109/TPWRD.2009.2036626

Chapter 9
Impact of Reward Crowdfunding on Entrepreneurial Nature of Crowdfunded Ventures:
Model of Client–Entrepreneur Relationship

Ivan D. Kotliarov
Peter the Great St. Petersburg Polytechnic University, Russia

ABSTRACT

The goal of the paper consists in identification of changes in entrepreneur-customer relations generated by the emergence of crowdfunding. The research is based on the concept of on-demand economy. It introduces taxonomy of rewards used in crowdfunding based on the type of value provided to backers and amount of this value. The analysis of reward crowdfunding demonstrates that backers actually finance ventures, not products. Crowdfunded venture is established in response to backers' needs and backers become customers of this venture. It means that such venture can be considered as venture on demand. Rise of this new type of ventures transforms the model of relations between customers and entrepreneurs. Customers become active stakeholders of business development and create demand for new ventures (not for new products). Thanks to this new model of relations customers can better satisfy their needs while entrepreneurs can reduce the level of risks. The paper contains a conceptual framework that describes this evolution of customers within the system of customer-entrepreneur relations.

DOI: 10.4018/979-8-3693-3423-2.ch009

1. INTRODUCTION

Crowdfunding is a novel tool of financial and social interaction that enables individuals and entrepreneurs to finance their needs by raising money from a relatively large group of people or a crowd (Langley, 2016; Mollick, 2014; Mollick, 2016). According to Kraus et al. (2016), crowdfunding is a part of crowdsourcing, a broader instrument that provides companies and individuals with an access to different resources. These resources are delivered by a crowd. As including but not limited to money, these resources substantially financial and social interaction. It can be noted that uber is an excellent example of a crowdsourcing-based organization.

Crowdfunding is also a part of fintech, a new wave in the development of the financial industry based on a symbiosis of financial services and digital technologies (Lee & Shin, 2018). In addition to this, crowdfunding is a part of the platform economy, as interactions between providers of financial resources (backers or funders) and funded projects are usually mediated by platforms which are either professional crowdfunding platforms or social media networks (Petruzzelli et al., 2019). As Langley (2016) rightfully states, "A defining feature of the crowdfunding economy is that dedicated public or private institutions are not the source of funding."

While crowdfunding can be used to finance both personal needs and entrepreneurial projects, only the latter will be analyzed in the present paper. The term "entrepreneurial" is used here in a broader sense to describe projects that create value not only for their initiators but also for external users. These projects include not only for-profit business ventures, but also non-commercial and social initiatives. Crowdfunding quickly gains popularity thanks to its apparent advantages in comparison with more traditional (and more formal) models of entrepreneurial finance. Following eminent advantages are stated here:

1. Crowdfunding can be used to finance ventures that can hardly get access to traditional sources of finance – novel products, innovation businesses at early stages, cultural and social projects and events (Mollick, 2014). Businesses facing financial problems is one such example (Josefy et al., 2017; Walthoff-Borm et al., 2018), Female entrepreneurship also makes good use of this incentive (Francesca et al., 2021). The COVID-19 pandemic adds a new crowdfunding option i.e., businesses in crisis caused by the lockdown (Farhoud et al., 2021).
2. Crowdfunding exists in many forms, ranging from pure donations to pre-sales to different types of debt or equity finance (Langley, 2016; Shneor, 2020). Unlike traditional sources of finance, a crowdfunding campaign may combine all these forms and propose different benefits for each form. The type and size of benefits depend on the type of backer participation. It makes crowdfunding flexible and easily adaptable to the various needs of entrepreneurs and backers.

3. Crowdfunding is more than just a source of finance. Backers do not only provide projects and ventures with money. They often participate in the promotion and development of projects, support projects with their expertise etc. (Ahsan et al., 2018; Andreas et al., 2019; Zhao & Ryu, 2020). It provides with social capital (Petruzzelli et al., 2019; Stanko & Henard, 2017). Moreover, crowdfunding is a tool of community building (Cai et al., 2021; Mollick, 2016). In addition to this, crowdfunding can inform entrepreneurs about the market potential of their venture (da Cruz, 2018).

It clearly indicates that despite all the obvious advantages of crowdfunding as a tool of finance, its potential goes far beyond. Crowdfunding profoundly transforms the system of interactions between entrepreneurs on one side, and customers and public on the other. The latter become actively involved in project development. This involvement leads to a shift in customers' and backers' participation in projects and transforms both the role of backers and the nature of crowdfunded businesses.

Surprisingly, these changes remain understudied in the current literature. Key research streams dedicated to relations between backers and entrepreneurs mainly analyze a) factors of success of crowdfunding campaigns (Allison et al., 2017; Butticè et al., 2017; Colombo et al., 2015; Crosetto & Regner, 2018; Dikaputra et al., 2019; Josefy et al., 2017; Liang et al., 2020; Mollick, 2014; Younkin & Kashkooli, 2016; Zhao & Ryu, 2020), b) optimal use of backers' resources (Butticè & Noonan, 2020), c) project's ability to fulfill its obligations towards backers such as its ability to deliver benefits (Mollick, 2014; Shneor & Munim, 2019; Zhao & Ryu, 2020), d) backers' motivation for participation in crowdfunding campaigns (Efrat et al., 2020; Huang, 2020; Inés & Moleskis, 2021; Ryu & Kim, 2016; Shneor & Munim, 2019; Zhang & Chen, 2019). New research streams emerge that cover the system of interactions between backers and businesses (Efrat & Gilboa, 2020; Petruzzelli et al., 2019). However, the transformation of the entrepreneurial model of crowdfunded businesses remains beyond the scope of the existing research.

The present paper represents an attempt to show the new structure of relations between entrepreneurs and customers created by crowdfunding. As crowdfunding has many forms, only reward crowdfunding will be analyzed. This choice can be explained by two reasons: a) in the case of reward, crowdfunding backers are also customers of the project which simplifies the analysis (Allison et al., 2017), b) reward crowdfunding is a prevalent form of crowdfunding (Kraus et al., 2016; Mollick, 2014).

Crowdfunding can be used to finance a project or a venture. The difference between them is that a project has a limited period of activity and normally produces a unique product such as a cultural event or a movie that will not be reproduced, while a venture is expected to be active for an unlimited period of time and can produce various products and services such as a museum, a theater, etc. Only ventures will be analyzed in the present paper as they are related to the long-term needs of customers.

Ventures on demand satisfy customers' long-term needs. Furthermore, customers cooperate with these ventures during a long period of time. This long-term cooperation substantially reduces the risks of cultural entrepreneurs. This research proposes that emergence of ventures on demand is one of the sides of the on-demand economy that is quickly growing now. This is an important development both for crowdfunding and for theory of entrepreneurship as it clarifies the change in entrepreneur-customer relations generated by crowdfunding.

An additional contribution of this paper comprises of a new taxonomy of rewards that crowdfunded ventures offer to their backers. This taxonomy helps to better understand the nature of rewards compared with existing classifications.

2. METHODS

This research combines desk research methods with a narrative literature review. It is based on the study of sources devoted to crowdfunding, specifically to reward crowdfunding. It does not use empirical data. The research draws on the concept of an on-demand economy (Frenken & Schor, 2017). Within an on-demand economy, users can get access to resources when they need them. The concept of an on-demand economy is complex and combines the following strategic models of access to resources. These models are not mutually exclusive. The models are stated as:

1. Users may immediately purchase the resources they need. This subset of the on-demand economy can be described as an economy of abundance—everything people may need is freely available on the market.
2. Resources may be temporarily rented to customers. While this model is also a part of an economy of abundance, it can also be described as an access economy, whereby people do not need to purchase resources if they want to use them, they may simply get temporary access to them.
3. Resources may be produced on demand. In this case users can get products that are not available on the market. These products are produced to satisfy specific requests.

The researcher further developed the concept of on-demand economy and demonstrated that reward crowdfunding helps to set up businesses on demand. This is discussed in detail in the following section.

3. TYPES OF REWARDS

It is integrally possible for new businesses to kickstart, or the current business to extend their influence by making use of reward crowdfunding. It appeals to larger group for support in return for a reward, for instance offering a specific product or project. Reward crowdfunding have a great impact on market economy. Rewards are duely offered in this model. The model of compensations that backers will receive is one of the key factors that determine backers' intention to participate in funding. New variations of this model have been proposed in the literature in order to specify different sub-types of crowdfunding, but the structure of the model basically remains the same (Paschen, 2017; Shneor, 2020). The most popular classification of crowdfunding reward models includes four elements (Kraus et al., 2016; Pietro, 2019).

1. Donation crowdfunding – no reward is offered to backers.
2. Reward crowdfunding – backers receive a non-monetary value. This form of crowdfunding exists in different variants depending on type of rewards proposed to backers (Meyskens & Bird, 2015; Zhao & Ryu, 2020). In reward crowdfunding, backers can receive items produced by crowdfunded ventures. Technically, this form of crowdfunding is very similar to pre-sales, as backers pay in advance for the product that will be delivered later when the venture starts operations (Allison et al., 2017; Short et al., 2017; Zhao & Ryu, 2020).
3. Debt crowdfunding, the interest – Backers' support represents a loan. Backers receive monetary compensation in the form of interest, whereby the loan itself is also paid back.
4. Equity crowdfunding – It is also known as investor model crowdfunding (Mollick, 2014). In equity crowdfunding, backers receive a share in project's profits or assets (Shneor, 2020).

Despite its popularity, this classification blurs some important aspects of models of rewarding. These are enumerated as follows:

1. It implicitly supposes that backers can receive either a full compensation of their support with extra profit i.e., reward, debt and equity crowdfunding or no compensation at all i.e., donation. However, these models represent just two poles of the continuum of compensations, which includes two basic models with different types of partial compensation (Habibi et al., 2017). As Zhao and Sun (2020) points out, value of compensation may be lower than the backer's contribution and this model can be described as a hybrid (hidden) donation. When studied in context of reward crowdfunding, all types of compensation such a financial, tangible, and intangible can imply hybrid donation. A con-

tinuum of rewards was proposed by Paschen (2017), but it does not cover all types of potential compensations and pays principal attention to the scheme of compensation such as equity and debt while not paying much attention to the nature of this compensation.

2. It does not include intangible compensations as a separate model of rewards. Intangible compensations are seen either as a part of donation crowdfunding or as an element of reward crowdfunding (Zhao & Ryu, 2020). However, while intangible compensations do not have material value, they may be valued by backers, so they do not belong to donation crowdfunding. On the other hand, intangible compensations are different from material objects (usually given as a reward in case of reward crowdfunding). So, this model of compensation should be excluded from reward crowdfunding.
3. It is difficult to clearly distinguish between reward crowdfunding and debt crowdfunding, as pre-sales are obviously debt-based (Paschen, 2017).
4. The distinction between debt crowdfunding and equity crowdfunding is not made clear enough. Equity crowdfunding includes a plethora of compensation models, and not all of them are equity-based (Mollick, 2014).

Considering the considerations above, we propose a two-dimensional classification of compensation models based on the following criteria (Fig. 1).

1. Nature of compensation: The nature of compensation could be intangible (non-material benefits), tangible, non-monetary (products and services supplied by the crowdfunded project), and monetary.
2. Value of compensation: The value of compensation could be pure donation (no compensation), partial donation (backers receive compensation, but its value is below the amount of financial support provided by backers), equal exchange (the value of compensation is equal to the value of funding) and positive compensation (the value of compensation is higher than the value of funding). It is essential to avoid confusion between the value of compensation and sufficiency of compensation. Even if the value of compensation is below the value of the backer's contribution, it may be seen as sufficient by backers.

Figure 1. Two-dimensional classification of models of compensation

Intangible value	None	Intangible token of gratitude	Intangible services	Intangible services
Tangible non-monetary value	None	Tangible token of gratitude	Standard product	Product with additional value
Monetary value		?	Interest-free loan	- Interests; - Profit sharing; - Dividends
	No compensation	Partial compensation	Full compensation	Positive compensation
	Pure donation	Partial donation	Equal exchange (swop crowdfunding)	Commercial transaction

The list of compensations given in Figure 1 is obviously not exhaustive. The features of Figure 1 are enumerated below:

1. One cell has been intentionally left blank (indicated by question mark). To the best of my knowledge, there are no crowdfunding projects that offer a partial monetary compensation to funders (which obviously does not mean that such projects may not exist). Indeed, from the financial point of view, it is not logical to offer backers less money than the amount they have paid. Partial compensation for donation mostly represents an expression of social value, which can hardly be achieved through money transfer. Projects based on partial donation generally use partial intangible or tangible non-material compensation.
2. In terms of funds, pure donation does not mean absolute freedom of use of money. Non-commercial ventures supported by backers typically publish reports with evidence of correct use of funds received via donation crowdfunding (Meyskens & Bird, 2015).
3. Both intangible value and tangible non-monetary value represent a reward (Zhao & Ryu, 2020). Thus, crowdfunding models based on these types of compensations can be described as reward crowdfunding.

4. Unlike tangible and monetary value, intangible value can be delivered to customers (at least partially) even before the crowdfunded venture becomes operational. A list of funders can be published on the project's website before the venture starts delivering the products and services it was set up for.
5. An intangible token of gratitude may be a "thank you" letter to the backer or a mention of the backer's name on the list of funders on the project's website (Zhao & Ryu, 2020). In other words, it conveys little material value. It is mostly an expression of social value. The same is true for tangible tokens of gratitude. They are small material objects which are valued for their social meaning. They commemorate the fact of participation in crowdfunding. If the entrepreneur chooses to use this model of compensation, he/she should thoroughly select a reward that will be valued by the community.
6. Intangible services represent actions that create value for backers. This value is intangible but can be a source of income (for example, promotion of backer's projects – the crowdfunded venture can put a link to the backer's website on its own internet page.
7. It is extremely difficult to make a precise distinction between an equal exchange and a commercial transaction in case of intangible compensation. This is why these two cells are merged into one.
8. The relationship between the entrepreneur and the backers in case of delivery of a standard product or a product with additional value represents a pre-order (Zhao & Ryu, 2020). Backers are customers who pay for the product that will be delivered in the future. This type of crowdfunding will be referred to as pre-sales crowdfunding (or simply pre-sales).
9. Standard product corresponds to a standard delivery without specific financial or non-financial terms. It means that backers have no privileges in comparison with other customers, the only advantage that the backers may receive is to be the first one to get the product. Other customers will be served only if there are some products left. Contrarily, in case of a product with additional value, backers get some benefits for example, a discount or a product with additional features etc.
10. A backer may get intangible, non-monetary and monetary compensations at the same time. It analyzes real benefits as more complicated. This combination of benefits is typical for reward crowdfunding, where tangible value is often supported by intangible ones in order to create additional value for backers. The goal of entrepreneurs is to design an effective combination of these types of compensations.
11. Different types of compensation can be used for different groups of backers within the same crowdfunding project (Meyskens & Bird, 2015; Tomczak & Brem, 2013). The entrepreneur has to pay attention to the composition of the

compensation portfolio in order to raise enough money and to provide backers with sufficient value without taking too high obligations. The term "hybrid crowdfunding" should be used to denote the models of crowdfunding described in pages 11 and 12.

It is important to highlight that the types of compensation described above are no more than just promises by project initiators. If a project announces a model of compensation for its backers, it does not mean that backers will indeed obtain this value as backers support projects that do not exist yet. Therefore, there is a risk of non-delivery of value (Renwick & Mossialos, 2017).

4. BUSINESSES ON DEMAND

The change in the model of financial interactions generated by crowdfunding inevitably leads to a profound transformation of relations between customers and businesses. However, the analysis of this phenomenon is missing from the literature.

Within the traditional model of customer-business relations customers are seen by entrepreneurs as a source of cash flows generated by more or less occasional purchases. Crowdfunding creates closer ties between customers and businesses and transforms customers into providers of different resources such as finance or expertise etc. Both backers and customers who do not provide direct financial support, participate in product development, project promotion etc. This new role of customers clearly demonstrates that crowdfunding is indeed a part of crowdsourcing as crowdfunding campaigns are often complemented by explicit or implicit crowdsourcing. Within this new model of business-customer relationship, customers are not just passive buyers anymore, they are active support providers for business ventures. However, the change in role of customers from the point of view of business is not the only change. The role of business from the point of view of customers also underwent a profound transformation, especially in the case of pre-sales crowdfunding. This shift was not analyzed in the literature on crowdfunding.

Due to pre-sales crowdfunding, customers declare their intention to buy the product before its actual production, and sometimes even before the company that will produce this product is set up. The customer confirms this intention by funding the project initiator. In many cases, these customers plan to continue their cooperation with this company after the initial delivery. They are ready to be constant users of this venture's product.

It means that customers do not just pre-order the product. They pre-order (if this term can be used in this situation) the company itself, as both the company and the product it plans to launch do not exist yet. Providing a project with financial support, customers initially participate in the organization of business and not just in product funding. The company is set up on demand to satisfy backers' needs in the product it promises to produce. Customers become active stakeholders in business ventures. This new role of backers is transparent in the case of cultural projects such as events, venues, movies, etc. As these projects are products at the same time; along with the feature that when the product is delivered, the project is over. When backers support a project, they simultaneously support its unique result. However, in the case of ventures this new relationship between entrepreneurs and backers is less visible because backers get a one-time reward while the venture is planned to exist for an unlimited period of time.

Backers in many cases plan to continue buying products of this venture as the venture meets their needs. This customer-venture relation continues, as other companies present on the market cannot satisfy their demands. So, the delivery of rewards does not mean that the cooperation between the venture and its backers stops. Furthermore, the delivery of rewards may be a long process, as in the case of educational ventures – courses provided as rewards may take a long period of time. Moreover, crowdfunding is often combined with donation crowdfunding (pure or hybrid). But even backers who donate in many cases, become customers of the venture when it is established. They could not contribute above the minimum level during the crowdfunding campaign for different reasons, but they are interested in using the venture's products and services.

Even if backers do not plan to continue buying products from the venture, the venture should be established in order for rewards to be delivered. This supports the view that the venture is set up on demand and simply continues servicing customers after the initial delivery of rewards.

As crowdfunding is often combined with crowdsourcing, customers not only support ventures financially but can also influence the project's product strategy to adapt the product's characteristics to their requirements and preferences (Butticè & Noonan, 2020). This influence further strengthens the role of pre-sales crowdfunding as a tool for setting up businesses on demand.

The concept of an on-demand economy is often used to analyze the contemporary model of consumption (Frenken & Schor, 2017). However, this concept is mostly applied to the possibility of instant access to products or services a customer is interested in, which is generally provided by digital platforms. I propose broadening the concept of an on-demand economy by including the idea of business on demand (Table 1). These considerations demonstrate that customers do not play a passive role in relations with businesses anymore. They are active resource providers and active stakeholders who participate in setting up businesses, whose products and services they are interested in (Figure 2).

Table 1. Structure of on-demand economy

	On-demand economy	
Type of on-demand economy	Product on demand	Business on demand
Nature	Instant access to products that customers are interested in (without actual ownership)	Possibility to support businesses that will supply products or services that customers are interested in
Organizational basis	Platform economy	Pre-order crowdfunding (for new businesses)

Figure 2. Evolution of customers within reward (pre-sales) crowdfunding

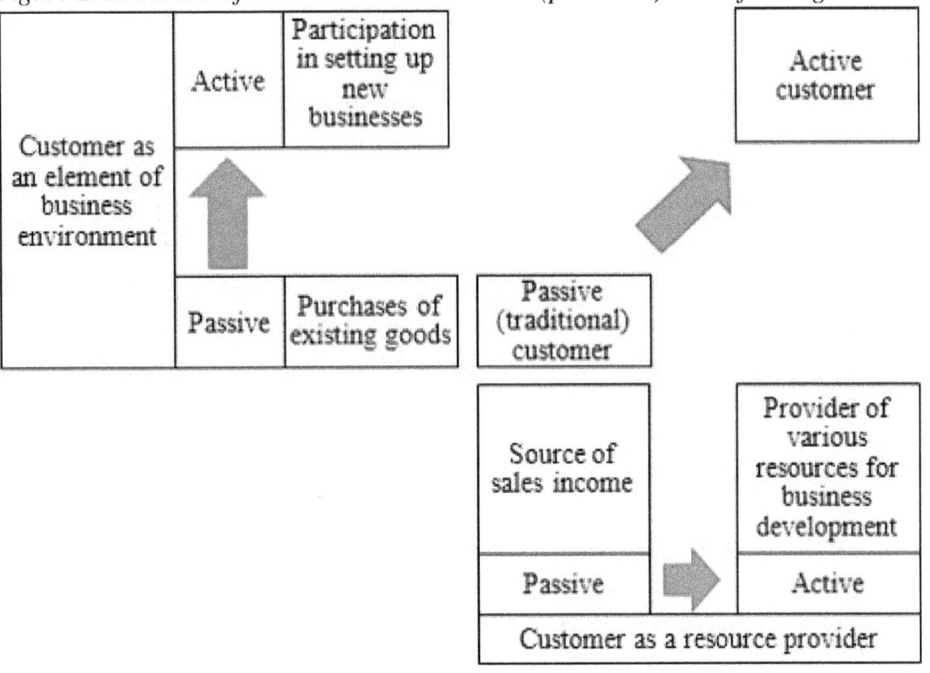

The role of customers within reward (pre-sales) crowdfunding evolves in two directions. One direction proceeds from passive purchases of products offered by existing companies to participation in setting up new ventures that will provide them with products they need. Second direction proceeds from being a source of cash flow from sales to becoming providers of resources for company's creation and for development of new products i.e., initial investments, expertise, or promotion etc. Within this new role customers have an active voice in development of the crowd-funded venture's products.

The active role of backers is confirmed by a fascinating phenomenon that can be described as inverse crowdfunding. Customers contact an existing venture and ask it to start a new project, promising their financial support for it i.e., customers express their readiness to become backers for this project. On the basis of these requests, businesses can identify projects that are worth launching. This phenomenon is primarily implicit and non-institutionalized. There are no places where backers can publish such requests, therefore they contact existing ventures personally. However, this phenomenon shows that customers can push businesses towards new projects and that they are ready to pay for them.

Within this approach, the coordinating nature of crowdfunded venture is apparent. These ventures do not simply act as producers, they function as private governments (Ménard, 2004). They coordinate backers' resources in order to satisfy backers' needs. One can say that backers vote financially for the election of the crowdfunded venture as an agent that will produce products or services for their needs. Backers are ready to provide this agent with additional resources beyond funding in order to ensure the compliance of the final product with their needs, expectations, and values (Butticè & Noonan, 2020). Backers will insist on taking their opinions into account, and ventures normally provide their backers with tools for that. Such backers-driven ventures can hardly be described as traditional capitalist firms as they bear strong similarities with consumers' cooperatives. This quasi-cooperative nature can be considered as a distinctive feature of crowdfunded businesses, particularly based on reward crowdfunding.

5. DISCUSSION

The concept of ventures on demand may have interesting implications for both theory and practice. Theoretical results represent a contribution to the growing literature on crowdfunding and a development of the theory of entrepreneurship. The crowdfunding theory, the concept of ventures on demand helps to identify one more distinctive feature that could help to differentiate reward (pre-order) crowdfunding from pre-order from existing companies. In the case of pre-order crowdfunding, backers finance the venture not the product. The contribution to the theory of entrepreneurship consists in the demonstration of the new model of relations between customers and entrepreneurs in the era of digitalization. Before the emergence of crowdfunding, entrepreneurs had to take two consecutive steps. First, they had to convince potential investors of viability of their future ventures. After this, when the venture starts operations entrepreneurs have to promote the product among the target audiences. These steps were separated over time, which increased entrepreneurial risks. Reward crowdfunding combines these two steps into one, as backers and customers are represented by the same community.

Entrepreneurs can now get information about the commercial prospects of their ventures along with financial support i.e., the support represented by pre-orders. So, entrepreneurs have to sell to customers (backers) not just the product itself but the concept of the venture they are pitching in. The appeal of their idea is directly represented to customers, and not to external investors. This entails that the venture is set up in response to customers' needs. This transformation of relations helps to better satisfy customers' demands. It is also beneficial for customers as this new model of relations substantially reduces entrepreneurial risks due to the absence of delimitation between two key stakeholders of the entrepreneurial ecosystem i.e., backers and customers. Furthermore, merging the two steps of venture development into one has eased and transformed the whole process of business initative and business marketing.

6. CONCLUSION

Crowdfunding became a game changer allowing projects and ventures with no access to traditional sources of finance to really take shape. This new form of financial relations in an era of digital technologies profoundly transformed the whole system of interactions between entrepreneurs and backers, leading to the emergence of the phenomenon of ventures on demand. Ventures on demand fill an important gap in the cultural entrepreneurial ecosystem, providing customers with a possibility of a higher level of satisfaction of their needs, while entrepreneurs can substantially reduce their risks. Possible directions of future research may include developing useful tools to regulate relations within venture on-demand projects so that both sides (backers and entrepreneurs) can collaborate efficiently.

REFERENCES

Ahsan, M., Cornelis, E. F. I., & Baker, A. (2018). Understanding backers' interactions with crowdfunding campaigns: Co-Innovators or consumers? *Journal of Research in Marketing and Entrepreneurship*, 20(2), 252–272. DOI: 10.1108/JRME-12-2016-0053

Allison, T. H., Davis, B. C., Webb, J. W., & Short, J. C. (2017). Persuasion in crowdfunding: An elaboration likelihood model of crowdfunding performance. *Journal of Business Venturing*, 32(6), 707–725. DOI: 10.1016/j.jbusvent.2017.09.002

Andreas, W., Holmesland, M., & Efrat, K. (2019). It is not all about money: Obtaining additional benefits through equity crowdfunding. *The Journal of Entrepreneurship*, 28(2), 270–294. DOI: 10.1177/0971355719851899

Butticè, V., Colombo, M. G., & Wright, M. (2017). Serial crowdfunding, social capital, and project success. *Entrepreneurship Theory and Practice*, 41(2), 183–207. DOI: 10.1111/etap.12271

Butticè, V., & Noonan, D. (2020). Active backers, product commercialisation and product quality after a crowdfunding campaign: A comparison between first-time and repeated entrepreneurs. *International Small Business Journal*, 38(2), 111–134. DOI: 10.1177/0266242619883984

Cai, W., Polzin, F., & Stam, E. (2021). Crowdfunding and social capital: A systematic review using a dynamic perspective. *Technological Forecasting and Social Change*, 162(1), 120412. DOI: 10.1016/j.techfore.2020.120412

Colombo, M. G., Franzoni, C., & Rossi-Lamastra, C. (2015). Internal social capital and the attraction of early contributions in crowdfunding. *Entrepreneurship Theory and Practice*, 39(1), 75–100. DOI: 10.1111/etap.12118

Crosetto, P., & Regner, T. (2018). It's never too late: Funding dynamics and self pledges in reward-based crowdfunding. *Research Policy*, 47(8), 1463–1477. DOI: 10.1016/j.respol.2018.04.020

Dikaputra, R., Sulung, L. A. K., & Kot, S. (2019). Analysis of success factors of reward-based crowdfunding campaigns using multi-theory approach in ASEAN-5 countries. *Social Sciences (Basel, Switzerland)*, 8(10), 293. DOI: 10.3390/socsci8100293

Efrat, K., & Gilboa, S. (2020). Relationship approach to crowdfunding: How creators and supporters interaction enhances projects' success. *Electronic Markets*, 30(4), 899–911. DOI: 10.1007/s12525-019-00391-6

Efrat, K., Gilboa, S., & Sherman, A. (2020). The role of supporter engagement in enhancing crowdfunding success. *Baltic Journal of Management*, 15(2), 199–213. DOI: 10.1108/BJM-09-2018-0337

Farhoud, M., Shah, S., Stenholm, P., Kibler, E., Renko, M., & Terjesen, S. (2021). Social enterprise crowdfunding in an acute crisis. *Journal of Business Venturing Insights*, 15(2), e00211. DOI: 10.1016/j.jbvi.2020.e00211

Francesca, B., Manganiello, M., & Ricci, O. (2021). Is equity crowdfunding the land of promise for female entrepreneurship? *PuntOorg International Journal*, 6(1), 12–36. DOI: 10.19245/25.05.pij.6.1.3

Frenken, K., & Schor, J. (2017). Putting the sharing economy into perspective. *Environmental Innovation and Societal Transitions*, 23, 3–10. DOI: 10.1016/j.eist.2017.01.003

Habibi, M. R., Davidson, A., & Laroche, M. (2017). What managers should know about the sharing economy. *Business Horizons*, 60(1), 113–121. DOI: 10.1016/j.bushor.2016.09.007

Huang, W. (2020). The study on the relationships among film fans' willingness to pay by film crowdfunding and their influencing factors. *Ekonomska Istrazivanja*, 33(1), 804–827. DOI: 10.1080/1331677X.2020.1734849

Inés, A., & Moleskis, M. (2021). Beyond financial motivations in crowdfunding: A systematic literature review of donations and rewards. *Voluntas*, 32(2), 276–287. DOI: 10.1007/s11266-019-00173-w

Josefy, M., Dean, T. J., Albert, L. S., & Fitza, M. A. (2017). The role of community in crowdfunding success: Evidence on cultural attributes in funding campaigns to "save the local theater.". *Entrepreneurship Theory and Practice*, 41(2), 161–182. DOI: 10.1111/etap.12263

Kraus, S., Richter, C., Brem, A., Cheng, C. F., & Chang, M. L. (2016). Strategies for reward-based crowdfunding campaigns. *Journal of Innovation and Knowledge*, 1(1), 13–23. DOI: 10.1016/j.jik.2016.01.010

Langley, P. (2016). Crowdfunding in the United Kingdom: A cultural economy. *Economic Geography*, 92(3), 301–321. DOI: 10.1080/00130095.2015.1133233

Lee, I., & Shin, Y. J. (2018). Fintech: Ecosystem, business models, investment decisions, and challenges. *Business Horizons*, 61(1), 35–46. DOI: 10.1016/j.bushor.2017.09.003

Liang, X., Hu, X., & Jiang, J. (2020). Research on the effects of information description on crowdfunding success within a sustainable economy—The perspective of information communication. *Sustainability (Basel)*, 12(2), 650. DOI: 10.3390/su12020650

Ménard, C. (2004). The economics of hybrid organizations. [JITE]. *Journal of Institutional and Theoretical Economics*, 160(3), 345–376. DOI: 10.1628/0932456041960605

Meyskens, M., & Bird, L. (2015). Crowdfunding and value creation. *Entrepreneurship Research Journal*, 5(2), 155–166. DOI: 10.1515/erj-2015-0007

Mollick, E. (2014). The dynamics of crowdfunding: An exploratory study. *Journal of Business Venturing*, 29(1), 1–16. DOI: 10.1016/j.jbusvent.2013.06.005

Mollick, E. (2016, April 21). The unique value of crowdfunding is not money — It's community. *Harvard Business Review*. https://hbr.org/

Paschen, J. (2017). Choose wisely: Crowdfunding through the stages of the startup life cycle. *Business Horizons*, 60(2), 179–188. DOI: 10.1016/j.bushor.2016.11.003

Petruzzelli, A. M., Natalicchio, A., Panniello, U., & Roma, P. (2019). Understanding the crowdfunding phenomenon and its implications for sustainability. *Technological Forecasting and Social Change*, 141, 138–148. DOI: 10.1016/j.techfore.2018.10.002

Pietro, F. D. (2019). Deciphering crowdfunding. In Lynn, T., Mooney, J., Rosati, P., & Cummins, M. (Eds.), *Disrupting Finance* [Series: Palgrave Studies in Digital Business and Enabling Technologies]. (pp. 1–14). Palgrave Pivot., DOI: 10.1007/978-3-030-02330-0_1

Renwick, M. J., & Mossialos, E. (2017). Crowdfunding our health: Economic risks and benefits. *Social Science & Medicine*, 191, 48–56. DOI: 10.1016/j.socscimed.2017.08.035 PMID: 28889030

Ryu, S., & Kim, Y. G. (2016). A typology of crowdfunding sponsors: Birds of a feather flock together? *Electronic Commerce Research and Applications*, 16, 43–54. DOI: 10.1016/j.elerap.2016.01.006

Shneor, R. (2020). Crowdfunding models, strategies, and choices between them. In Shneor, R., Zhao, L., & Flåten, B. T. (Eds.), *Advances in Crowdfunding* (pp. 21–42). Palgrave Macmillan., DOI: 10.1007/978-3-030-46309-0_2

Shneor, R., & Munim, Z. H. (2019). Reward crowdfunding contribution as planned behaviour: An extended framework. *Journal of Business Research*, 103(2), 56–70. DOI: 10.1016/j.jbusres.2019.06.013

Short, J. C., Ketchen, D. J.Jr, McKenny, A. F., Allison, T. H., & Ireland, R. D. (2017). Research on crowdfunding: Reviewing the (very recent) past and celebrating the present. *Entrepreneurship Theory and Practice*, 41(2), 149–160. DOI: 10.1111/etap.12270

Stanko, M. A., & Henard, D. H. (2017). Toward a better understanding of crowdfunding, openness and the consequences for innovation. *Research Policy*, 46(4), 784–798. DOI: 10.1016/j.respol.2017.02.003

Tomczak, A., & Brem, A. (2013). A conceptualized investment model of crowdfunding. *Venture Capital, 15*(4), 335–59. da Cruz, J. V. (2018). Beyond financing: Crowdfunding as an informational mechanism. *Journal of Business Venturing, 33*(3), 371–93. https://doi.org/DOI: 10.1080/13691066.2013.847614

Walthoff-Borm, X., Schwienbacher, A., & Vanacker, T. (2018). Equity crowdfunding: First resort or last resort? *Journal of Business Venturing*, 33(4), 513–533. DOI: 10.1016/j.jbusvent.2018.04.001

Younkin, P., & Kashkooli, K. (2016). What problems does crowdfunding solve? *California Management Review*, 58(2), 20–43. DOI: 10.1525/cmr.2016.58.2.20

Zhang, H., & Chen, W. (2019). Backer motivation in crowdfunding new product ideas: Is it about you or is it about me? *Journal of Product Innovation Management*, 36(2), 241–262. DOI: 10.1111/jpim.12477

Zhao, L., & Ryu, S. (2020). Reward-based crowdfunding research and practice. In Shneor, R., Zhao, L., & Flåten, B. T. (Eds.), *Advances in Crowdfunding* (pp. 119–143). Palgrave Macmillan., DOI: 10.1007/978-3-030-46309-0_6

Zhao, L., & Sun, Z. (2020). Pure donation or hybrid donation crowdfunding: Which model is more conducive to prosocial campaign success? *Baltic Journal of Management*, 15(2), 237–260. DOI: 10.1108/BJM-02-2019-0076

Chapter 10
Improving the Efficiency of Using Equity Captial in Oil and Gas Companies of Russia:
The Study of the Financial Performance Indicators

Irina Filimonova

Institute of Economics and Industrial Engineering SB RAS, Novosibirsk State University, Russia

Anna Komarova

Institute of Economics and Industrial Engineering SB RAS, Novosibirsk State University, Russia

Irina Provornaya

Novosibirsk State University, Russia

Timofey Mikheev

Novosibirsk State University, Russia

ABSTRACT

The paper is devoted to the study of the indicators of the financial performance of the Russian oil and gas companies in the era of the digital transformation of markets. Nowadays oil and gas companies operate against the background of the financial and technological sanctions, the global trends for decarbonization, volatile oil prices, as well as challenges imposed by the changes in the structure and

DOI: 10.4018/979-8-3693-3423-2.ch010

quality of the resource base. The paper aims to propose the decomposition of the coefficient of the sustainability of economic growth for the petrolium companies. The calculations were carried out for the major Russian petroleum companies in 2012-2020. It was revealed that there are two main drivers for the changes in the coefficient mainly connected with the capital and profit ratios. The results of the study justify the complex evaluation of the factors of the companies' development, as well as the formation of suitable managerial decisions. It is possible to increase the coefficient of sustainability of economic growth within the framework of management activities in different ways.

NOVELTY STATEMENT

This chapter introduces an innovative approach to decomposing the sustainability of economic growth (SEG) in the oil and gas sector using the chain substitution method. The analysis, applied to major Russian oil and gas companies, offers a new diagnostic framework for enhancing the efficient use of equity capital. This chapter, by identifying key performance drivers, offers actionable insights for enhancing financial strategies and lays the foundation for automated evaluation tools applicable across industries. This method's applicability and practical recommendations contribute to both academic literature and industry practices.

1. INTRODUCTION

As many markets are currently undergoing a digitalization process, this presents new problems and challenges for industries and society. As the technical and technological side of digitalization advances, there is an increasing demand for precise algorithms that would help to assess a company's performance in various spheres. Oil and gas businesses operate as one of Russia's most influential sectors. The necessity to solve the issues of the oil and gas companies is defined by the high dependence of the country's economy on the oil and gas revenues, the presence of financial and technological sanctions, as well as the need to meet the requirements of the decarbonization process (Filimonova et al., 2020). The assessment of the performance of companies operating in the oil and gas industry allows for the identification of major trends in their evolution, as well as key opportunities and challenges. The statements of financial performance indicate the responses of enterprises to the ongoing fluctuations in the economic environment (Dubolazov et al., 2021). This approach employs a variety of economic indicators, coefficients, and methods to analyze these results, allowing for the identification of problematic patterns resulting from companies' lack of preparedness for specific market alterations. Thus, smart company management activities not only help to achieve the goal of minimizing the negative impact of the problem, but also to build specific scenarios in case of a repeat of the changes that have taken place in the market.

Economic indicators allow for the assessment of the performance of mining companies (Kozlov et al., 2017). With their assistance, it is possible to review the enterprise's results for a certain period in general, as well as the performance of its elements (Abushova et al., 2017). Each company chooses its indicators based on operational experience and industry specifics. The wrong choice can lead to strategically unfavorable consequences, such as lack of growth prospects, bankruptcy, etc. (Sokolitsyn et al., 2019). To avoid such situations, it is necessary to understand the characteristics of each indicator. Furthermore, the environmental policies of various countries have presented oil and gas companies with new challenges in recent decades (Orazalin & Mahmood, 2018). The optimal capital structure of a company has long been studied. A certain share of borrowed capital in the overall structure allows companies to grow faster; therefore, in those years, the main research interest was the question of the share of borrowed capital (McConnell & Muscarella, 1985). Over time, researchers began to distinguish other components in the structure, particularly those associated with the emergence of new financial instruments (Wanke et al., 2019).

The capital structure of companies in the oil and gas industry is a topic of growing interest among researchers (Ahmed & Sabah, 2021). This structure reflects the company's activities, policies, and other operational decisions. It serves as a key indicator of sustainable development, providing business owners and investors with insights into the organization's performance (Warszawski, 1996). In the oil and gas sector, key factors influencing capital structure formation include the size of the enterprise and sustainable development parameters (Hashmi et al., 2020). This relationship arises because major players in the oil and gas market tend to have lower bankruptcy risks, making them highly attractive to lenders (Hamam et al., 2020). Additionally, large oil and gas companies often benefit from state support and implement favorable dividend policies, which further increase their access to borrowed funds (Kim & Choi, 2019). Russian oil and gas companies exhibit unique characteristics due to the structure of the national economy (Ishuk et al., 2015). Companies frequently use decomposition analysis to assess the contribution of specific factors to their efficiency. More recently, Kim and Patel (2017) decomposition has been applied to study the impact of employee stock ownership on firm performance and the effect of subsidiaries on overall company outcomes, among other applications (Ma et al., 2013). This chapter aims to propose and calculate the decomposition of the coefficient of economic growth sustainability, with a focus on Russian oil and gas companies' development. This analysis will allow for a comprehensive evaluation of the factors driving these companies' growth and support the formation of informed managerial decisions.

1.1 Capital Structure in the Oil and Gas Industry

Financial capital is perhaps one of the most influential factors in the oil and gas industry, determining companies' stability, growth, and overall sustainability (Kim & Choi, 2019). Capital structure, in simple terms, is the proportion of debt and equity that a firm employs in its venture. It is crucial to seek the most appropriate proportion of the two components for a company's financial stability. The degree to which this balance holds varies depending on the size of the company, the market forces, risk profiles, as well as access to external capital equity in the oil and gas sector (Liu et al., 2023). The financial data analysis revealed that oil and gas companies heavily depend on debt, likely due to the high capital intensity of this industry. Exploration, extraction, and infrastructural investments are capex-intensive, and to fund these projects, many firms turn to debt. However, the high use of debt may lead to increased financial risk, especially when market conditions are unfavorable, such as low oil prices. Therefore, it is crucial for companies to keep close track of their debt-to-equity ratios so as not to over-leverage themselves and face problems with solvency (AlimoradiJaghdari et al., 2020). Equity capital, on the other hand, helps to minimize financial risk because it enables business organizations to finance their operations without obtaining loans from financial intermediaries. To recap, effective equity financing has the key advantage of increasing a company's financial performance by decreasing its capital cost (Ahmed & Sabah, 2021). Furthermore, investors and creditors tend to view firms with higher equity levels favorably, as they can readily raise additional capital from the market when necessary. This is especially true in the oil and gas sector, where fluctuating underlying commodity prices often result in huge hurdles.

Sustainable development is a fourth dimension that has an impact on the capital structure of the oil and gas companies. Over the past few years, there has been a growing concern about the evolution of business models based on ESG concepts. Thus, a number of firms worldwide have included the sustainable development goals in their financial planning. According to Villarón-Peramato et al. (2018), a company's capital structure plays a crucial role in determining its trajectory towards sustainable development. It serves as a benchmark for determining whether the company is headed in the right direction and sets the highest standards for future sustainability. Thus, it can be pointed out that the factors influencing the elements of capital structure are also influenced by specific economic conditions of the Russian oil and gas industry. According to Hashmi et al. (2020), larger Russian oil and gas firms receive state aid, which cuts their risk of going bankrupt and makes them less risky for creditors. Such government support, especially regarding the dividend policy and the magnitude of these enterprises, stabilizes the companies' capital structure and equity financing (Kim & Choi, 2019).

1.2 Unique Features of Russian Oil and Gas Companies

Some characteristics set the Russian oil and gas industry apart from its international counterparts. The state heavily invests in the major Russian oil and gas companies, highlighting the sector's importance in the Russian economy (Al-Shaiba et al., 2019). Such state backing enables these firms to secure a financial base and avoid bankruptcy issues, making them more appealing to both domestic and international investors. State-owned banks and government promotion of financing conditions facilitate the easier acquisition of equity capital (Hamam et al., 2020). Also, thanks to the optimal policies, Russian oil and gas companies provide quite high dividends, which guarantee stable revenues to shareholders, thus improving investors' attitudes (Kim & Choi, 2019).

A third source of competitive advantage for Russian oil and gas companies lies in their size and structure of operation. Some of these companies are vertically integrated, meaning they participate in all aspects of a project, ranging from exploration and production to processing and marketing of the product. It helps in overcoming operational risks on a large scale and increases efficiency, giving these organizations a competitive advantage in global markets (Khoso et al., 2022). Furthermore, because these companies play a significant role in the national economy, many of them establish strong relationships with financial institutions. These institutions provide them with favorable lending rates and other financial conditions that other industries, particularly small and mid-sized ones or those less significant to the economy, may find unattainable (Ishuk et al., 2015). The political influence of Russia, the world's largest supplier and exporter of oil and natural gas, also contributes to the capital structure of these firms. State participation, along with the industry's export orientation, ensures that Russian oil and gas companies have access to both equity and debt funds, which are key sectors. These aspects of Russian oil and gas companies are somewhat understandable, as they help the latter remain financially sound and are an integral part of the national economy, making them investors' darlings.

2. LITERATURE REVIEW

The capital structure of oil and gas companies, particularly in the context of equity capital, is a widely studied subject due to the unique financial and operational challenges these companies face. In the broader context, the literature on capital structure is deeply rooted in theories of corporate finance, especially the trade-off theory, pecking order theory, and agency cost theory, which explain how companies decide on the mix of debt and equity. This section reviews the existing literature on capital structure in the oil and gas industry, focusing on the factors that influence financial performance, the role of state support and governance, and the implications of sustainable development on financial strategy.

2.1 Factors Influencing Capital Structure in the Oil and Gas Industry

Hence, the capital structure of the oil and gas companies depends on certain factors that define the credit and non-credit balance of the financing. The industry is a high-risk industry primarily because it works in a capital-intensive structure and requires large-scale capital outlay for exploration and extraction, infrastructure development, and technology upgrade (Kim & Choi, 2019). Therefore, the selection of the appropriate capital structure may be complicated and specific for the oil and gas companies. Some of the factors that affect the capital structure in these firms include market factors, size of the firms, their risk characteristics, tax effects/impact, and external finance constraints. The fluctuating price of oil and gas is one of the influential causes that impact the capital structure decisions of the oil and gas companies (Liu et al., 2023). The market for oil is very sensitive to changes in supply and demand forces, political risk factors, and economic factors. If the price of oil rises, then it is expected that companies will continue with their operations and activities and also increase borrowing in order to finance their investments. Nonetheless, during periods of declining prices, the companies with high levels of debt become vulnerable to some extent to financial risk as the companies' capacity to pay off such debts decreases in proportion to the declining revenues; therefore, in the operation of oil and gas companies, the debt-to-equity ratio should be controlled in order to avoid high financial risks due to fluctuations in prices. This means that there is stability in the capital structure so as to reduce the instances of insolvency, especially during a period of unstable markets.

Another factor that may influence the capital structure of the firms operating in the oil and gas industry is the size of the company. Larger firms, especially those with vertically integrated structures and a large asset base, are better placed in terms of market access and cost of capital than their counterparts in the small firms. Such large oil and gas companies are also seen as less risky credit risks in the eyes of lending institutions as well as investors due to their size, market power, and diversification of revenue sources. Hashmi et al. (2020) also identify that bigger firms have the least probability of bankruptcy and further improve access to credit at less cost and better leverage ratios. Also, these firms are characterized by economies of scale, which reduce costs and increase profitability, thus enhancing a good balance between debt and equity financing. Another important factor that determines the capital structure of the oil and gas companies is tax factors. Interest on debt is tax-allowable in many countries, which implies interest taxes act as a shield that reduces the overall tax expenses for companies with large debts. Khoso et al. (2024) noted that this tax-induced advantage means that firms deploy debt to finance their working capital needs since the cost of capital for such firms is cut down. But at the same time, I should note that the same debt can be a source of meaningful tax shields and that even operational leverage can be rather dangerous in a period of low profitability. Oil and gas companies have to, therefore, balance the tax shield against the implications of higher leverage.

Another strategic factor that is evaluated when determining the capital structure is risk management. Market risk, operational risk, environmental risk, and geopolitical risks are some of the risks that are associated with the oil and gas industry. It was carried further to discuss that high-risk firms prefer equity financing rather than debt to avoid financial distress in bad macroeconomic conditions. Ahmed and Sabah (2021) further posited that equity capital can act as a financial buffer that can absorb any unfavorable market situation since it does not contain a fixed date for repayment like the debt capital. On the other hand, risk-lower companies may go for more leverage by opting to adopt high debt levels due to secure cash flows to maximize returns on investment. Last but not least is the availability of external financing, which is also relevant to the capital structure. It is important for oil and gas companies to obtain a large amount of capital for capital-intensive projects such as drilling more wells or constructing pipelines; this can be in the form of debt or equity, which is determined by the financial health status, credit rating, and relationship with financial institutions. This is consistent with the assertion that companies in a good financial position and high credit rating have easy access to debt at lower costs and hence are able to sustain high leverage. On the other hand, firms, which have weak access to debt instruments, tend to use equity instruments as the means of financing the capital requirements (Hamam et al., 2020).

2.2 State Support and Corporate Governance in Russian Oil and Gas Companies

The two main areas that affect the choice of capital structures in oil and gas companies include state support and corporate governance, with particular reference to the Russian oil and gas industry. The main sector of the Russian economy that is receiving large state support is the oil and gas industry, and this support enhances the financial robustness of the firms and the minimum risk of their bankruptcy. Such state support can be in various forms, such as preferential credit from state-owned banks, subsidies, and other forms of favorable regulatory backdrop (Shiobara, 2006). A Russian company gets equity capital more easily through state-backed financing on account of the fact that they are sheltered from equalities of market volatility. However, there are other factors as well in the determination of the capital structure of the oil and gas industries, some of which include corporate governance and state support. Effective corporate governance approaches, for instance, accurate financial disclosures, an efficient board of directors, and the implementation of powerful risk management systems can foster a corporation's performance and capacity when it comes to attracting investors (Thurner & Proskuryakova, 2014). That is why companies with efficient governance policies have a higher potential of keeping the proper capital structure, which is regarded as a significant factor affecting the company's financial risks. Further, efficient firms can mitigate agency costs and hence enhance their capital markets and the cost of financing (Buck, 2003).

That is why the role of state support and its influence on the system of corporate governance is especially acute in the situation with Russian oil and gas companies. Yakovlev (2008) highlighted that the industry is largely filled with state-owned enterprises; he stated that, as a result, Russian corporate governance in oil and gas industries is highly dependent on government involvement. This type of governance structure can therefore have a direct impact on the capital structure, bringing either benefits or otherwise. On the one hand, it is evident that government intervention makes available financial resources that do not expose firms to the high risk of bankruptcy. On the other hand, the high extent of state control may reduce the freedom that these firms have in making their own financial decisions, which may result in inefficient capital structures. Like the Western majors, Russian oil and gas companies also reap bonuses of steady dividend support from the state. These policies render the companies more appealing to equity investors since they are associated with reliable and certain cash flows. Based on the work of Adachi (2010), the ability of Russian oil and gas firms to depend on state support and the statutory policies that encourage favorable dividends ensure the firms have a sound capital structure that employs most of the equity financing. This stability is especially relevant for the companies working in the oil and gas (O&G) sector, which is known for its fluctuations.

2.3 Impact of Sustainable Development on Capital Structure

Specifically, sustainable development has emerged as a significant factor concerning capital management of oil and gas organizations since such entities have to adapt to several dominant ESG factors (Adachi, 2010). The shift toward sustainable development on the international level, especially in the aspect of decreasing greenhouse emissions and switching to green energy, has a major impact on the business planning strategies of the oil and gas industries (Feng et al., 2018). This shift has forced a change in perspective regarding the level of debt and equity financing those firms will be able to undertake since achieving sustainable development goals entails the need to invest in costlier new technologies and structures. Arguably, the oil and gas sector is capital-intensive, and this complicates one's ability to fully appreciate the notion of sustainable development in this particular domain (Bogan, 2012). To enhance sustainability performance, such as providing clean energy, decreasing carbon emission rates, and increasing energy density, firms are mandated to incur high capital expenditures. Such investments require the careful regulation of the capital structure; firms need to ensure they have access to enough capital, but at the same time it should be sustainable. Villarón-Peramato et al. (2018) also state that capital structure can also act as a reference for sustainable development because companies with the right balance ratio of debt to equity sources have the advantage when targeting investors who take into account CSR factors. Research shows that those firms that engage themselves in reporting on sustainability are in a better position to access equity financing from socially conscious investors who are always on the lookout for stable and healthy investments.

The change of focus to sustainability has also impacted the cost of capital for the oil and gas companies. Thus, firms with well-developed ESG profiles are expected to have a lower cost of borrowing and better funding accessibility since they present less risk for investors and financiers. Kim and Choi (2019) see that such sustainably oriented companies are valued more and can attract funds more effectively, including both debt and equity funds. This has important implications for the capital structure decisions because the companies that follow a sustainable development strategy can decrease their reliance on debt and increase the equity, thus reducing the overall financial risks associated with the increases in leverage (Aibar-Guzmán et al., 2022). But the process of attaining these sustainability objectives alongside the management of capital structure can be very complex for most oil and gas firms. Debt financing has for long been embraced by the industry considering the fact that exploration, production, and infrastructure investment often involve large sums of money. On the one hand, debt can be advantageous in that it offers tax-shield benefits, which is a type of operating cash flow, and often overrides the cash flow disadvantage derived from excessive leverage. On the other hand, a high level of debt leads to financial risk and constrains a company's ability to invest in sustainable development. AlimoradiJaghdari et al. (2020) argue that for companies in the oil and gas sector, there is the need to find the right proportions of debt to equity so that the company can have the required funding to finance long-term projects on sustainability without having to load its balance sheet with too many liabilities.

However, besides the financial obstacles, there is the regulatory burden, which forces the oil and gas companies to consider other, more sustainable means of operation. To curb emissions of greenhouse gases and encourage the use of clean energy sources, governments and other regulatory agencies in different parts of the world are tightening their environmental standards. This sort of regulatory climate has put pressure on oil and gas industries to develop and procure new technologies as well as infrastructure that most of the time entails large amounts of capital outlay. According to Ahmed and Sabah (2021), affordable capital, appropriately channeled to address these new regulations, and a stable capital structure should be achieved for organizations' sustainable growth. In addition, expectations from the investors are also placed more on sustainability since the majority of the investors invest in businesses that have incorporated the ESG factors in their operations. In response to this changing investor attitude, the aforementioned industries have now directed their attention to equity funding since equity funders are more receptive to future-oriented investments in sustainability projects than debt funders, who are fixated on immediate returns. The possibility to mobilize equity capital from investors' eager on sustainable investments is instrumental for the oil and gas organizations, which are interested in financing the transition to the low-carbon business model as well as decreasing the environmental footprint.

3. MATERIAL AND METHODS

The coefficient of sustainability of economic growth (SEG) is an indicator reflecting the rate of increase in equity capital due to the enterprise's financial results. For oil and gas companies, this indicator is the most relevant since there is an objective need to increase the return on the use of equity capital obtained from capitalized profits and reduce dependence on borrowed funds. Consideration of various methods of the factor analysis made it possible to choose in favor of the method of chain substitutions, which refers to a deterministic model (Qiu et al., 2015). This method allows for the determination of the degree of influence of individual factors on the change in the value of the effective indicator by step-by-step replacement of the value of each factor indicator (driver) by the actual value (Boyer & Filion, 2007). The results obtained in this way take into account the change in one, then two, three, etc. subsequent factors, while not changing the indicators of other factors.

The model proposed for consideration is an eight-factor model of the sustainability of economic growth coefficient calculated according to formula (1).

$$K_{seg} = \frac{RE}{WACE} = \frac{\overline{Cap}}{WACE} \times \frac{\overline{Cap_a}}{\overline{Cap}} \times \frac{\overline{Cap_r}}{\overline{Cap_a}} \times \frac{Rev}{\overline{Cap_r}} \times \frac{SP}{Rev} \times \frac{EBT}{SP} \times \frac{NP}{EBT} \times \frac{RE}{NP} = R_1 \times S_1 \times S_2 \times R_2 \times Prof \times R_3 \times R_4 \times S_3 \quad (1)$$

where RE – retained earnings; WACE – weighted average cost of equity; $\overline{Cap}$ – average total capital; $\overline{Cap_a}$ – average capital advanced into assets; $\overline{Cap_r}$ – average amount of real capital advanced in assets used for entrepreneurial purposes; Rev – operating revenue; SP – sales profit; EBT – earnings before tax; NP – net profit; R_1 – ratio of total capital to equity capital; S_1 – share of capital actually advanced into assets in total capital; S_2 – share of capital advanced into assets, for entrepreneurial purposes, in real capital; R_2 – the rate of return of real capital advanced into assets used for entrepreneurial purposes; Prof – profitability of sales; R_3 – the ratio of profit before tax to profit from sales; R_4 – the ratio of net profit to profit before tax; S_3 – share of retained net profit in total net profit. The chain substitutions were further transformed into eight indicators of the company's financial and operational performance. The calculations were based on data from the financial statements and yearly reports of the seven Russian major oil and gas companies, Rosneft, LUKOIL, Surgutneftegaz, Gazpromneft, Tatneft, NOVATEK, and Bashneft, from 2012 through 2020.

Based on the proposed approach and described data, the SEG coefficient was calculated for the major Russian oil and gas companies in dynamics for 2012–2020 as shown in Table 1. It was revealed that most of the companies follow general trends, which are usually defined by external challenges. Critically small coefficient values for all companies fell in 2020—the period of a pandemic: the minimum level of oil and gas production over the past 10 years, a drop in demand, a strong decline in oil and gas prices. Negative values of the indicator in some periods until 2020 indicate that the companies Surgutneftegaz, Tatneft, and NOVATEK are loyal to their shareholders and are not inclined to fully capitalize on their net profit.

Table 1. The SEG coefficient for the Russian oil and gas companies

Company	2012	2013	2014	2015	2016	2017	2018	2019	2020
Rosneft	0.13	0.17	0.07	0.09	0.02	0.04	0.08	0.09	-0.02
LUKOIL	-	-	0.12	0.06	0.02	0.08	0.12	0.11	-0.12
Surgutneftegaz	0.17	0.13	0.34	0.21	-0.04	0.05	0.20	0.02	-
Gazpromneft	0.17	0.13	0.08	0.07	0.15	0.14	0.18	0.09	0.00
Tatneft	0.14	0.12	0.15	0.13	0.12	0.02	0.15	-0.03	0.05
NOVATEK	0.17	0.26	0.02	0.09	0.41	0.17	0.16	0.62	0.01

Company	2012	2013	2014	2015	2016	2017	2018	2019	2020
Bashneft	0.19	0.02	0.03	0.18	0.10	0.36	0.17	0.10	-0.05
Industry overall	0.23	0.23	0.22	0.18	0.11	0.14	0.23	0.22	0.05

Also noteworthy is the period from 2014 to 2016, in which many companies show a steady downward trend in the coefficient. This is due to the imposition of sectoral sanctions on Russian oil and gas companies and the decline in oil and gas prices (Saitova et al., 2022). Furthermore, the main drivers were calculated based on the proposed approach of the chain substitutions of the decomposition of the SEG coefficient as shown in Table 2.

Table 2. Changes of the elements of the SEG coefficient decomposition of the Russian oil and gas companies in 2020

Company	R_1	S_1	S_2	R_2	Prof	R_3	R_4	S_3
Rosneft	1.64	-0.09	-17.02	0.22	0.02	5.84	-0.24	1.00
LUKOIL	0.51	0.18	16.68	0.44	0.03	-2.59	0.30	7.34
Surgutneftegaz	0.00	-	3.22	0.21	0.11	7.92	0.00	-
Gazpromneft	0.77	0.10	23.56	0.23	0.03	5.49	0.06	1.00
Tatneft	0.59	-0.20	-13.68	0.27	0.17	1.29	0.58	1.00
NOVATEK	0.21	0.89	6.43	0.16	0.06	14.27	0.06	1.00
Bashneft	0.65	0.41	6.28	0.36	-0.04	4.49	0.13	3.59

For almost all research companies, the largest share of the change of the SEG coefficient is explained by the share of capital advancing into assets in total capital and the ratio of profit before tax to profit from sales.

4. DISCUSSION

The proposed approach can serve as a diagnostic tool for the problem areas that allow for the identification of the drivers of economic growth for each company. Moreover, the impact on such drivers will support the improvement of the efficiency of using equity capital. It is possible to increase the coefficient of sustainability of economic growth within the framework of management activities in different ways, such as an increase of the reinvestment ratio, improvement of the level of production management and sales of manufactured products, implementation of competent control over the financial, labor, and material resources of the company, improving the use of fixed assets, and increasing the level of financial independence of the organization. Thus, the following specific proposals for improving the stability of the functioning of each studied oil and gas company can be highlighted. For Rosneft, it is advisable to influence financial leverage by reducing the share of borrowed funds in the capital structure.

To improve the efficiency of using equity capital, LUKOIL's management should take measures to increase productivity in the area of the core activities and reduce costs by enhancing technologies for the extraction and processing of oil and gas. The sustainable development of Surgutneftegaz will ensure the impact on the production profitability driver by reducing the cost of production and improving its quality (improving production technologies). Companies Tatneft and Bashneft should consider revising the dividend policy, which will have an impact on the driver of the reinvestment ratio.

NOVATEK has the most significant number of levers. To increase the efficiency of using equity capital, the company's management should pay attention to parameters such as dividend policy and production cost. Thus, to improve the efficiency of using equity capital, oil and gas companies can usually be recommended to reduce dividend payments in favor of reinvesting profits. Overall, the proposed approach can be applied to the analysis of the company's operations in different industries. It can serve as a methodic basis for the creation of automatic evaluation tools. However, it is advisable to combine it with other tools of performance assessment to make the results more robust. Moreover, the regular reevaluation of the parameters will increase the flexibility and efficiency of the managerial decisions, and due to time-consuming calculation, that can be achieved with the application of information technologies.

5. CONCLUSION

This study fills a research gap by utilizing an approach to break down sustainability indicators for economic growth, which aids in company comparisons. This study aids in elucidating the rational investment requirements of oil and gas companies, such as the financial resources required for the latest technological advancements in field development or comprehensive digitalization. Furthermore, it facilitates the assessment of competitive advantages, leading to a more efficient use of equity capital. l. By identifying potential issues that may arise due to dynamic changes in the economy, senior management can enhance an existing algorithm to highlight areas of concern. This would in turn assist in identifying competitive strengths and directing the right financial investment for a particular line of business to match the investment with strategic goals for optimized returns.

The oil and gas industry in Russia heavily relies on the recent liberal yet unpredictable economic, political, and environmental climate, necessitating the preservation of substantial managerial flexibility. The use of equity capital fosters this flexibility, enabling companies to quickly adapt to environmental changes, recover any losses, and maximize benefits when necessary. To obtain a multi-dimensional picture of the company's performance, new approaches to efficiency and competitiveness measurement are necessary, complementing existing solutions. If properly implemented, these significant advancements will assist oil and gas firms in operating efficiently and fundamentally in response to external pressures, thereby enhancing the realization of steady and sustainable growth in the uncertain oil and gas market.

REFERENCES

Abushova, E., Burova, E., Suloeva, S., & Shcheglova, A. (2017). Complex approach to selecting priority lines of business by an enterprise. In *Procedings of the 2017 6th International Conference on Reliability, Infocom Technologies and Optimization (Trends and Future Directions)(ICRITO)* (pp. 565-569). IEEE. DOI: DOI: 10.1109/ICRITO.2017.8342491

Adachi, Y. (2010). *Building Big Business in Russia: The Impact of Informal Corporate Governance Practices* (1st ed.). Routledge., DOI: 10.4324/9780203856000

Ahmed, I. E. S., & Sabah, A. (2021). The determinants of capital structure of the GCC oil and gas companies. *International Journal of Energy Economics and Policy*, 11(2), 30–39. DOI: 10.32479/ijeep.10570

Aibar-Guzmán, B., García-Sánchez, I. M., Aibar-Guzmán, C., & Hussain, N. (2022). Sustainable product innovation in agri-food industry: Do ownership structure and capital structure matter? *Journal of Innovation & Knowledge*, 7(1), 1–10. DOI: 10.1016/j.jik.2021.100160

Al-Shaiba, A. S., Al-Ghamdi, S. G., & Koc, M. (2019). Comparative review and analysis of organizational (in) efficiency indicators in Qatar. *Sustainability (Basel)*, 11(23), 6566. DOI: 10.3390/su11236566

Alimoradi Jaghdari, S., Mehrabanpour, M. R., & Najafi Moghadam, A. (2020). Determining the effective factors on financing the optimal capital structure in oil and gas companies. *Petroleum Business Review*, 4(3), 1–20. DOI: 10.22050/pbr.2020.255708.1133

Bogan, V. L. (2012). Capital structure and sustainability: An empirical study of microfinance institutions. *The Review of Economics and Statistics*, 94(4), 1045–1058. DOI: 10.1162/REST_a_00223

Boyer, M. M., & Filion, D. (2007). Common and fundamental factors in stock returns of Canadian oil and gas companies. *Energy Economics*, 29(3), 428–453. DOI: 10.1016/j.eneco.2005.12.003

Buck, T. (2003). Modern Russian corporate governance: Convergent forces or product of Russia's history? *Journal of World Business*, 38(4), 299–313. DOI: 10.1016/j.jwb.2003.08.017

Druzhinin, A., & Alekseeva, N. (2020). Efficiency assessment and research of the internal risks in the innovation-active industrial cluster. In *SPBPU IDE '20:Proceedings of the 2nd International Scientific Conference on Innovations in Digital Economy* (pp. 1-7). ACM Digital Library. DOI: DOI: 10.1145/3444465.3444501

Dubolazov, V., Simakova, Z., Leicht, O., & Shchelkonogov, A. (2021). The impact of digitalization on production structures and management in industrial enterprises and complexes. In Schaumburg, H., Korablev, V., & Ungvari, L. (Eds.), *Technological transformation: A new role for humans, machines, and management: TT-2020* (pp. 39–47). Springer., DOI: 10.1007/978-3-030-64430-7_4

Ezu, G. (2020). Effect of capital structure on financial performance of oil and gas companies quoted on the Nigerian Stock Exchange. *International Journal of Business and Management*, 8(4), 293–305. DOI: 10.24940/theijbm/2020/v8/i4/BM2004-046

Feng, Y., Chen, H. H., & Tang, J. (2018). The impacts of social responsibility and ownership structure on sustainable financial development of China's energy industry. *Sustainability (Basel)*, 10(2), 1–15. DOI: 10.3390/su10020301

Filimonova, I. V., Komarova, A. V., Provornaya, I. V., Dzyuba, Y. A., & Link, A. E. (2020). Efficiency of oil companies in Russia in the context of energy and sustainable development. *Energy Reports*, 6, 498–504. DOI: 10.1016/j.egyr.2020.09.027

Hamam, M. D., Layyinaturrobaniyah, L., & Herwany, A. (2020). Capital Structure and Firm's Growth in Relations to Firm Value at Oil and Gas Companies Listed in Indonesia Stock Exchange. *Journal of Accounting Auditing and Business*, 3(1), 14–28. DOI: 10.24198/jaab.v3i1.24760

Hashmi, M. H. A., Khan, M., & Ajmal, M. M. (2020). The impact of internal and external factors on sustainable procurement: A case study of oil and gas companies. *International Journal of Procurement Management*, 13(1), 42–62. DOI: 10.1504/IJPM.2020.105189

Ishuk, T., Ulyanova, O., & Savchitz, V. (2015). Approaches of Russian oil companies to optimal capital structure. In *Proceeding of the IOP Conference Series: Earth and Environmental Science* (pp.1-5). IOP Publishing. DOI: DOI: 10.1088/1755-1315/27/1/012066

Khoso, A. K., Darazi, M. A., Mahesar, K. A., Memon, M. A., & Nawaz, F. (2022). The impact of ESL teachers' emotional intelligence on ESL Students academic engagement, reading and writing proficiency: Mediating role of ESL students motivation. *International Journal of Early Childhood Special Education*, 14(1), 3267–3280. DOI: 10.9756/INT-JECSE/V14I1.393

Khoso, A. K., Khurram, S., & Chachar, Z. A. (2024). Exploring the Effects of Embeddedness-Emanation Feminist Identity on Language Learning Anxiety: A Case Study of Female English as A Foreign Language (EFL) Learners in Higher Education Institutions of Karachi. *International Journal of Contemporary Issues in Social Sciences*, 3(1), 1277–1290.

Kim, K. Y., & Patel, P. C. (2017). Employee ownership and firm performance: A variance decomposition analysis of European firms. *Journal of Business Research*, 70, 248–254. DOI: 10.1016/j.jbusres.2016.08.014

Kim, S. T., & Choi, B. (2019). Price risk management and capital structure of oil and gas project companies: Difference between upstream and downstream industries. *Energy Economics*, 83, 361–374. DOI: 10.1016/j.eneco.2019.07.008

Komarova, A., Filimonova, I., Nemov, V. Y., & Provornaya, I. (2020). *Integrated indicators of companies' efficiency in the petroleum industry*. In *Proceedings of the 20th. International Multidisciplinary Scientific GeoConference SGEM*, 2020, 311–318. DOI: 10.5593/sgem2020/5.2/s21.038

Kozlov, A. V., Teslya, A. B., & Zhang, X. (2017). Principles of assessment and management Approaches to innovation potential of coal industry enterprises. *Записки Горного института, 223* (англ.), 131-138. DOI: DOI: 10.18454/PMI.2017.1.131

Lieberson, S., & O'Connor, J. F. (1972). Leadership and organizational performance: A study of large corporations. *American Sociological Review*, 37(2), 117–130. DOI: 10.2307/2094020 PMID: 5020209

Liu, H., Zhu, Q., Khoso, W. M., & Khoso, A. K. (2023). Spatial pattern and the development of green finance trends in China. *Renewable Energy*, 211(C), 370–378. DOI: 10.1016/j.renene.2023.05.014

Ma, X., Tong, T. W., & Fitza, M. (2013). How much does subnational region matter to foreign subsidiary performance? Evidence from Fortune Global 500 Corporations' investment in China. *Journal of International Business Studies*, 44(1), 66–87. DOI: 10.1057/jibs.2012.32

McConnell, J. J., & Muscarella, C. J. (1985). Corporate capital expenditure decisions and the market value of the firm. *Journal of Financial Economics*, 14(3), 399–422. DOI: 10.1016/0304-405X(85)90006-6

Orazalin, N., & Mahmood, M. (2018). Economic, environmental, and social performance indicators of sustainability reporting: Evidence from the Russian oil and gas industry. *Energy Policy*, 121, 70–79. DOI: 10.1016/j.enpol.2018.06.015

Pavlov, N., Kalyazina, S., Bagaeva, I., & Iliashenko, V. (2021). Key digital technologies for national business environment. In Murgul, V., & Pukhkal, V. (Eds.), *Energy management of municipal facilities and sustainable energy technologies EMMFT 2019. Advances in Intelligent Systems and Computing* (pp. 151–158). Springer., DOI: 10.1007/978-3-030-57453-6_12

Qiu, X. H., Wang, Z., & Xue, Q. (2015). Investment in deepwater oil and gas exploration projects: A multi-factor analysis with a real options model. *Petroleum Science*, 12(3), 525–533. DOI: 10.1007/s12182-015-0039-4

Saitova, A. A., Ilyinsky, A. A., & Fadeev, A. M. (2022). Scenarios for the development of oil and gas companies in Russia in the context of international economic sanctions and the decarbonization of the energy sector. *Sever i Rynok: Formirovanie Ekonomicheskogo Poryadka*, 25(3), 134–143. DOI: 10.37614/2220-802X.3.2022.77.009

Shiobara, T. (2006). Oversights in Russia's corporate governance: The case of the oil and gas industry. In Tabata, S. (Ed.), *Dependent on oil and gas: Russia's integration into the world economy* (pp. 85–114). Slavic Research Center.

Sokolitsyn, A., Ivanov, M., Sokolitsyna, N., & Babkin, I. (2019). Analytical model of the Interrelation between Enterprise's Activity and its Financial Stability. In *Proceedings of the IOP Conference Series: Materials Science and Engineering* (pp.1-11). IOP Publishing Ltd. DOI: DOI: 10.1088/1757-899X/618/1/012086

Thurner, T., & Proskuryakova, L. N. (2014). Out of the cold–the rising importance of environmental management in the corporate governance of Russian oil and gas producers. *Business Strategy and the Environment*, 23(5), 318–332. DOI: 10.1002/bse.1787

Villarón-Peramato, O., García-Sánchez, I. M., & Martínez-Ferrero, J. (2018). Capital structure as a control mechanism of a CSR entrenchment strategy. *European Business Review*, 30(3), 340–371. DOI: 10.1108/EBR-03-2017-0056

Wanke, P., Azad, M. A. K., Emrouznejad, A., & Antunes, J. (2019). A dynamic network DEA model for accounting and financial indicators: A case of efficiency in MENA banking. *International Review of Economics & Finance*, 61, 52–68. DOI: 10.1016/j.iref.2019.01.004

Warszawski, A. (1996). Strategic planning in construction companies. *Journal of Construction Engineering and Management*, 122(2), 133–140. DOI: 10.1061/(ASCE)0733-9364(1996)122:2(133)

Yakovlev, A. A. (2008). *State-business relations and improvement of corporate governance in Russia* (BOFIT Discussion Paper No. 26/2008). Bank of Finland. https://ssrn.com/abstract=1324548

Chapter 11
Involvement of Participation in Network Communication as a Factor of Its Sustainability:
The Public Policy and Conditions of Its Formation

Anna M. Kuzmina
St. Petersburg State University, Russia

Alexey E. Kuzmin
The Russian Presidential Academy of National Economy and Public Administration, Russia

ABSTRACT

The paper deals with the phenomenon of public policy and conditions of its formation through the contractual choice. Purpose of the research is to characterize the process of involving the target audience in the communication of non-profit organizations in social networks from the point of view of digitalization tasks. Research objectives are to determine a set of methods and tools in the media communication practice of working with the audience in social networks to increase the degree of involvement of the target audience in communication; to characterize parameters of audience reflection in various digital systems for its impact on engagement. Authors prove the hypothesis that network communication is necessary for providing the conventional basis for public policy formation and determines the sustainability of

DOI: 10.4018/979-8-3693-3423-2.ch011

Copyright © 2025, IGI Global. Copying or distributing in print or electronic forms without written permission of IGI Global is prohibited.

network interactions through the prism of contract choice. In addition, the authors determine the need to support the contract choice within the framework of communication of public sphere actors by the mechanism of their involvement in the digital communication process.

DESCRIPTION

This study presents an innovative approach by exploring the crucial role of participation in network communication as a determinant of sustainability within public policy frameworks. Unlike previous research that primarily focused on technological or infrastructural aspects, this study uniquely emphasizes the social dimension, particularly the participatory dynamics that contribute to the long-term viability of network communication systems. By integrating contemporary theories of public policy formation, the research identifies vital conditions that enhance sustainability, providing valuable insights for policymakers and stakeholders in the evolving digital landscape.

1. INTRODUCTION

Network communication sustainability is an urgent and essential issue that addresses the modern needs of digital society's functioning, as it impacts the stability of information exchange between various entities in a globalized world (Nielsen & Thomsen, 2011). Economic transactions, interpersonal communication, and governance processes increasingly depend on digital networks; thus, robust communication systems must be constructed to maintain uninterrupted operations (Dlouhá et al., 2018). Sustainable network communication requires the ability of the network platform originations, hardware, software, and human resources to adapt to new demands without necessarily raising new demands on the network's functionality and response (Zutshi & Creed, 2018).

Further, the sustainability of network communication cannot be solely technical; it should also follow economic and social parameters. Some of the critical objectives of economic sustainability involve efficiently managing resources to ensure that networks can run effectively without compromising their utilization of resources, such that costs are being cut and numerous environmental effects are visible (Mamonov et al., 2016). While technical sustainability deals with making the network technology sustainable for communication and Internet Service Providers, social sustainability is concerned with facilitating access to communications networks by all society players to the technology, ensuring that the digital divide is closed.

These aspects are essential because vagaries in entry can give birth to substantial socio-economic divides, especially in less developed areas.

Public policy plays a pivotal role in promoting the sustainability of network communication by setting standards, providing incentives for innovation, and regulating practices that could harm the long-term viability of communication systems (Laužikas & Miliūtė, 2020). Policies that encourage investment in sustainable technologies and practices are essential for maintaining the integrity and resilience of network infrastructures over time. As the digital landscape continues to evolve, the importance of sustainable network communication will only increase, necessitating ongoing attention from policymakers, industry leaders, and the public. Public policy significantly influences the sustainability of network communication mainly through the formulation of policies on sustainable communication standards, encouragement of innovation and policies against practices that are detrimental to the achievement of sustainable communication (Dlouhá et al., 2013). Support from policies that foster the usage of sustainable assets and methods is vital in ensuring that the twenty-seven networks' infrastructures are sturdy and sustainable in the long run (Laužikas & Miliūtė, 2020). Hence, conveying concepts of sustainable network communication will be critical as flexibility in the digital space grows, prompting stakeholders such as policymakers, business professionals, and society to pay more attention to the issue.

1.2 The Role of Public Policy in Network Communication

Public policy is also very significant in the development and implementation of the networks of communication, especially in aspects like the improvements in the technology of the communication network, aspects of accessibility and equity of the networks of communication, as well as the sustainability of the networks of communication (Bauer, 2010). The governments of nations across the globe have realized that proper communication is a crucial element in any nation's security and economic development and to make sure that the society in question is well united and thus have put in place policies aiming at proper development and management of these channels. Communication regulation forms a large part of the interface policy in the networking sector. It is imperative because regulatory frameworks exist for properly functioning communication networks, equal opportunities for the companies involved, and banning monopolistic access to specific channels (Spulber & Yoo, 2005). For instance, net neutrality policies are implemented to make it impossible for Internet service providers to act in ways that limit access to content or particular services, thereby preserving the competitive dynamics of the Internet (Park et al.,

2016). They make innovative solutions possible and enable all immediate users of the resources to have equal opportunities for their use.

Other public policy areas have contributed to supporting the development and establishment of advanced communication networks. Governments usually subsidize and encourage installing broadband-speed Internet and other technologies, especially in remote or rural areas (Whitt, 2003). These policies are essential for reducing the digital divide and enabling all of its citizens to have the appropriate technologies that will allow them to engage in the digital market. Another way that it impacts the network is through the setting of policies and frameworks that determine standards and standard practices. The world governments and the international bodies set the formalisms for different networks to be compatible and secure. These standards are essential for international communication processes and for users' privacy protection and secure data transmission (Koch-Baumgarten & Voltmer, 2010).

However, another area of interest in the public policy of network communication is how it helps society deal with future issues like cybercrimes or data protection. With the growth of complexity in the communication networks, there is also the growth of the network's susceptibility to cyber threats and data incursions (Crozier, 2007). Government intervention is vital since it creates policies that provide guidelines for protecting essential infrastructures and personal data and builds confidence in digital communication. Some policies that governments have set for security are the policies that require companies to conduct security assessments frequently and put efficient encryption mechanisms into practice, which can be considered as an example of how governments attempt to protect network communication (Howlett, 2009). Besides, it is essential to acknowledge that public policy is crucial in increasing cooperation between the public and the private domain, especially in developing various activities and projects. The difficulties that come with network communication have further been acknowledged by many governments, meaning that it will only be in the capability of the public sector to handle it with others. Thus, the search for assistance from the private sector has been taken (Nielsen & Thomsen, 2011). Outsourcing is a tool that empowers innovations and increases the quality and reliability of the services provided by the networks during different crises.

1.3 Importance of Participation in Public Policy Formation

Engagement in policymaking is the defining feature of democracy and the process of making laws relevant, efficient, acceptable, and feasible (Holman & Dutton, 1977). Laypeople and stakeholders must be included in network communication flow policies since their consequences impact society. This strategy helps ensure that the policies adopted are responsive to the public's needs, issues, and vision (Kathlene & Martin, 1991). The two discussed concepts both point to an increased improvement in the exercise of transparency and accountability in the governance processes when participatory policy formation is encouraged. Thus, all stakeholders, including the citizens, industry and civil society organizations (CSOs), have a ready force to influence policies to ensure they are accepted as legitimate and trustworthy. This sense of legitimacy is essential when it comes to the politics of the networks, which, among other things, include matters touching on data privacy, rights in cyberspace, and equal access (Peng & Tao, 2022). Through engaging the citizens in the process, policymakers ensure that the people have confidence in the policies that shall be put in place, hence the need to enable communication policies to take root and exist for a long time.

In addition, being involved in policy formulation enables policymakers to include local knowledge and facts, which are very important when developing policies suitable for the local community. It is crucial in network communication because the needs and challenges are heterogeneous across various geographical regions or socio-economic statuses (Liu et al., 2020). For example, when it comes to formulating policies for the provision of actual physical infrastructure regarding the development of broadband, consultation with communities will help formulate policies that will address problems of relevant sectors in specific areas that lack such access. These localized efforts assist in reducing the digital divide, thus increasing other people's access to means of communication. Policymaking with the participation of the stakeholders also helps support innovation significantly. Due to the engagement of stakeholders from different contexts, such as the governmental, industrial, academic and public sectors, aspects of participatory processes enable the formation of new ideas and solutions that may not be developed in the manner of a more bureaucratic approach to process a solution (Moon, 2020). It applies even more to network communication as it constantly develops, and new challenges and opportunities are bound to surface as time passes. For instance, inclusive public consultations and stakeholder workshops are convenient platforms for radical ideas in cybersecurity, digital governance, or infrastructure advancement (Borgia, 2014).

Diversity inclusion also helps in minimizing the consequences of policy failure since the decision involves input from all the parties. Hence, the policies and decisions formulated without involving various stakeholders are likely to face opposition from other individuals and are bound to have low success rates. On the other hand, participatory processes help the policymakers foresee the likely implementation problems, shortcomings and other unforeseen issues associated with a policy, hence improving policies that would better fit the changing landscape (Berkvens et al., 2024). In the case of network communication, given the rapid changes in technology and the high levels of interest attached to many policies, this capacity proves essential to making interventions sustainable and effective.

1.4 Challenges in Achieving Sustainable Network Communication

Organizing sustainable network communication raises several issues, and most arise from constant advancements in technology in the field of communication. There is still the issue of the digital divide, which results in inequalities in communication network access among the deprived groups. This gap is further magnified by other socio-economic disparities where the deprived, the poor, and the countryside are prone to lack adequate access to broadband speed or any other digital items (Imoize et al., 2021). Therefore, the above groups cannot engage in the communication network and the related policymaking, which affects the emergence of a sustainable and effective communication network (Nižetić et al., 2020). Other factors, such as environmental issues, are significant challenges that affect network communication sustainability. The growth of the data centers as a part of information technologies' infrastructure has also contributed to the overall energy demand and, thus – to the emissions. Indeed, as the usage of digital services increases at exponential rates, the environmental impacts of these infrastructures are a worrying factor. The task is to regulate the current energy consumption and think about the future environmental consequences of an increasingly growing society using digital technologies (Ahmad et al., 2021). Thirdly, there is a very high rate of technological product change, which has resulted in the creation of electronic wastes which are difficult to recycle, hence causing many setbacks in the management of the technological lifecycle in an environmentally effective manner (Singh et al., 2020).

This layer of complexity is underpinned by the increasing sophistication of 'Network Communication Systems' that are inherently pervasively vulnerable. Depending on the complexity of the systems, these reinforce and expand their vulnerability to cybercrimes that hamper their stability and dependable nature. Protecting these networks demands considerable resources, which proves to be a severe issue in pursuing stable, secure lines of communication. Besides, because technological advancement is increasingly thrilling and markedly faster than the capacity of governance systems to change, governance challenges can thwart the appropriate orientation of innovation toward sustainable development (Mondejar et al., 2021).

The second source of complexity stems from the numerous organization members' shouldering interconnectivity: network communication. A government has its self-interests and the interest of the citizens it serves, while business and CSOs have their agendas, which sometimes are opposite, all pulling different ways. Thus, reaching a specific consensus about sustainable policies among these diverse groups takes work. It reveals a need to coordinate these objectives toward achieving a sustainable digital future, although it is pretty challenging because establishing this coordination entails a lot of bargaining and compromise (Ahmad et al., 2021). The challenges mentioned above exemplify that the sustainability of community network communication takes work. Solving them requires a complex strategy that will focus simultaneously on using technology, protecting the environment, providing equal access to opportunities, and reliable regulation. Only in this way can an integrated and optimized structural design for network communication systems guarantee a secure future.

1.5 The Role of Technology in Enhancing Participation

Technological systems, specifically in the digital age, have considerably changed how individuals and corporations engage with others, especially in policymaking processes and decision-making systems (Harahap et al., 2023). In this case, using technology to encourage mass and real-time participation is one of the primary benefits of using technology. For example, users interact in online forums and social media and engage in e-governance and many other platforms that have provided a broader spectrum of participants access to knowledge and a broader spectrum of participants in deliberative arenas that define public policy (Haleem et al., 2022). Also, technology has made it easy to get involved with small or high risks due to the reduced hurdles. According to Sandi et al. (2021), the use of smartphones, business meetings online, and other related digital tools enable individuals to participate in policy-related discourse regardless of their location and status compared to a few years ago. It has been especially relevant in ensuring that groups that do not usually get the chance to participate in policy-making due to their status of being vulnerable or minority groups can have that chance. The availability of these technologies has also contributed to the increase of more frequent, effective, and protracted participation as compared to the previous methods whereby people could only participate at certain times and places since the ideas could be submitted at the convenience of the owners (Lihua et al., 2020).

Moreover, using technology in the participation process makes the whole process more transparent and accountable. One realizes the clarity and openness when meeting discussions, decisions, and policy drafts are recorded electronically. The record allows stakeholders to track the form and weigh in their policymaking process (Thunberg & Arnell, 2022). It enables stakeholders to be actively involved without having to doubt the process through which their input will be included or if their concerns will be met. AI and big data analytics have increased participation by allowing comprehensive input analysis from varied stakeholders. He argues that these technologies can detect patterns of interest, preference, and concern that could be undetected with conventional pattern analysis (Kricorian et al., 2020). In this way, the overall stakeholder interest before the policymaker is more accurately represented than if a particular stake exerts undue influence. Automated tools also help in individualized feedback and engagement, particularly in increasing the relevance of participation calls (Kricorian et al., 2020).

However, important issues that must be considered when using technology to increase participation include digital literacy, privacy and the risk of creating a digital divide (Kricorian et al., 2020). To increase the effectiveness of technological participation, it is essential that all participants be adequately trained for their use, and other problems associated with data security should be premised (Morris, 2023). Thus, although technology can improve participation, it has to be adopted wisely and combined with equal opportunities to impact network communication's sustainability positively.

2. LITERATURE REVIEW

2.1 Theoretical Foundations of Network Communication Sustainability

Network communication sustainability is linked to the general concept of sustainable development, which refers to improving human well-being without detriment to the future generations' quality of life while considering the social, environmental, and economic objectives (Ashrafi et al., 2020). This concept has also been adopted for communication networks' sustainability, which concerns the sturdiness and effectiveness of the technology or communication networks and its contribution towards enhancing long-term social and environmental objectives (Lihua et al., 2020). A crucial theoretical lens in this regard is socio-technical systems, where communication networks are realized as socio-technical systems of flow characterized by technical and social elements. This approach underlines that technology and society are co-dependent and that sustainability of network communication cannot occur solely through technological advancements but might require the engagement of multiple, more or less obvious social factors such as regulators, industrialists, or the society (Haleem et al., 2022). In this respect, the socio-technical approach also attaches significance to the governance element, as the coordination mechanisms and rules impact the progression of the technological environment and the distribution of access to the means of communication (Ahmad et al., 2021).

Another valuable theoretical assumption is rooted in the line of study known as environmental communication, which explores how communication practices and technologies shape people's environmental concerns and actions (Singh et al., 2020). This body of work posits that sustainable communication networks should reduce their detrimental impacts on the environment and facilitate people's responsibilities towards the environment (Kantabutra & Ketprapakorn, 2020). Ecological modernization expands the notion by claiming that combining environmental and economic objectives through technological progress is possible, given that the proper policies and institutional framework are established. Thus, it is possible to conclude that the theoretical framework for network communication sustainability is based on various socio-disciplinary approaches. It considers the insights from studying socio-technical networks, environmental communication, and ecological modernization perspectives. Such views imply that sustainability must incorporate technology, society, and the environment into a systemic approach.

2.2 The Role of Public Policy in Shaping Network Communication

Thus, public policy is a critical determinant of the sustainability and effectiveness of networked communications systems, especially in today's digital society, where networking technologies are essential to social organization. Public policy includes rules, resources, and perceptions within societies that define network communication systems' design, governance, and usage (Robinson, 2020). Policies affect networks most evidently, where the government of a country plays the central role in delivering regulation interventions. For instance, the basic policies defining the procedures for distributing the spectrums are central in guaranteeing that the communication networks will be free from interference. This is especially the case with high population density zones such as urban centers where the need for wireless communication is most felt (Weible et al., 2020). Sustainable spectrum management policies help improve a communications network's ability and dependency, thus minimizing network overcrowding and service interruptions.

Furthermore, it is essential to note that equal communication over a network is an outcome of public policy. Thus, the policy measures for overcoming the digital divide, for example, subsidies for the connection of households in areas with limited access to the Internet, are instrumental in opening up and developing the possibilities of using digital communication technologies for all population groups (Tsoy et al., 2021). Such policies underpin principles of social justice and support the conversation's sustainable development by encouraging equal opportunities in the digital context. Another parameter of the formation of network communication is economic stimuli resulting from the activity of public policy. The government should provide financial incentives such as tax exemptions, subsidies and other incentives to support investments in emerging new communication technologies and facilities (Straßheim, 2020). For example, some policies envisage bringing fiber-optic networks or 5G technology into an organization that will improve the rate of the network communication system in that organization due to the improvement in the physical structures (Dunlop et al., 2020). Such policies fuel technological advancement needs and guarantee that networking communication systems effectively meet future needs.

Besides the formal and the economic levers, the public policies for the networks are also responsible for defining the ethical and legal boundaries of communication. Rules of privacy, data protection, and cybersecurity are necessary for building confidence in Autonomous system numbers (ASNs) based on the technique of digital communication (Balog-Way et al., 2020). These values indicate that users need to increase the employability of network communication technologies and the ability to optimize their sustainability; however, weak legal protection may prevent users from actively utilizing these technologies. Moreover, it is necessary to add that public policy can affect the utilization of green technology in network communication, minimizing the adverse environmental effects of these systems. Measures that encourage using energy-proficient data centers, encourage proper disposal and recycling of electronic waste, and support using renewable energy sources are essential in lowering the carbon footprint of network communication. By introducing environmental suggestions into policy regulations, governments can prevent the extension of networking communication systems from hurting the environment.

2.3 Participation and Stakeholder Involvement in Network Communication

Engagement and funding of those concerned are other critical aspects that boundary the sustainability of network communication systems. Therefore, the participation of numerous stakeholders, hence government entities, private sector organizations, civil society organizations and the public, is vital to formulating policies and practices that would ensure that the usage of the networks is fair and responsive to the needs of every user. Participative decision-making is thus in consonance with the tenets of democratic governance, which suggests that decision-making processes should be executed openly and involve citizens in the decision-making processes (De Luca et al., 2022). The study also revealed that jointly set policies are more efficient and stable than those formulated exclusively by the leaders. It is because they cater to the interests and requirements of the stakeholders, they have the most impact. For example, Troise and Camilleri (2021) in their research noted that when the stakeholders are involved in the development of the network communication policies, then the policies that are being formulated are more likely to solve some of the existing problems, such as the provision of access to information, individuals' rights to data protection, among other. This way, the overall sustainability improvement of the network is promoted, and a positive impact on the public interest is guaranteed.

Among the conditions which can cause actual participation in network communication to have little meaning, the most important one is the conflicting interests of various stakeholders. The government institutions' objectives may be compliance with the law and security of the nation, whereas the company/organization may have profit-making and market domination as their motive (Schmidt et al., 2020). While the state has interests in security, justice, social order, and the public good, civil society organizations may hold contentious concerns, such as digital rights, privacy and fairness. Addressing such diverse and often competing interests is essential, making the process more of a political nature where stakeholders seek to find points where everyone can agree and be satisfied with the outcome, which is the network's sustainability (Kassen, 2020). The digital divide is another hindering reason for efficient participation; it includes a difference between those connected to digital tools and those without connection. This bifurcation can stifle the participation of various groups, including people with low incomes, minorities, and those in rural areas, in network communication and policy formulation. More specific efforts to address this issue include increasing the availability of high-speed internet connections, raising the populations' digital literacy levels and engaging minorities in policy decision-making processes (De Luca et al., 2022).

Another exciting area of research explored in the literature is the application of technologies to enhance stakeholders' engagement in network communication. Using technology, communication media and developments, such as social networking sites, broader and more representative participation can be achieved since stakeholders can participate in discussions, provide feedback and contribute to the discussion and formulation of decisions from remote centers (Kassen, 2020). However, the applicability of these tools depends on the level of users' engagement, tool and content accessibility, interface friendliness, and credibility (Franklin & Roberts, 2018). For the improvement and sustenance of network communication, these platforms must, therefore, be secure, free from prejudice, and developed in a way that can put the consumer first. People's involvement in the network communication system is crucial for sustaining these networks. Thus, if the client and multiple other interests are presented and all stakeholders are encouraged to participate in the process of policymaking, then communication networks that are appropriate, dependable and fair are likely to be created. Further research should aim to find other strategies for involving stakeholders and reducing the barriers that hinder stakeholder engagement in network communication (Balog-Way et al., 2020).

2.4 Technological Advancements and Their Impact on Network Sustainability

Information technology has impacted the sustainability of network communication in the following ways due to its innovations. Current phenomena like 5G, sophisticated cloud computing, and the Internet of Things (IoT) have created new ideas about boosting network standards and availability. However, these advancements also meet many problems that need solving to be sustainable in the long run. Another one of the revolutionary technologies is 5G, which is expected to provide much higher download speed, much shorter wavelengths, and the simultaneous connectivity of countless devices (Abid et al., 2022). 5G has virtues since it may support sustainable development by improving smart cities, energy efficiency, and intelligent health care, but there are vices. There are questions regarding the environmental influence of the infrastructure that 5G implies, such as the deployment of multiple small cells used to relay the signals and the energy demands that would be required.

Cloud computing is another technological advancement that has played a massive role in determining the sustainability of networks. Availability of online services and, more particularly, the adoption of cloud technologies has improved ways of using resources since storage and Data processing have been migrated towards the cloud, improving scalability (Martín & Fernández, 2022). However, the energy consumption of these servers, especially the data centers that act as the central receivers of cloud services, has now become a cause for concern. Data centers use significant amounts of electricity, and their consumption is projected to rise as organizations turn to the cloud. Scholars have called for efficient data center architectures and designs incorporating green energy to lessen carbon emissions (Popkova et al., 2022). It has also escalated the introduction of a new form of network, the Internet of Things (IoT), with its new dimensions for network sustainability. The current number of IoT devices is around 20 billion, and according to experts, this figure will grow to 25 billion by 2030; this means that IoT devices constantly collect data that needs to be processed and transmitted through networks. Despite the recent discussions linking IoT to the enhancement of sustainable practices in industries, including agriculture and transport and energy ministries of different countries, there are problems regarding the amount of data to be handled in different IoT applications and the energy required for handling them (Abid et al., 2022). Some scholars have suggested creating IoT reference models focusing on energy consumption, security, and environmental friendliness in this context.

In addition, there is an opportunity to make a network more sustainable by adopting AI technology in network management. This invention helps improve network performance, determine the failures, and, thus, decrease energy usage (Roblek et al., 2020). The deployment of AI in the operation of modern complex networks can assist in automating activities and optimizing the use of resources, hence minimizing the effects of network operations on the environment.

2.5 Environmental and Social Considerations in Network Communication

Environmental and social factors now form an important part of the discussion in network publicity due to their impact on sustainability. With the growth of the globally interconnected platform, the emissions of energy consumption, carbon footprint and electronic waste concerning communication networks' provision have been an area of concern (Thunberg & Arnell, 2022). The number of data centers needed to meet the new demands of digital communication continues to increase, and it has a vast impact on energy consumption. Current research shows that data center consumes around 0. 1% of the global electricity, and the usage is expected to increase as the world becomes smart (Kricorian et al., 2020). Measures to reduce the effects of network communication towards the environment have escalated to integrating green computing techniques and energy-saving devices. For example, new functionalities of server virtualization and the effectiveness of cooling systems in data centers have minimized energy use (Mondejar et al., 2021). Also, using renewable energy by leading tech giants like Google and Apple is a positive sign, denoting the low carbon impact of digital communication technologies.

Another significant environmental concern linked to network communication is electronic waste, abbreviated as e-waste. Due to the extremely high rate of technological advancement, there is an orderly dumping of devices, which creates a high demand for electronics recycling (Robinson, 2020). E-waste consists of elements such as lead, mercury, cadmium, and beryllium, which negatively impact the environment and human health if disposed of appropriately. Awareness campaigns, take-back and recycling programs, and the circular economy regarding disposal, reuse, and refurbishment of electronic devices, especially in developing countries, are viable solutions to the e-waste issue. Network communication plays a crucial role on the social front as it influences the overall society, for example, concerning accessibility and egalitarianism. The respondents identified that the digital divide or the difference between 'haves' and 'have-nots' in terms of access to technology worsens social inequalities and reduces the chances of participation in the digital ecosystem (Weible et al., 2020). For instance, non-urban areas or impoverished groups may experience a notable challenge in gaining high-speed broadband connections, thus preventing them from harnessing the opportunities provided through digital communication.

Access to network communication is more effective and supports better social conditions for all society groups without considering economic class, gender or region, which also forms part of social sustainability. It is imperative to note that efforts to mitigate the digital gap are crucial for enhancing social equity in society, including government-backed broadband connectivity projects and digital literacy education (Thunberg & Arnell, 2022). Furthermore, data protection is vital for building confidence in networked communication because abuses in this area are more sensitive and harm specific individuals, especially vulnerable ones.

3.0 SUMMARY

In the introduction, the author identifies the topicality of the problem and underscores that independence is critical in the modern world of network communication – this pertains to participation and public policy. It increases the understanding of the necessity of integrated thinking, which requires the application of technological, social, economic, and environmental perspectives. Public policy is critical in creating sustainable Communication and Social Media Networks (CSMN), specifically emphasizing mobilizing stakeholders. The literature review builds upon these threads to discuss the theory regarding networks' sustainability, policy issues, stakeholders, and technology advancements. This paper also analyses broader aspects of network communication, including the interaction between people and the physical environment, energy usage, electronic waste disposal, and the digital gap. The review suggests that cleaning up the environment and giving a facelift to the social landscape of communication networks should go hand in hand with technological advancement. This system approach offers a holistic view of the issues and prospects of attaining sustainable network communication, which could be helpful to policymakers, business people and academics.

4.0 CONCLUSION

In conclusion, network communication sustainability is a rather delicate topic that cannot be solved using technological advancement alone, the intervention of policymakers, environmental conservationists, and social justice activists. Thus, the study emphasizes the function of public policies in providing germinal foundations for forming sustainability for the communication networks, including the policy of participation in which possible stakeholders are included in the process. Thus, efficient public policy not only promotes technological solutions and equal access to technology but also serves as a solution to environmental issues regarding network communication, for example, energy consumption and e-waste management. The most pressing concerns when building communication networks are the policies that will foster the development of green technologies and enable the provision of an all-inclusive digital platform while protecting users' privacy.

Moreover, it is also important to consider that the social elements of network communications sustainability are significant since they define how digital technologies affect various groups of people in society. The digital divide issue shall be improved by precisely defining and designing purposeful projects to increase the usage of digital technologies and levels of digital competencies among the socially vulnerable population. According to the research, network communication can only be sustainable when an organization assumes a systems approach to the organization's operations, environment, and technology. Thus, thanks to such a multifaceted strategy, all stakeholders can guarantee that digital communication's positive outcomes are maximally pro-aggressively distributed, and the world becomes more humane and perspective-driven. This conclusion stresses the need for the politics of policymaking, business management, and civil engagement to work together to foster solutions for the complex issues of the sustainability of network communication.

REFERENCES

Abid, N., Marchesani, F., Ceci, F., Masciarelli, F., & Ahmad, F. (2022). Cities trajectories in the digital era: Exploring the impact of technological advancement and institutional quality on environmental and social sustainability. *Journal of Cleaner Production*, 377, 134378. DOI: 10.1016/j.jclepro.2022.134378

Ahmad, T., Zhang, D., Huang, C., Zhang, H., Dai, N., Song, Y., & Chen, H. (2021). Artificial intelligence in sustainable energy industry: Status Quo, challenges and opportunities. *Journal of Cleaner Production*, 289, 125834. DOI: 10.1016/j.jclepro.2021.125834

Ashrafi, M., Magnan, G. M., Adams, M., & Walker, T. R. (2020). Understanding the conceptual evolutionary path and theoretical underpinnings of corporate social responsibility and corporate sustainability. *Sustainability (Basel)*, 12(3), 760. DOI: 10.3390/su12030760

Balog-Way, D., McComas, K., & Besley, J. (2020). The evolving field of risk communication. *Risk Analysis*, 40(S1), 2240–2262. DOI: 10.1111/risa.13615 PMID: 33084114

Bauer, J. M. (2010). Regulation, public policy, and investment in communications infrastructure. *Telecommunications Policy*, 34(1-2), 65–79. DOI: 10.1016/j.telpol.2009.11.011

Berkvens, L., Roets, A., Haesevoets, T., & Verschuere, B. (2024). Local civil society organizations' appreciation of different local policy decision-making instruments. *Local Government Studies*, 50(3), 545–572. DOI: 10.1080/03003930.2023.2227592

Borgia, E. (2014). The Internet of Things vision: Key features, applications and open issues. *Computer Communications*, 54(1), 1–31. DOI: 10.1016/j.comcom.2014.09.008

Crozier, M. (2007). Recursive governance: Contemporary political communication and public policy. *Political Communication*, 24(1), 1–18. DOI: 10.1080/10584600601128382

De Luca, F., Iaia, L., Mehmood, A., & Vrontis, D. (2022). Can social media improve stakeholder engagement and communication of Sustainable Development Goals? A cross-country analysis. *Technological Forecasting and Social Change*, 177(1), 121525. DOI: 10.1016/j.techfore.2022.121525

Dlouhá, J., Barton, A., Janoušková, S., & Dlouhý, J. (2013). Social learning indicators in sustainability-oriented regional learning networks. *Journal of Cleaner Production*, 49, 64–73. DOI: 10.1016/j.jclepro.2012.07.023

Dlouhá, J., Henderson, L., Kapitulčinová, D., & Mader, C. (2018). Sustainability-oriented higher education networks: Characteristics and achievements in the context of the UN DESD. *Journal of Cleaner Production*, 172, 4263–4276. DOI: 10.1016/j.jclepro.2017.07.239

Dunlop, C. A., Ongaro, E., & Baker, K. (2020). Researching COVID-19: A research agenda for public policy and administration scholars. *Public Policy and Administration*, 35(4), 365–383. DOI: 10.1177/0952076720939631

Haleem, A., Javaid, M., Qadri, M. A., & Suman, R. (2022). Understanding the role of digital technologies in education: A review. *Sustainable operations and computers*, *3(4)*, 275-285. DOI: DOI: 10.1016/j.susoc.2022.05.004

Harahap, M. A. K., Kraugusteeliana, K., Pramono, S. A., Jian, O. Z., & Ausat, A. M. A. (2023). The role of information technology in improving urban governance. *Jurnal Minfo Polgan*, 12(1), 371–379. DOI: 10.33395/jmp.v12i1.12405

Holman, H. R., & Dutton, D. B. (1977). A case for public participation in science policy formation and practice. *Southern California Law Review*, 51(6), 1505–1534. 10822/715792 PMID: 11661665

Howlett, M. (2009). Government communication as a policy tool: A framework for analysis. *Canadian Political Science Review*, 3(2), 23–37. DOI: 10.24124/c677/2009134

Imoize, A. L., Adedeji, O., Tandiya, N., & Shetty, S. (2021). 6G enabled smart infrastructure for sustainable society: Opportunities, challenges, and research roadmap. *Sensors (Basel)*, 21(5), 1709. DOI: 10.3390/s21051709 PMID: 33801302

Kantabutra, S., & Ketprapakorn, N. (2020). Toward a theory of corporate sustainability: A theoretical integration and exploration. *Journal of Cleaner Production*, 270(4), 122292. DOI: 10.1016/j.jclepro.2020.122292

Kassen, M. (2020). E-participation actors: Understanding roles, connections, partnerships. *Knowledge Management Research and Practice*, 18(1), 16–37. DOI: 10.1080/14778238.2018.1547252

Kathlene, L., & Martin, J. A. (1991). Enhancing citizen participation: Panel designs, perspectives, and policy formation. *Journal of Policy Analysis and Management*, 10(1), 46–63. DOI: 10.2307/3325512

Koch-Baumgarten, S., & Voltmer, K. (Eds.). (2010). *Public policy and the mass media: The interplay of mass communication and political decision making* (Vol. 66). Routledge. DOI: 10.4324/9780203858493

Kricorian, K., Seu, M., Lopez, D., Ureta, E., & Equils, O. (2020). Factors influencing participation of underrepresented students in STEM fields: Matched mentors and mindsets. *International Journal of STEM Education*, 7(1), 1–9. DOI: 10.1186/s40594-020-00219-2

Laužikas, M., & Miliūtė, A. (2020). Impacts of modern technologies on sustainable communication of civil service organizations. *Entrepreneurship and sustainability issues, 7*(3), 2494-2509. DOI: DOI: 10.9770/jesi.2020.7.3(69)

Lihua, W. U., Tianshu, M. A., Yuanchao, B. I. A. N., Sijia, L. I., & Zhaoqiang, Y. I. (2020). Improvement of regional environmental quality: Government environmental governance and public participation. *The Science of the Total Environment*, 717(5), 137265. DOI: 10.1016/j.scitotenv.2020.137265 PMID: 32092810

Liu, C., Dou, X., Li, J., & Cai, L. A. (2020). Analyzing government role in rural tourism development: An empirical investigation from China. *Journal of Rural Studies*, 79, 177–188. DOI: 10.1016/j.jrurstud.2020.08.046

Mamonov, S., Koufaris, M., & Benbunan-Fich, R. (2016). The role of the sense of community in the sustainability of social network sites. *International Journal of Electronic Commerce*, 20(4), 470–498. DOI: 10.1080/10864415.2016.1171974

Martín, J. M. M., & Fernández, J. A. S. (2022). The effects of technological improvements in the train network on tourism sustainability. An approach focused on seasonality. *Sustainable Technology and Entrepreneurship*, 1(1), 100005. DOI: 10.1016/j.stae.2022.100005

Mondejar, M. E., Avtar, R., Diaz, H. L. B., Dubey, R. K., Esteban, J., Gómez-Morales, A., Hallam, B., Mbungu, N. T., Okolo, C. C., Prasad, K. A., She, Q., & Garcia-Segura, S. (2021). Digitalization to achieve sustainable development goals: Steps towards a Smart Green Planet. *The Science of the Total Environment*, 794, 148539. DOI: 10.1016/j.scitotenv.2021.148539 PMID: 34323742

Moon, M. J. (2020). Fighting COVID-19 with agility, transparency, and participation: Wicked policy problems and new governance challenges. *Public Administration Review*, 80(4), 651–656. DOI: 10.1111/puar.13214 PMID: 32836434

Nielsen, A. E., & Thomsen, C. (2011). Sustainable development: The role of network communication. *Corporate Social Responsibility and Environmental Management*, 18(1), 1–10. DOI: 10.1002/csr.221

Nižetić, S., Šolić, P., Gonzalez-De, D. L. D. I., & Patrono, L. (2020). Internet of Things (IoT): Opportunities, issues and challenges towards a smart and sustainable future. *Journal of Cleaner Production*, 274(5), 122877. DOI: 10.1016/j.jclepro.2020.122877 PMID: 32834567

Park, M. J., Kang, D., Rho, J. J., & Lee, D. H. (2016). Policy role of social media in developing public trust: Twitter communication with government leaders. *Public Management Review*, 18(9), 1265–1288. DOI: 10.1080/14719037.2015.1066418

Peng, Y., & Tao, C. (2022). Can digital transformation promote enterprise performance?—From the perspective of public policy and innovation. *Journal of Innovation & Knowledge*, 7(3), 100198. DOI: 10.1016/j.jik.2022.100198

Popkova, E. G., De Bernardi, P., Tyurina, Y. G., & Sergi, B. S. (2022). A theory of digital technology advancement to address the grand challenges of sustainable development. *Technology in Society*, 68(4), 101831. DOI: 10.1016/j.techsoc.2021.101831

Robinson, S. C. (2020). Trust, transparency, and openness: How inclusion of cultural values shapes Nordic national public policy strategies for artificial intelligence (AI). *Technology in Society*, 63, 101421. DOI: 10.1016/j.techsoc.2020.101421

Roblek, V., Meško, M., Bach, M. P., Thorpe, O., & Šprajc, P. (2020). The interaction between internet, sustainable development, and emergence of society 5.0. *Data*, 5(3), 80. DOI: 10.3390/data5030080

Sandi, H., Afni Yunita, N., Heikal, M., Nur Ilham, R., & Sinta, I. (2021). Relationship Between Budget Participation, Job Characteristics, Emotional Intelligence and Work Motivation As Mediator Variables to Strengthening User Power Performance: An Emperical Evidence From Indonesia Government. *Morfai Journal*, 1(1), 36–48. DOI: 10.54443/morfai.v1i1.14

Schmidt, L., Falk, T., Siegmund-Schultze, M., & Spangenberg, J. H. (2020). The objectives of stakeholder involvement in transdisciplinary research. A conceptual framework for a reflective and reflexive practise. *Ecological Economics*, 176, 106751. DOI: 10.1016/j.ecolecon.2020.106751

Singh, S., Sharma, P. K., Yoon, B., Shojafar, M., Cho, G. H., & Ra, I. H. (2020). Convergence of blockchain and artificial intelligence in IoT network for the sustainable smart city. *Sustainable Cities and Society*, 63(10), 102364. DOI: 10.1016/j.scs.2020.102364

Spulber, D. F., & Yoo, C. S. (2005). Network regulation: The many faces of access. *Journal of Competition Law & Economics*, 1(4), 635–678. DOI: 10.1093/joclec/nhi024

Straßheim, H. (2020). The rise and spread of behavioral public policy: An opportunity for critical research and self-reflection. *International Review of Public Policy, 2*(2: 1), 115-128. DOI: DOI: 10.4000/irpp.897

Thunberg, S., & Arnell, L. (2022). Pioneering the use of technologies in qualitative research–A research review of the use of digital interviews. *International Journal of Social Research Methodology*, 25(6), 757–768. DOI: 10.1080/13645579.2021.1935565

Troise, C., & Camilleri, M. A. (2021). The Use of Digital Media for Marketing, CSR Communication and Stakeholder Engagement. In Camilleri, M. A. (Ed.), *Strategic Corporate Communication in the Digital Age* (pp. 161–174). Emerald Publishing Limited., DOI: 10.1108/978-1-80071-264-520211010

Tsoy, D., Tirasawasdichai, T., & Kurpayanidi, K. I. (2021). Role of social media in shaping public risk perception during COVID-19 pandemic: A theoretical review. *International Journal of Management Science and Business Administration*, 7(2), 35–41. DOI: 10.18775/ijmsba.1849-5664-5419.2014.72.1005

Weible, C. M., Nohrstedt, D., Cairney, P., Carter, D. P., Crow, D. A., Durnová, A. P., Heikkila, T., Ingold, K., McConnell, A., & Stone, D. (2020). COVID-19 and the policy sciences: Initial reactions and perspectives. *Policy Sciences*, 53(4), 225–241. DOI: 10.1007/s11077-020-09381-4 PMID: 32313308

Whitt, R. S. (2004). A Horizontal Leap Forward: Formulating a New Communications Public Policy Framework Based on the Network Layers Model. *Federal Communications Law Journal*, 56(3), 587–673.

Zutshi, A., & Creed, A. (2018). Declaring Talloires: Profile of sustainability communications in Australian signatory universities. *Journal of Cleaner Production*, 187, 687–698. DOI: 10.1016/j.jclepro.2018.03.225

Chapter 12
Management of the Petroleum Projects on the Russian Arctic Shelf:
Project Management Aspects

Anna Komarova
Trofimuk Institute of Petroleum Geology and Geophysics SB RAS, Novosibirsk State University, Russia

Irina Filimonova
Trofimuk Institute of Petroleum Geology and Geophysics SB RAS, Novosibirsk State University, Russia

Vasily Nemov
Trofimuk Institute of Petroleum Geology and Geophysics SB RAS, Novosibirsk State University, Russia

Mohamed M. Adel A. M.
Novosibirsk State University, Russia

ABSTRACT

The study is devoted to a relevant problem of management and evaluation of the petroleum projetcs in Russian Arctic shelf. The significance of the region is supported by a number of factors, such as a vast resource base, important geopolitical location, and others. The Arctic region presents unique logistics, transportation, production and market challenges. The aim of the research is the development of the multi-criteria decision-making approach for the evaluation of the Arctic petroleum projects. Both qualitative and quantitative indicators (economic, technological, logistical) of the efficiency of the project implementation were considered and the

DOI: 10.4018/979-8-3693-3423-2.ch012

difference in their evaluation was studied. Results of the evaluation of the forteen petroleum projects showed the influence of the qualitative criteria on the overall problem which in turn pointed out their importance when evaluating projects in the Arctic region as financial evaluation does not have the sufficient efficiency. For the given set of objects, the preliminary sequence of development was determined under current conditions.

1. INTRODUCTION

The Russian Arctic shelf is considered the richest in oil and gas resources, containing over 50% of the region's total oil and gas resources. However, most of these resources remain undiscovered, and only one field is currently in production (Kryukov & Moe, 2018; Chanysheva and Ilinova, 2021).

The oil and gas industry is capital-intensive, and the offshore part intensifies it significantly. Thus, it is essential to break projects into investment phases such as exploration, drilling, production and logistics (Filimonova et al., 2020; Iliinskij et al., 2020). Offshore drilling has the same principle as its onshore counterpart and the addition of the challenge of stability and limited available space. Offshore drilling has the same principle as its onshore counterpart and the addition of the challenge of stability and limited available space (Konakhina et al., 2019). The second phase of field transportation infrastructure involves the design, construction and manufacturing facility system. Usually, the product would be transported from the field through pipelines to the shore and fed into the local pipeline system or exported directly (Merkulov, 2020). Shuttle tankers will require the facility to have storage and offloading capability (Fadeev et al., 2021). The Arctic shelf has a lot of logistics challenges defined by the lack of infrastructures and pack ice, which limits traversing the waters in the region to only using ice-strengthen tankers or ice-breaker escorts (Max and Johnson, 2019). The Arctic region is environmentally fragile. Therefore, the minimization of oil spills is a pressing issue (Nemov et al., 2020).

The Foundations of State Policy of the Russian Federation in the Arctic in the period to 2035 is a strategic policy introduced By the Russian government in 2020. It outlines the long-term objective of the development strategy in terms of economic growth, national security and environmental protection for the Arctic zone of the Russian Federation. Particularly, plans to upgrade the icebreaker fleet in four regional airports and to construct transport infrastructure (railways and ports) to support the Arctic region's social and economic development and uplifting of the region (Kikkas 2020). However, the government introduced a tax break strategy to promote petroleum projects in the region with the key provision of reducing the Mineral Extraction Tax (MET) to 5% for offshore oil production and 1% for gas production for the first 15 years. For liquefied natural gas (LNG) projects and gas-to-chemicals production, zero MET was implemented for 12 years, gradually increasing to a high rate between 13 to 17 years (Vatansever, 2020).

In addition to policies and taxation, western sanctions played a part in halting or delaying all development plans. These sanctions target certain companies and personnel in specific industries in a way that prohibits Western companies from supplying technologies and finance (Kozlov et al., 2020). Despite such challenges, Russia's focus on domestic development, with state policies supporting Arctic exploration and production, though long-term development depends on global market conditions and political dynamics (Shapovalova et al., 2020).

Due to the wide range of variables and criteria that need to be taken into account while evaluating and ranking petroleum projects on the Arctic shelf, a comprehensive approach like the Multi-Criteria Decision-Making (MCDM) approach is necessary to assess all the considerations systematically while evaluating and ranking different fields (Lukashevich et al., 2019). The Multi-criteria decision-making (MCDM) approach is a comprehensive framework that extends beyond merely comparing financial aspects when evaluating alternatives. It includes technical performance, environmental impact, feasibility and other relevant factors, not solely focusing on cost or profitability (Wei et al., 2021). Moreover, to handle the uncertainty and imprecision associated with decision-making in the supplier section, the MCDM technique with fuzzy and intuitionistic fuzzy Technique for Ordered Preference by Similarity to the Ideal Solution (TOPSIS) methods, combined with flexible entropy, provides a more suitable and accurate assessment compared to traditional methods. (Wood, 2016).

Wang et al., (2020) present Multi-Criteria Decesion-Making (MCDM) model that integrates decision-making techniques, including the Analytical Hierarchy Process (AHP) and the techniques for order preference by similarity to the ideal solution (TOPSIS) to asses suppliers based on multiple criteria such as cost, quality and delivery. These enhance decision-making efficiency by providing a structural approach and addressing quantitative and qualitative factors. At the same time, Kolios et al., (2016) focused on energy-related problems by utilizing the MCDM approach under stochastic inputs. It highlights how to handle uncertainty and different criteria successfully through TOPSIS and PROMETHEE methods. Fadeev and Tsukerman (2020) evaluate the technical and economic potential to asses the Western Arctic shelf development. They determine the most promising crucial developmental fields by using comparative analysis of various factors, such as infrastructure, financial viability and technical challenges. It also highlighted the importance of economic and technical factors for the success of Arctic petroleum projects.

Besides the complexity of Arctic projects due to environmental, technical and economic risk factors, the MCDM methods for the same decision-making problem can help by evaluating and balancing these factors and choosing the optimal choice development strategies under uncertain conditions (Zolotuhin and Stepin, 2019). Pangsri (2015) investigates the application of MCDM methods for project selection in construction companies. As a result, all methods like AHP and TOPSIS enhance decision-making by considering qualitative and quantitative factors, ultimately improving project management outcomes.

Thus, the main factor explaining the low level of activity in the region is the high risk of the project implementation, which the Russian government has been trying to address in recent years. The institutional changes in 2020 introduced several tax breaks and other incentives for hydrocarbon development in the Arctic region. However, it might not be enough due to falling oil prices and a lack of infrastructure. Due to the changing framework of the development of the Arctic, it is essential to introduce complex assessment methods for the project's efficiency. Therefore, the research aims to develop the MCDM approach for evaluating the portfolio of Arctic petroleum projects.

2. LITERATURE REVIEW

Petroleum resources are abundant on the Russian Arctic shelf and are essential to the nation's economic development and energy security. However, the severe Arctic climate, geopolitical and environmental issues, resource distribution, and the possibility of further exploration and extraction create distinct project management obstacles (Gautier et al., 2009). This literature review investigates the crucial aspects of managing petroleum projects in the Russian Arctic, focusing on risk management, technological innovation, environmental sustainability, and stakeholder engagement.

2.1 Risk Management in Arctic Petroleum Projects

The Arctic region is home to untapped oil and gas reserves, but access and operational hazards are increased by harsh weather and ice cover, which makes it hard to locate and get these resources (Gautier et al., 2009). In contrast, Samimi (2020) examines risk management practices in oil and gas refineries. It outlines various strategies for identifying, assessing, and mitigating risks associated with refinery operations. It emphasizes the importance of the oil and gas sector in implementing robust risk management frameworks to enhance safety, operational efficiency, and regulatory compliance.

2.2 Environmental and Operational Risks

Gautier et al., (2009) assess the potential for undiscovered oil and gas resources in the Arctic, estimating significant quantities of both. It highlights the Arctic's importance for future energy supplies while noting the region's challenging conditions, such as harsh weather and ice cover, which pose significant operational risks and difficulties for resource exploration and extraction.

Bashkin et al., (2017) assess geo-environmental risks associated with the oil and gas industry in the Russian Arctic. It evaluates the environmental impacts in areas affected by oil and gas operations, focusing on potential risks and the degradation of polar ecosystems. It emphasizes the need for effective management strategies to mitigate environmental damage and manage pollution in Arctic regions.

2.3 Geopolitical Risks

The effective governance of the Arctic Ocean is essential due to the increasing regional geopolitical, environmental, and economic activities. There is a need to focus on collaborative international frameworks to address challenges such as resource extraction, environmental protection, and territorial disputes in the rapidly changing Arctic, along with geopolitical risk in the Arctic and its influence over regulatory frameworks and international relations, adding complexity for project managers (Young, 2016).

2.2 Technological Innovations for Arctic Projects

The main idea of Bambulyak and Frantzen (2009) is that oil transport from the Russian part of the Barents Region is crucial for the growth of Arctic resources. The report examines the status of oil transport infrastructure, challenges related to harsh environmental conditions, and the strategic significance of Russia's energy sector.

2.2.1 Ice-Resistant Infrastructure

The development of ice-resistant platforms and sub-sea technologies has been critical in addressing the challenges posed by the Arctic environment. The assessment of oil transport infrastructure in the Russian Barents Region as of January (2009) focuses on its importance for Arctic resource development and the challenges posed by the region's harsh environmental conditions. These innovations ensure the stability and durability of infrastructure, even in harsh conditions. It also emphasizes the strategic role of the Barents Region in Russia's energy sector (Bambulyak & Frantzen, 2009).

2.2 Remote Monitoring Systems

Petrov (2016) examines the Arctic's emerging "other economies," focusing on knowledge, creativity, and innovation as drivers of economic development beyond traditional resource extraction. The study explores how new economic activities, such as creative industries, scientific research, and cultural initiatives, can reshape the region's economic landscape. It also highlights the significance of human capital and local knowledge in fostering sustainable growth and suggests that these non-extractive sectors offer new opportunities for Arctic communities. Moreover, it also emphasizes the need for policy support to enhance these alternative economic models, positioning the Arctic as a frontier for innovation and creativity.

Heleniak (2021) examines demographic trends in the Arctic, focusing on population decline, ageing, and migration. These shifts pose challenges for regional socio-economic development, including labour shortages and increased dependency on older populations. It highlights the need for policies to retain residents and adapt to demographic changes, emphasizing economic diversification as key to the Arctic's future resilience.

2.3 Environmental Sustainability and Regulatory Compliance

Curtis and Kaufman (2020) explore the responsive decision made by protection officers in the case of environmental management. The study reveals that these officers rely heavily on auditory information and informal cues rather than solely on visible data. It also highlights the importance of understanding the decision-making processes and the role of subjective factors in shaping environmental protection strategies.

2.3.1 Environmental Impact Assessments

Agarkov et al., (2018) address the environmental impacts of energy resource development in the Arctic region. They focus on such activities affecting the Arctic environment, highlighting the need for effective measures to mitigate adverse ecological effects and ensure sustainable practices in energy extraction. At the same time, Kruk et al., (2018) examine the environmental and economic damage caused by oil and gas field development in the Arctic shelf of the Russian Federation. They analyse the adverse ecological impacts and the economic costs associated with these developments, underscoring the need for improved management practices to balance resource extraction with environmental protection.

2.3.2 Compliance with Environmental Regulations

Humpert and Raspotnik (2012) analyze the future of Arctic shipping along the Transpolar Sea Route. They discussed the potential for increased shipping activity due to melting ice and improved navigation conditions while addressing the associated challenges, including environmental risks, geopolitical tensions, and the need for international cooperation to ensure safe and sustainable shipping practices in the Arctic.

Schmidt (2011) examines the environmental health concerns associated with Arctic oil drilling plans, highlighting risks such as oil spills, pollution, and the potential harm to fragile ecosystems. It emphasizes the challenges of mitigating these risks due to the Arctic's harsh conditions and remote location, raising questions about the region's readiness for large-scale oil exploration.

2.4 Stakeholder Engagement and Strategic Planning

The complex relationship between Arctic communities and extractive industries emphasises the need to move beyond a purely extractivist approach. The consequence of considering local perspectives and cosmologies favours greater engagement with indigenous communities and alternative development models to address the social and environmental uncertainties posed by resource extraction in the Arctic (Wilson and Stammler, 2016).

2.4.1 Engaging Local Communities

Lawrence and Larsen (2017) discuss the political challenges and impacts of mining on Sami lands. They highlight that current planning processes frequently marginalize Sami perspectives, prioritizing industrial development over indigenous rights and environmental protection. The paper emphasizes the need for improved planning practices that incorporate Sami views and address the socio-environmental consequences of mining activities to ensure more equitable and sustainable outcomes.

2.4.2 Strategic Planning and Adaptive Management

Carayannis et al., (2021) examined the development of Arctic offshore energy resources and the importance of strategic management. They emphasized the importance of strategic planning and management in addressing the unique challenges of Arctic offshore projects, including environmental risks, technological demands, and geopolitical factors. It outlines how effective strategic management can enhance project sustainability and success in the Arctic energy sector.

Dmitrieva et al., (2021) define and develop a conceptual framework for the strategic sustainability of offshore Arctic oil and gas projects. In order to guarantee that Arctic offshore projects may controlled in a manner that strikes a balance between financial gains and environmental and social sustainability, they provide necessary guidelines and tactics. It emphasizes the inclusion of sustainable considerations into strategic planning and decision-making processes for Arctic resource development.

The oversight of petroleum projects on the Russian Arctic shelf requires addressing various complex challenges. Effective risk management, technological innovation, environmental sustainability, and stakeholder engagement are essential for successful project execution. Adaptive management techniques and sustainable practices will be crucial to balancing resource development with environmental preservation and geopolitical stability as Arctic petroleum exploration expands.

3. MATERIALS AND METHODS

The primary methodological approach used in the chapter is Mult-Criteria Decision Making (MCDM), which can be in various forms and can adapt to different criteria. A specific form of MCDM, known as the Analytical Hierarchy Process (APH), was formulated by Thomas Saaty in 1980. AHP divides the MCDM problem into different levels of hierarchies, making it easier to analyze. It then uses the $m \times n$ decision matrix to assess the relative importance of alternatives concerning each relation to each criterion (Triantaphyllou, 2000). AHP is a well-established method used in decision-making tasks. It is effectively applied to solve complex problems of decision-making, such as determining priorities and resource allocation (Dhiman et al., 2020). AHP uses the following formula:

$$A^*_{AHP} = \max \sum_{j=1}^{n} a_{ij} w_j$$

or

$$i = 1, 2, 3, \ldots, m \qquad (1)$$

Where A^*_{AHP} is the best alternative for AHP, a_{ij} is the actual value of alternative i concering criterion j and w_j is the weight of criterion j, m is the number of criteria. AHP uses pairwise comparison to determine the relative importance or weight of each alternative in terms of each criterion. This method is used by Saaty in AHP in order to calculate criteria weight and it consists of an $n \times n$ reciprocal matrix that compares each criterion with the other. The idea of pairwise here is comparing each criterion with the other; for example, criteria A is more important or of the same importance as criteria B, and so on. The main challenge with this method is to quantify the linguistic choices of these criteria in the pairwise comparison. The widely used option is for pairwise comparisons to be quantified by a linear scale. This scale is a one-to-one mapping between the criteria with a set of numbers that imitates the linguistic choices of preferences taken by the decision maker. Thus, the values assigned according to the following: 1 means both criteria are equally important; 3 means one criterion has weak importance over another; 5 means one criterion is essential; 7 means demonstrated importance; 9 means absolute importance; 2, 4, 6, 8 means intermediate values used for in-between judgments.

The TOPSIS evaluates the alternatives by calculating their Euclidean distance to two ideal points: the positive Ideal Solution (PIS) represents the most favourable scenario even though the Negative Ideal Solution (NIS) represents the least favourable scenario. According to this, the optimal solution is the one with the shortest distance to PIS and the longest distance from NIS (Dhiman et al., 2020). After constructing the decision matrix, the TOPSIS model proceeds to normalize the elements of the matrix to convert them into non-dimensional with the following formula:

$$r_{ij} = \frac{a_{ij}}{\sqrt{\sum_{i=1}^{m} a_{ij}^2}} \quad (2)$$

Where r_{ij} is the normalized element in the matrix, a_{ij} is the original value for alternative i and criterion j.

After normalizing the matrix in MCDM, the entropy method calculated the criteria weights by assessing the amount of information each criterion provided. Its objective weight is a parameter that describes the value of a criterion in an MCDM problem. The entropy method measures the variability in the adat for each criterion. Criteria with lower entropy (more consistent data) receive higher weights, while those with higher entropy (more variability) get minimum weights. After normalizing the decision matrix when implementing one of the methods discussed before, we proceed with calculating the entropy and divergence, which in turn is used in calculating the criteria weights with the equations below.

$$e_j = -\frac{1}{\log m} \sum_{i=1}^{m} r_{ij} \log r_{ij} \quad (3)$$

$$d_j = |1 - e_j| \quad (4)$$

$$w_j = \frac{d_j}{\sum_{j=1}^{n} d_j} \quad (5)$$

Where e_j is the entropy, d_j is the divergence, m is the number of alternatives, w_j is the criterion weight. The third step here, is a unique comparision to the previous methods, is to calculate the ideal and negative-ideal solutions. For each criterion, we identify and highlight the best and worst value among all alternatives, and then we calculate the Euclidean distance for each alternative. Lastly, from the Euclidean distance, we calculate the relative closeness to the ideal solution, which is the performance score of each alternative (*Pi*), then rank the alternatives where the best alternative would be the one with the shortest distance to the ideal solution (Triantaphyllou, 2000).

$$S_i^+ = \left[\sum_{j=1}^{n}(V_{ij} - V_j^+)^2\right]^{0.5} \quad (6)$$

$$S_i^- = \left[\sum_{j=1}^{n}(V_{ij} - V_j^-)^2\right]^{0.5} \quad (7)$$

$$P_i = \frac{S_i^-}{S_i^+ + S_i^-} \quad (8)$$

Where *S* is the Euclidean distance between each alternative and the ideal (positive) and worst (negative). Vij is the weighted-normalized value representing an alternative's performance w.r.t. criteria n, *V+* and *V-* denote the highest and lowest values among the set of alternatives in each criterion, respectively. *Pi* is the performance score derived from these distances, indicating how close an alternative is to the ideal solution.

Database for the evaluation

Based on the data from the Russian International Affairs Council (RIAC) and other open sources, the database of petroleum offshore fields (see Table 1) shows minimal exploration activity in the eastern Arctic shelf, with companies focusing on the western shelf due to better infrastructure, including that builds for the Yamal peninsula and nearby ports. In contrast, the eastern region lacks the necessary infrastructure, making it less attractive for development. The average water depth in the region is 88 meters, and oil discoveries are spread equally among the four water areas.

Table 1. Fields database in the Russian Arctic shelf

Water Area	Field	Content type	Distance from shore	Average water depth (m)
Barents Sea	Ludlovskoye	NG	240km west of North Island. 200km north of Shtokman field	220
	Ice	NG	70km northeast of Shtokman field	240
	Murmansk	NG	200km from Murmansk	100
Pechora Sea	Severo-Gulyaevskoye	~75% NG	100km northwest of Varandey	20
	Dolginskoye	Oil	120km south of Novaya Zemlya. 110km north offshore	45
	Medynskoye-sea	Oil	45km northeast of Varandey	15
	Varandey-Sea	Oil	11km northwest of Varandey	16
South Kara Sea	Leningradskoye	NG	100km offshore	120
	Rusanovskoye	NG	70km north of Leningradskoye	75
	Victory	~75% NG	250km offshore	85
	Beloostrovskoye	NG	East of Rusanovskoye	75
Ob Bay	Chugoryakhinskoye	~70% NG	24 km offshore	12
	Severo-Kamennomysskoye	NG	50 km from Yamburgskoye field	13
	Ob	NG	20 km northwest of Yamburg	12

The proposed approach of the evaluation was applied to all forteen field in the region.

4. RESULTS

During the first stage of the research, the criteria and indicators were researched and organized into a threefold classification, considering the economic, technological, and logistical aspects. Each section includes multiple criteria that significantly affect the performance of projects in the region. A total of eight criteria are considered four criteria for both the quantitative and qualitative. The distribution among the sections is as follows- CAPEX, Taxes, and Revenue are the qualitative criteria in the economic section, and Structure Type (quantitative) and Sea-ice Conditions (qualitative) are in the technological section. The three qualitative criteria under the logistical section are Market Availability, Transportation, and Infrastructure Proximity (see Table 2).

Table 2. Developed criteria and their indicators

Section	Criteria	Indicators
Economic	CAPEX	estimated drilling costs according to estimated number of wells and average daily rate of contractors
	Taxes	MET calculation taking into consideration tax incentives for arctic oil and gas
	Revenue	oil and gas price forecasts, estimated revenue from sales
Technological	Structure Type	water depth, nature of operations
	Sea-ice conditions	geographical distribution of sea-ice, ice-free season
Logistical	Market Availability	potential markets, foreign energy strategies, consumption/production rates and trends
	Transportation	geographical location, mode of transportation
	Infrastructure Proximity	geographical study of the field's location relative to the required infrastructure

There are two groups of criteria: quantitative and qualitative groups. The quantitative criteria are CAPEX, Taxes, Revenue, and Structure type. CAPEX based on estimated drilling costs, calculated from average day rates of both jack-ups and drillships released by IHS Markit in 2020 and project life estimates from the production profile of each field in line with the design reference of both Prirazlomnoye and Snohvit. Revenue criterion will be based on the estimated product sales, taking Brent Crude as a reference due to its similarity to Arctic oil (ARCO) and long-term price forecasts from the EIA for oil and natural gas. The last economic criterion (Taxes) will be based on the payable amount of mineral extraction tax (MET) during the project life, accounting for the tax incentives introduced as a part of the Russian government's Arctic development strategy in 2020 and take the released tax records of the Russian Federal Tax Service as a reference. Lastly, the Structure Type criterion estimates facility cost per field, using Prirazlomnoye (Russia) and Goliat (Norway) to be taken as cost references for GBS and FPSO platforms, adjusting the inflation rates accordingly.

Qualitative criteria are analyzed and ranked for each field. In order to determine the ice-free duration for each field, data on the sea-ice conditions from the National Snow & Ice Data Center (NSIDC) were examined on the sea-ice extent in 2020 every month. Market Availability uses the BP Statistical Review of 2020 to analyze the potential markets and their energy strategies, ranking them with the help of Russia's export strategy towards these markets. Transportation methods combine the proximity of the fields to offloading points with the help of a constructed map with the field content, using the map to determine the product handling method. The last qualitative criterion is infrastructure proximity, which uses a constructed map that pinpoints the field's locations and existing infrastructure in the Arctic region. Each field is studied to identify the required infrastructure for its development. Finally, all the fields are ranked using the Analytical Hierarchy Process (AHP) model.

The Multi-Criteria Decision-Making (MCDM) model aims to rank a group of alternatives in terms of potential according to the chosen criteria while identifying the best alternative among the group. It presents a performance score calculated using the models combining all criteria. The evaluation includes the ranking of quantitative and qualitative criteria, separately and combined, as shown in the following table (see Table 3).

The final results revealed that the Chugoryakhinskoye field has the highest potential, which lies in the Ob Bay, and the Ice field in the northern Barents Sea has the lowest potential. It also shows that the closest field to Yamal has the highest performance score and high ranking due to the region's development. At the same time, fields in the Barents Sea region have the lowest ranks.

Table 3. MCDM evaluation

Field	Rank (Quantitative)	Rank (Qualitative)	Rank (Combined)
Ludlovskoye	11	13	12
Ice	13	13	14
Murmansk	12	11	11
Severo-Gulyaevskoye	8	12	13
Dolginskoye	4	6	10
Medynskoye-sea	5	6	9
Varandey-Sea	10	6	8
Leningradskoye	14	9	7
Rusanovskoye	1	9	6
Victory	3	2	3
Beloostrovskoye	2	2	2
Chugoryakhinskoye	7	1	1
Severo-Kamennomysskoye	6	4	4
Ob	9	4	5

DISCUSSION

One of the research results is the comparison of the rankings based on qualitative and quantitative criteria. There are differences and similarities between the two sets of criteria for the selected criterion. Both criteria yielded a similar understanding of the Barents Sea, but the situation is quite different for the remaining water areas. The quantitative ranking has a clear preference for Kara Sea fields, followed by the large fields in the Pechora Sea, derived from the rankings of each criterion stated previously. On the other hand, qualitative ranking prefers the fields in the Northern section of the Kara Sea and Ob Bay. The combined results showed the influence of the qualitative criteria on the overall problem, especially in the Chugoryakhinskoye field, having the highest ranking. It highlights the importance of qualitative factors in sensitive regions like the Arctic shelf, like Rusanovskoye, which ranked 1st on quantitative criteria but dropped to 6th level overall, while Chugoryakhinskoye rose from 7th to the highest potential.

The Arctic region's vast size compels experts to divide it into separate zones, subsequently evaluated differently. As seen during the research, each water area has its unique condition, whether from a nature/environmental perspective similar to accessibility issues or the proximity to infrastructure, influencing project success beyond traditional financial criteria. The results acquired from this research based on the current situation in the region and undergoing development indicate that the fields surrounding the Yamal Peninsula have the highest potential for development from different perspectives. The first perspective is the content of natural gas as its consumption globally is increasing rapidly due to the current agenda of emission reduction—Second, there are issues with the proximity of required infrastructure and product handling. Finally, China's market potential is higher than Europe's due to energy policies and the current political strain. The results also show that projects in the Barents Sea are inadvisable due to the near impossibility from a financial and technical point of view, even with infrastructure in the Murmansk area. The best solution for the Barents Sea is the construction of infrastructure in the Novaya Zemlya archipelago, which will give access to the whole of the northern Barents region.

5. CONCLUSION AND RECOMMENDATIONS

The Arctic shelf is a crucial oil production region that is mostly untapped. Several factors, such as a vast resource base, significant geopolitical location, and growing attention from the Russian government, supported the regional significance. The full and effective utilization of the Arctic shelf hydrocarbon resources has been a goal of the Russian Federation for a long time. Due to the depletion of conventional onshore resources, recent taxation incentives, and the continuous melting of Arctic ice that makes the region accessible, Russia is now closer than ever to achieving this goal. The development process and choice of the objects require precise decisions from the management of the companies and the state. The number of discovered fields on the shelf and the undiscovered potential compels us to approach the region differently for evaluation to avoid complications and obstacles like those met with Shtokman and Prirazlomnoye, which led to the use of the Multi-Criteria Decision-Making (MCDM). In this chapter, we used MCDM to rank the discovered oil and gas fields on the Russian Arctic shelf and identify those with the highest potential in the region. The final results showed the influence of the qualitative criteria on the overall problem, which pointed out their importance when evaluating projects in the Arctic region, as financial evaluation does not have sufficient evaluation efficiency. In conclusion, for the given set of objects, a preliminary sequence of development was determined that highlighted where to proceed and where to hold off. However, re-evaluation should take place regularly based on new information and regional development.

ACKNOWLEDGEMENTS

The chapter was prepared with the financial support of the Russian Foundation for Basic Research in the framework of the research project N° 20-010-00699.

REFERENCES

Agarkov, S., Motina, T., & Matviishin, D. (2018, August). The environmental impactcaused by developing energy resources in the Arctic region. In IOP Conference Series: Earth and Environmental Science (Vol. 180, No. 1, p. 012007). IOP Publishing. DOI: 10.1088/1755-1315/180/1/012007

Bambulyak, A., & Frantzen, B. (2009). Oil transport from the Russian part of the Barents Region. Status; Riksorgan för Sveriges Lungsjuka, (January), 107.

Carayannis, E. G., Ilinova, A., & Cherepovitsyn, A. (2021). The future of energy and the case of the arctic offshore: The role of strategic management. Journal of Marine Science and Engineering, 9(2), 134. DOI: 10.3390/jmse9020134

Chanysheva, A., & Ilinova, A. (2021). The future of Russian arctic oil and gas projects: Problems of assessing the prospects. Journal of Marine Science and Engineering, 9(5), 528. DOI: 10.3390/jmse9050528

Curtis, J., & Kaufman, S. (2020). "It's not what you see but what you hear…": Understanding environment protection officers' responsive decision making. Journal of Environmental Management, 262, 110336. DOI: 10.1016/j.jenvman.2020.110336 PMID: 32250813

S. Dhiman, H., Deb, D., S. Dhiman, H., & Deb, D. (2020). Multi-criteria decision-making: An overview. Decision and control in hybrid wind farms, 19-36.

Dmitrieva, D., Cherepovitsyna, A., Stroykov, G., & Solovyova, V. (2021). Strategic sustainability of offshore arctic oil and gas projects: Definition, principles, and conceptual framework. Journal of Marine Science and Engineering, 10(1), 23. DOI: 10.3390/jmse10010023

Fadeev, A., Kalyazina, S., Levina, A., & Dubgorn, A. (2021). Requirements for transport support of offshore production in the Arctic zone. Transportation Research Procedia, 54, 883–889. DOI: 10.1016/j.trpro.2021.02.143

Fadeev, A. M., & Tsukerman, V. A. (2020, April). Assessment of the prospects of developing the western arctic shelf fields on the basis of evaluation of the technical-economic potential of fields. In IOP Conference Series: Earth and Environmental Science (Vol. 459, No. 6, p. 062024). IOP Publishing. DOI DOI: 10.1088/1755-1315/459/6/062024

Filimonova, I. V., Komarova, A. V., Provornaya, I. V., & Mishenin, M. V. (2020). Changes in the structure of the oil raw material base as a factor determining federal budget revenues. Mining Journal, (4), 30–36. DOI: 10.17580/gzh.2020.04.06

Gautier, D. L., Bird, K. J., Charpentier, R. R., Grantz, A., Houseknecht, D. W., Klett, T. R., Moore, T. E., Pitman, J. K., Schenk, C. J., Schuenemeyer, J. H., Sørensen, K., Tennyson, M. E., Valin, Z. C., & Wandrey, C. J. (2009). Assessment of undiscovered oil and gas in the Arctic. Science, 324(5931), 1175–1179. DOI: 10.1126/science.1169467 PMID: 19478178

Heleniak, T. (2021). The future of the Arctic populations. Polar Geography, 44(2), 136–152. DOI: 10.1080/1088937X.2019.1707316

Humpert, M., & Raspotnik, A. (2012). The future of Arctic shipping along the transpolar sea route. Arctic Yearbook, 2012(1), 281–307.

Iliinskij, A., Afanasiev, M., Wei, T. X., Ishel, B., & Metkin, D. (2020, November). Organizational and management model of smart field technology on the arctic shelf. In Proceedings of the International Scientific Conference-Digital Transformation on Manufacturing, Infrastructure and Service (pp. 1-5). https://doi.org/DOI: 10.1145/3446434.3446510

Kikkas, K. N. (2020, September). Analysis of social development in the Arctic countries. In IOP Conference Series: Materials Science and Engineering (Vol. 940, No. 1, p. 012115). IOP Publishing. DOI: 10.1088/1757-899X/940/1/012115

Kolios, A., Mytilinou, V., Lozano-Minguez, E., & Salonitis, K. (2016). A comparative study of multiple-criteria decision-making methods under stochastic inputs. Energies, 9(7), 566. DOI: 10.3390/en9070566

Konakhina, N. A., Baranova, T. A., Mokhorov, D. A., & Prischepa, A. S. (2019, July). Innovative and technological development of the Russian Arctic. In IOP Conference Series: Earth and Environmental Science (Vol. 302, No. 1, p. 012101). IOP Publishing. DOI: 10.1088/1755-1315/302/1/012101

Kozlov, A., Kankovskaya, A., & Teslya, A. (2020, July). Digital infrastructure as the factor of economic and industrial development: case of Arctic regions of Russian North-West. In IOP Conference Series: Earth and Environmental Science (Vol. 539, No. 1, p. 012061). IOP Publishing. DOI: 10.1088/1755-1315/539/1/012061

Kruk, M., Semenov, A., Cherepovitsyn, A., & Nikulina, A. Y. (2018). Environmental and economic damage from the development of oil and gas fields in the Arctic shelf of the Russian Federation. https://www.um.edu.mt/library/oar//handle/123456789/37958

Kryukov, V., & Moe, A. (2018). Does Russian unconventional oil have a future? Energy Policy, 119, 41–50. DOI: 10.1016/j.enpol.2018.04.021

Lawrence, R., & Larsen, R. K. (2017). The politics of planning: Assessing the impacts of mining on Sami lands. Third World Quarterly, 38(5), 1164–1180. DOI: 10.1080/01436597.2016.1257909

Lukashevich, N., Garanin, D., & Konnikov, E. (2019). Modeling of the parameters of the investment project based on the information-statistical approach. In Proceedings of the 33rd International Business Information Management Association Conference, IBIMA 2019: Education Excellence and Innovation Management through. Vision (Basel), 2020, 8743–8752.

Merkulov, V. I. (2020, September). Analysis of Russian Arctic LNG projects and their development prospects. In IOP Conference Series: Materials Science and Engineering (Vol. 940, No. 1, p. 012114). IOP Publishing. DOI: 10.1088/1757-899X/940/1/012114

Nemov, V. Y., Filimonova, I. V., & Komarova, A. V. (2020, April). Assessment of the Mutual Influence of Energy Intensity of the Economy and Pollutant Emissions. In IOP Conference Series: Earth and Environmental Science (Vol. 459, No. 6, p. 062025). IOP Publishing. DOI: 10.1088/1755-1315/459/6/062025

Pangsri, P. (2015). Application of the multi criteria decision making methods for project selection. Universal Journal of Management, 3(1), 15–20. DOI: 10.13189/ujm.2015.030103

Petrov, A. N. (2016). Exploring the Arctic's "other economies": Knowledge, creativity and the new frontier. The Polar Journal, 6(1), 51–68. DOI: 10.1080/2154896X.2016.1171007

Samimi, A. (2020). Risk management in oil and gas refineries. Progress in Chemical and Biochemical Research, 3(2), 140–146. DOI: 10.33945/SAMI/PCBR.2020.2.8

Schmidt, C. W. (2011). Arctic oil drilling plans raise environmental health concerns. https://doi.org/DOI: 10.1289/ehp.119-a116

Shapovalova, D., Galimullin, E., & Grushevenko, E. (2020). Russian Arctic offshore petroleum governance: The effects of western sanctions and outlook for northern development. Energy Policy, 146, 111753. DOI: 10.1016/j.enpol.2020.111753

Triantaphyllou, E., & Triantaphyllou, E. (2000). Multi-criteria decision making methods. Springer Us., DOI: 10.1007/978-1-4757-3157-6_2

Vatansever, A. (2020). Put over a barrel? "Smart" sanctions, petroleum and statecraft in Russia. Energy Research & Social Science, 69, 101607. DOI: 10.1016/j.erss.2020.101607

Wang, C. N., Tsai, H. T., Ho, T. P., Nguyen, V. T., & Huang, Y. F. (2020). Multi-criteria decision making (MCDM) model for supplier evaluation and selection for oil production projects in Vietnam. Processes (Basel, Switzerland), 8(2), 134. DOI: 10.3390/pr8020134

Wei, Z., Zhu, S., Dai, X., Wang, X., Yapanto, L. M., & Raupov, I. R. (2021). Multi-criteria decision making approaches to select appropriate enhanced oil recovery techniques in petroleum industries. Energy Reports, 7, 2751–2758. DOI: 10.1016/j.egyr.2021.05.002

Wilson, E., & Stammler, F. (2016). Beyond extractivism and alternative cosmologies: Arctic communities and extractive industries in uncertain times. The Extractive Industries and Society, 3(1), 1–8. DOI: 10.1016/j.exis.2015.12.001

Wood, D. A. (2016). Supplier selection for development of petroleum industry facilities, applying multi-criteria decision making techniques including fuzzy and intuitionistic fuzzy TOPSIS with flexible entropy weighting. Journal of Natural Gas Science and Engineering, 28, 594–612. DOI: 10.1016/j.jngse.2015.12.021

Young, O. R. (2016). Governing the arctic ocean. Marine Policy, 72, 271–277. DOI: 10.1016/j.marpol.2016.04.038

Zolotuhin, A. B., & Stepin, Y. P. (2019, November). Problem and models of multi-criteria decision making and risk assessment of the arctic offshore oil and gas field development. In IOP Conference Series: Materials Science and Engineering (Vol. 700, No. 1, p. 012050). IOP Publishing. DOI: 10.1088/1757-899X/700/1/012050

Chapter 13
Prospects of Human-Centric Technology's Influence on Economic and Non-Economic Aspects of Public Life:
The Influence of Modern Technologies on Economic and Non-Economic Life

Oksana Pirogova
Peter the Great St. Petersburg Polytechnic University, Russia

Vladimir Plotnikov
 https://orcid.org/0000-0002-3784-6195
St. Petersburg State University of Economics, Russia

Sergey Barykin
 https://orcid.org/0000-0002-9048-009X
Peter the Great St. Petersburg Polytechnic University, Russia

Anna Karmanova
Peter the Great St. Petersburg Polytechnic University, Russia

Irina Kapustina
Peter the Great St. Petersburg Polytechnic University, Russia

Natalya Golubetskaya
St. Petersburg University of Management Technologies and Economics, Russia

Alexander Igorevich Puchkov
St. Petersburg University of Management Technologies and Economics, Russia

DOI: 10.4018/979-8-3693-3423-2.ch013

Anna Sedyakina
Peter the Great St. Petersburg Polytechnic University, Russia

Ahdi Hassan
Global Institute for Research Education and Scholarship, The Netherlands

Mikhail Loubochkin
Saint Petersburg State University, Russia

ABSTRACT

Human-centric technologies are one of the emerging trends in technological progress. The problems of its formation have not been studied well enough at the moment, which opens up a field for scientific research. In particular, the question of the impact of human-centric technologies on the economic and non-economic life of society remains open. Empirical methods of scientific research are used, in particular, to give disclosure the phenomena of scientific and technological progress at the present stage in the context of the study of human-centric technologies, and include observation, description, and measurement. As a result, it was possible to form and apply a basic approach to assessing the impact of human-centric technologies on economic and social life. The characteristics of their basic directions of development are given. The authors substantiate and finally formulate the conclusion that the formation of human-centric technologies is based on subjective and economically supported objective factors.

Novelty: The research provides a novel framework for understanding these technologies' influence. It reveals that they offer promising improvements in societal well-being and sustainable growth despite potential challenges to existing economic arrangements. It introduces a balanced evaluation of immediate setbacks versus long-term benefits, with a unique focus on the specific benefits for small and medium-sized enterprises (SMEs). Additionally, the study explores how these technologies might transform traditional industrial economies and contribute to sustainable growth, offering fresh insights into their potential to reshape economic and social structures.

1. INTRODUCTION

The basic concept of human-centric design in the circumstances of human-machine interaction and focused on prioritizing human needs, capabilities, and limitations in system creation, which is more natural, safe, and effective (Boy, 2017). Further, promote user-involved approaches that incorporate feedback at every stage of development, ensuring that the final product aligns with human requirements and increases overall user experiences. Such an approach bridges a gap between human factors and technology, making the system serve users more efficiently and effectively. The evolution of human-centric manufacturing as a core component of Industry 5.0. It focuses on the integration of advanced technologies with human expertise to create manufacturing systems that are not only efficient but also personalized and sustainable. Moreover, there has been a shift towards collaborative robots (cobots), AI-driven decision-making, and the emphasis on human well-being in industrial processes. It outlines future research directions and challenges in achieving this human-centric approach within Industry 5.0 (Lu et al., 2022).

The emerging concept of Industry 5.0 emphasizes human-centric approaches that integrate advanced technologies with human creativity and skills. The main idea focuses on how Industry 5.0 seeks to harmonize the collaboration between humans and machines, moving beyond automation to enhance human well-being. It identifies key challenges, such as ethical concerns and the need for workforce upskilling (Adel, 2022). Munsamy and Telukdarie (2019) present a Human Resource Management (HRM) model designed to optimize processes by adopting Industry 4.0 technologies. They focus on the significant benefits of integrating digital tools such as AI, data analytics, and automation into HRM. This integration can enhance efficiency, streamline operations, and improve decision-making in industrial settings. They highlighted the potential benefits of this model, including better workforce management and process optimization, and provided a framework for implementing these technologies effectively.

Azarenko et al., (2020) present a model for developing human capital with innovative characteristics in the context of the digital economy. The main idea is that the digital economy requires new approaches to human capital development, emphasizing innovation, continuous learning, and digital skills. His model outlines strategies for enhancing human capital by integrating advanced technologies and fostering creativity, aiming to prepare the workforce for the challenges and opportunities of the digital era. Pirogova and Makarevich (2020) discuss the formation of human capital within enterprises in the context of digitalization. They emphasize developing digital skills and competencies to enhance workforce efficiency and adaptability. Moreover, it outlines strategies for integrating digital technologies into human capital development, highlighting the role of continuous learning and innovation in maintaining competitiveness in the digital economy.

Mullin et al,. (2021) address the crucial need for organizations to recognize their responsibility in fostering inclusion, diversity, equity, and accessibility (IDEA) and developing leadership skills that effectively promote these principles. The main idea is that embedding IDEA into leadership practices can enhance organizational culture and improve healthcare outcomes by ensuring a more equitable and inclusive environment.

The main idea behind enhancing cultural competency in the archival profession in the United States is that implementing practices focused on equity, diversity, and inclusion (EDI) can improve the representation and accessibility of archival materials. The framework proposed aims to guide archivists in developing skills and strategies that address cultural biases and ensure more inclusive archival practices, ultimately fostering a more equitable profession (Engseth, 2018). Kline (2001) explores technological determinism, believing technology is the primary driver of societal change and shapes human behavior and culture. It examines this theory's origins, development, and criticisms, arguing that technological determinism oversimplifies the relationship between technology and society. It also highlights the importance of considering social, economic, and cultural factors in understanding how technology influences and is influenced by society.

Plotnikov and Pirogova (2018) examine how identifying and developing essential organizational competencies is crucial for enhancing its value and competitive advantage. They argue that these competencies can be used as a tool for enterprise value management. It outlines a framework for leveraging these competencies to optimize business processes, improve decision-making, and drive sustainable growth, positioning key competencies as central to effective enterprise management.

The chosen research topic is highly relevant, addressing emerging trends that may soon become prevalent. The study focuses on human-centric technologies and their interaction with society's economic and non-economic life. This chapter aims to develop an approach for assessing the impact of these technologies on various societal aspects and to determine their prospects. The key objectives are

1. Identify the influence of human-centric technologies on the global economy,
2. Analyze their impact on public life and
3. Highlight promising directions for their development in both economic and non-economic contexts.

Future research could explore technical guidelines for these technologies and predict their economic impact on specific industries.

LITERATURE REVIEW

Economics Aspect of Human-Centric Technology:

1. Labour Market and Employment:

AI can displaces jobs, However, it also creates new employment opportunities by increasing demand for products and services and examines the relationship between AI-driven productivity gains and employment trends (Crawford & Calo, 2026). It analyses the demand side on the effect of AI to identify whether increased productivity from AI leads to job creation or further displacement, providing insights into the conditions under which AI-driven demand can favourably impact employment. Brynjolfsson and McAfee (2014) explore rapid advancements in digital technologies, particularly AI and automation, reshaping the economy and society. They argue that such technologies significantly impact economic growth and productivity and create challenges, including job displacement and increased inequality. They highlight the need for policies to address these challenges, emphasizing education, innovation, and social safety nets to broadly ensure the benefits of technological progress.

Lamba and Subramanian (2020) examine India's economic trajectory, highlighting a paradox of high economic dynamism coupled with uneven development. They argue that despite rapid GDP growth, the nation needs help to achieve comparable social indicators such as health, education, and infrastructure improvements. They highlight imbalances due to weak governance, underinvestment in public goods, and regional disparities. They also emphasize the need for policy reforms to align economic growth with more inclusive and sustainable development outcomes in India.

2. Economic Growth and Productivity

Brynjolfsson and McAfee (2014) analyse the transformative impact of digital technologies, particularly AI and automation, on the economy. They argue that these innovations boost productivity and economic growth, exacerbate income inequality, and disrupt Labor markets by displacing specific jobs. It focused on policies that enhance education, foster innovation, and provide social safety nets to ensure the benefits of technological progress are widely shared and mitigate the adverse effects on workers. Bughin et al., (2018) highlighted that while demand for technological, social, and higher cognitive skills will increase, routine manual and basic cognitive skills will decline. Workers must adapt through reskilling and upskilling to remain competitive in the evolving job market. The authors emphasize the importance of proactive education and training strategies to address the growing skill gaps, ensuring the workforce is prepared for automation-driven changes.

Mokyr et al., (2015) explores historical concerns about technological change and its impact on jobs and economic growth. They argue that although technological advancements have consistently sparked widespread unemployment, the emergence of new industries and jobs has unfounded such fears. They suggest that current advances in AI and automation may heighten anxiety. The long-term effects mirror past trends, leading to economic growth and new opportunities rather than sustained job loss. Autor (2015) examines fears of widespread job loss due to automation; employment remains robust. He argues that automation often complements rather than replaces human Labor, leading to new tasks and jobs that machines cannot perform. While automation can reduce demand for routine jobs, it also increases productivity and demand for more complex tasks, creating new employment opportunities. Autor emphasizes the need for policies that support workers adapting to technological changes to ensure continued job growth.

3. Innovation and Market Dynamics

Cusumano et al., (2019) analyse the rise and dominance of digital platforms like Amazon, Google, and Facebook. They explore that these platforms create value by facilitating user interactions, driving innovation, and leveraging network effects. They also discuss the challenges and risks associated with platform dominance, such as regulatory scrutiny, market concentration, and the ethical implications of data usage. It offers strategic insights for businesses navigating the competitive and rapidly evolving platform economy.

Zuboff (2019) explores that tech companies exploit personal data to predict and influence behaviour, creating a new economic system termed "surveillance capitalism." It argues that this system prioritizes profit over individual autonomy and privacy, leading to unprecedented corporate power and social control. It also warns of the profound implications for democracy and human rights. It urges society to resist this encroachment on freedom and advocate for stronger regulations to protect individual rights in the digital age.

Non-Economic Aspects of Human-Centric Technologies

1. Social Structures and Cultural Norms

Crawford and Calo (2016) argue that AI research overlooks AI technologies' social and ethical implications. They highlighted the need for a multidisciplinary approach that includes social scientists, ethicists, and affected communities to address issues like bias, fairness, and accountability in AI systems. The authors emphasize that without considering these broader impacts, AI developments risk perpetuating inequality and harm, and they call for more inclusive research practices to ensure AI can be developed in any way that is equitable, transparent, and beneficial to society as a whole.

Meanwhile, Hampton and Wellman (2003) discovered that high-speed internet access improves local social capital, bolsters community relationships, and promotes neighbourly exchanges. They show that internet communication enriches and complements offline social networks in suburban areas where physical distance could usually limit social contact. They conclude that opening up new opportunities for social interaction and Internet connectivity can significantly contribute to the upkeep and revitalization of local communities.

2. Governance and Public Policy

Janssen and Estevez (2013) discuss that governments can increase efficiency by adopting lean principles and platform-based governance models. They argue that governments can provide more effective public services with fewer resources by streamlining processes, reducing waste, and leveraging digital platforms for service delivery. It also highlights the potential of these approaches to improve responsiveness, transparency, and citizen engagement while addressing challenges such as resistance to change and the need for cross-sector collaboration.

Mergel et al., (2019) explore the concept of digital transformation in the public sector through insights gathered from expert interviews. They found that digital transformation is broadly understood as integrating digital technologies into all government areas, fundamentally changing how public services are delivered and how government operates. They highlighted vital themes such as the need for cultural change, the importance of leadership, and the challenges of balancing innovation with public accountability. They provide a nuanced definition of digital transformation, emphasizing its complexity and the need for strategic, long-term planning.

Eubanks (2018) examines the use of digital technologies and data analytics in social welfare systems, often to the detriment of marginalized populations. It argues that these technologies can reinforce existing inequalities by profiling, monitoring, and unfairly targeting low-income individuals. It highlights such practices' ethical and social implications, advocating for greater scrutiny and reforms to ensure that technology uplifts rather than penalizes the disadvantaged.

3. Environmental Sustainability

Forti et al., (2020) provide a comprehensive analysis of global electronic waste (e-waste) trends and reveal that e-waste reached a record 53.6 million metric tons in 2019, with only 17.4% adequately recycled. The paper highlights improper disposal's environmental and health risks and emphasizes the need for improved e-waste management and recycling practices. Further, it advocates for adopting circular economy principles to reduce e-waste generation, recover valuable materials, and mitigate the negative impact on the environment and human health.

Gungor et al., (2011) discuss the various communication technologies and essential standards for implementing innovative grid technologies. Such communication technologies include wired and wireless systems that enable real-time data exchange within intelligent grids. They highlight the essential standards and protocols supporting the interoperability and integration of different components within the smart grid. By thorough examination, the study aims to underscore their role in enhancing the efficiency and reliability of energy distribution and management in modern intelligent grid systems.

3. RESULTS

Human-centric technologies replace specific current solutions, including those from top vendors, while complementing contemporary fundamental methods to treat numerous challenges arising from human activity. Contemporary basic techniques aim to automate as much as possible (Fierro et al., 2022). Automation like this lower and simplifies the cost of mass production while creating new, hitherto unattainable opportunities in the service sector. With the use of hardware and software systems, calculations for marketing, business, and non-economic fields like biology, chemistry, and medicine can now be completed that could not be completed manually or with less sophisticated machinery. For the first time in human history, humanity has the opportunity to touch the very foundations of the universe thanks to the ability to calculate models with a vast number of influencing factors and analyze big data. This includes learning in-depth details about the workings of the human body and, if necessary, connecting to its various processes; deciphering even the most minor details of physical and chemical bonds in substances to develop materials with the desired properties; and computing socioeconomic processes and interactions to maximize efficiency and improve people's lives.

Technological advancement is reflected in the international economy's unparalleled expansion within social and economic contexts, particularly regarding stability (de and Gaillard, 2022). With the formation of a solid middle class, humanity has improved the well-being of most people and conquered hunger and epidemics in many parts of the world. It even turned out to have learned over time how to smooth out the decline phases during long-term economic cycles gradually: today's most significant national and international crises are far less destructive than they were a century or more ago, mainly due to the sophisticated instruments of economic authorities based on scientific discoveries and research (Sufi and Taylor, 2022).

However, the fundamental strategies for technological advancement currently in use have some drawbacks, particularly when combined with a flawed socioeconomic structure (Eum and Lee, 2022). Modern innovations ruin humans, their infrastructure, and the environment in conflicts due to a lack of mutual understanding and effective interaction methods across different world regions (de and Rodríguez, 2022). A typical serial product replaces an individual after the shift to mass automated production, making individualization costly due to economic inefficiencies. Ultimately, in the modern world, jobs are being created at a noticeably lower rate than their disappearance. In those that have arisen, most workers are merely cheap parts of the assembly line's mechanisms, not creative or intellectual components (Rigger et al., 2022).

Middle-level employees are frequently deprived of employment due to the modern economic reality, which frequently supports the automation of middle-level creative and intellectual jobs rather than regular, challenging ones. The quality of work produced by neural networks trained to draw drawings upon request continuously improves (Chavlis and Poirazi 2021). Based on data analysis results, marketing goods will be more accurate than an experienced manager advising you on which advertising campaigns to prioritize, where to allocate budget, and which ones to discontinue (Khan et al., 2022). To develop these solutions, considerable resources and teams with experience may be needed. However, their implementation is nearly free compared to a complex automated production line or, for instance, a handyman robot for the construction industry. The latter's development seems improbable given the variations in conditions at each construction site and the associated risk that a misguided marketing robot will reduce the marketing budget's effectiveness. In contrast, a handyman robot will cause accidents, fatalities, and facility destruction (Iturralde et al., 2022).

So, it would seem valid to ask: Is using human-centric technology on a large scale in life practical and feasible? Despite the preceding, significant circumstances facilitate an affirmative response to this inquiry.

First, high-tech consumer goods and services can be introduced in addition to production technologies (Zhang et al., 2022). Additionally, as these technologies advance, many middle-class suppliers and wealthy clients are targeted to consume customized, human-centric solutions (Xin and Jin, 2022). For instance, the demand for personalized treatment is long overdue. Plans for its availability range from routine monitoring of all parameters by the world's top doctors using cutting-edge equipment to automatic monitoring with essential devices that alert the attending physician as soon as alarm signals are received, allowing them to identify the impending crisis before it materializes (Jabbar et al., 2022).

Second, there is a specific issue with corporate automation of intellectual tasks, which is only sometimes feasible. Furthermore, a fundamental weakness of contemporary approaches to developing "smart" systems is that they rely solely on mathematical models and techniques, making it impossible to describe social and economic systems in all of their complexity and adjust to changes in them. Among the most well-known and sophisticated intelligent systems are neural networks, machine learning, and extensive data analysis. Their foundations are models that, upon initial observation, have a resemblance to the model of the human brain: neurons and the connections that form between them (Kumar et al., 2022).

According to Uzquiano and Arlotta (2022), the human brain consists of approximately 80 billion neurons, each of which can create multiple connections with other neurons. The creation of artificial intelligence appears to be a straightforward task: to produce intelligence that is several times stronger than that of a human, one must have enough computer power to generate 800 billion neurons and the necessary number of connections between them. Everything is actually far more intricate.

The human brain is divided into numerous departments, each with a focus. Many work concurrently, sometimes sending out contradictory signals, to calculate the information received and build a solution (Vachha and Middlebrooks, 2022). In addition, the brain is a component of a giant organism. The peripheral nervous system in this creature receives inputs from external sources and transmits them to the brain. The newly formed brain would have nothing to process without it.

The entire system is nevertheless far more complex because, besides the creature processing the data, there is an outside environment where the signals originate. The brain interprets and processes signals and information based on a person's biological makeup, the structure of the outside world, which is made up of numerous subatomic particles and several fundamental forces, and the essential need that all living things have in order to survive and procreate.

For instance, displaying a baby a gazillion picture of couches and audio to help them comprehend what they are will not help them understand what a sofa is. Through touching, pushing, and other manipulations, as well as observation of adults, the infant learns what a sofa is by viewing different sofas repeatedly from hundreds and millions of different postures and angles in various contexts.

The child's biological makeup and matter's physical makeup provide accurate comprehension. They are designed to make sitting and lying on relatively soft, specially-shaped things more convenient for an individual. Sitting on a stone surface for an extended period can cause harm; however, if the shape is fundamentally different, this can also disrupt blood flow. As a result, the child's mental image of a sofa is formed by its practicality, which is derived from both its biological makeup and surroundings. Now that they know what a sofa is and how it works, the toddler will not mistake it for a fire hose or a storefront window.

The resulting robot will perceive things differently, including what is hot or cold, comfortable to sit on, and convenient and inconvenient, even if a software and hardware complex is constructed with enough sophistication. It will be a part of the environment, but our understanding of it will be entirely different. They will thereby develop a conceptual apparatus of their own that is incompatible with the human one. Furthermore, attempts to grow the number of neurons and program billions of words, images, and sounds into their networks cannot, in theory, create intelligence with the same level of power as a human. The question of building such a complex system has yet to be addressed, even in the far future.

Thus, only straightforward creative and intellectual tasks, often mundane and generally repetitive, can be automated now or shortly. The human-centric approach to innovation is the most successful for integrating intricate, unique creative and intellectual parts into automated systems that are put into production. Implementing this policy is contingent upon economic factors, which are economic growth and efficiency.

4. DISCUSSION

The influence of modern technologies and their accelerating development at the international level is very prominent in economic and non-economic life. Understandably, such influences are not only in the short to medium term but also in the long term. The development of human-centric technologies also belongs to such areas, a phenomenon still in its early stages but with enormous potential. Economic growth is one of the most critical factors for its implementation. (Bondarev & Krysiak, 2021).

Economic growth is intensive and qualitative by nature. The development and implementation of new technologies almost entirely provide it. Therefore, if two workers produce 30 units of output per day on an outdated machine, one worker could produce 100 units of output in the same unit of time on current equipment. This leads to intensive economic growth and productivity rises due to increased efficiency. At the same time, the total number of resources, including labour, land, capital, and entrepreneurial skills, remains unchanged or even falls. We are discussing technological and innovative economic growth.

However, large-scale growth is inherently quantitative. For example, if two workers work 30 hours daily on one machine, four on two identical machines will work 60 hours daily. Owing to the rule of decreasing returns to scale, 200 workers using 100 machines may likely produce 2500 pieces per day.

Digitalization and new technologies have a significant role in social life, extensive and intensive economic growth, and overall improvements in general quality of life. Even though intensive growth is more similar to exponential growth and less susceptible to the law of diminishing returns because the number of production factors remains constant, whereas extensive growth is more linear and subject to the law of diminishing returns to scale, it is still necessary because it is frequently much more accessible to provide. The number of production components involved in economic usage can be increased to new technologies, which enable the development of new areas (deserts, swamps to be drained, territories to be reclaimed, and water resources to be used), growing population, which simultaneously increases labour resources, entrepreneurial spirit, and consumer spending; more structures, machinery, and equipment are being built.

The primary contribution of human-centric technology to intensive economic growth can be precisely described as longer-term, complicated, but also more efficient, nearly exponential, as opposed to linear. Human-centric technologies suggest integrating people as creative and intellectual contributors into systems where they are most helpful.

We are discussing a minor modification of the current conception of technological advancement. However, we are also discussing significant capital expenditures and a modification of the current production machinery, including in large-scale mass production, where human-centric technologies are utilized sparingly, mainly because of the historical circumstances surrounding their development. We are discussing a shift in the dominant companies in the market, specifically, a sharp decline in their market share.

Instead, small- and medium-sized business representatives will mostly take their place since they possess a notably superior ability to create business and production processes almost from the ground up. One such factory is a furniture company that uses CNC machines to autonomously generate carved parts, especially particularly intricate ones, based on 3D models entered into the machines. The company makes solid wood furniture (Stepputat et al., 2021). To varying degrees, a person carries out specific tasks, including production. For example, he finalizes parts, feeds them, and maintains quality control on machine tools in a semi-automated mode. It also serves as the system's focal point, producing 3D models that users may enjoy within the editor. Naturally, some tasks in the 3D editor can be automated, such as creating standard patterns and pieces. However, the professional will still be responsible for carrying out the primary tasks and making most of the decisions. A milling machine with the described functionality is shown in Figure 1.

Figure 1. Woodworking CNC router

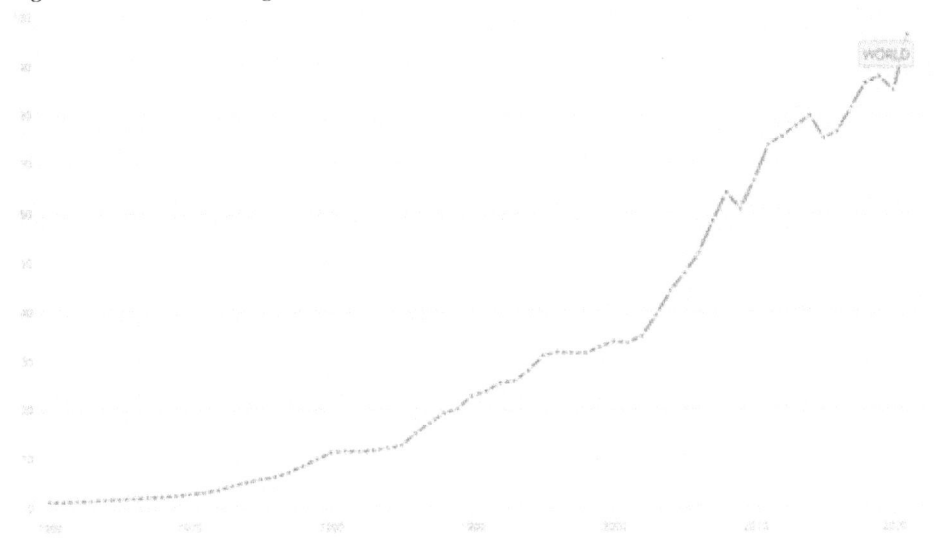

Traded at a price slightly higher than that of large mass production, high-quality products with excellent environmental characteristics are possible, given the low cost of such equipment and even the ability to assemble it yourself with sufficient knowledge of microcontrollers and programming.

5. CONCLUSIONS AND IMPLICATIONS

Human-centric technologies are a somewhat controversial phenomenon. As in many other cases, the principal contradiction lies in the opposition of long-term benefits to short-term losses from their implementation and at the level of individual enterprises and the national and global economies.

To assess the impact of human-centric technologies, it is essential to understand the factors that both enable and obstruct their implementation. When these technologies are adopted, some factors will inevitably shape their outcomes. The main non-economic driver behind human-centric technologies lies in their capacity to offer tangible benefits and enhance the quality of life that resonate deeply with people and rephrase beyond mere monetary valuation. Economically, these technologies promise to accelerate growth and intensify competition in the long run. Whereas, in the short to medium term, their adoption may disrupt large international businesses, as smaller local enterprises deliver superior, more tailored products. This shift may cause a crisis among dominant players, potentially leading to a broader economic downturn until a new economic order is established.

Funding

The research is partially funded by the Ministry of Science and Higher Education of the Russian Federation under the strategic academic leadership program "Priority 2030" (Agreement 075-15-2021-1333 dated 30.09.2021).

ACKNOWLEDGEMENT

The development of human-centric technologies for small and medium-sized businesses seems inevitable, given that they are frequently affordable or even created and used by enthusiasts. Unless large corporations successfully lobby public authorities on behalf of their interests and secure the implementation of crippling or permanent restrictions on the conduct of small and medium-sized businesses similar to those imposed in certain nations and industries in response to the COVID-19 pandemic.

REFERENCES

Adel, A. (2022). Future of industry 5.0 in society: Human-centric solutions, challenges and prospective research areas. *Journal of Cloud Computing (Heidelberg, Germany)*, 11(1), 40. DOI: 10.1186/s13677-022-00314-5 PMID: 36101900

Autor, D. H. (2015). Why are there still so many jobs? The history and future of workplace automation. *The Journal of Economic Perspectives*, 29(3), 3–30. DOI: 10.1257/jep.29.3.3

Azarenko, N., Kazakov, O., Kulagina, N., & Rodionov, D. (2020, September). The model of human capital development with innovative characteristics in digital economy. []. IOP Publishing.]. *IOP Conference Series. Materials Science and Engineering*, 940(1), 012032. DOI: 10.1088/1757-899X/940/1/012032

Bondarev, A., & Krysiak, F. C. (2021). Economic development and the structure of cross-technology interactions. *European Economic Review*, 132, 103628. DOI: 10.1016/j.euroecorev.2020.103628

Boy, G. A. (2017). A human-centered design approach. In *The handbook of human-machine interaction* (pp. 1–20). CRC Press. DOI: 10.1201/9781315557380-1

Brynjolfsson, E. (2014). The second machine age: Work, progress, and prosperity in a time of brilliant technologies.

Bughin, J., Hazan, E., Lund, S., Dahlström, P., Wiesinger, A., & Subramaniam, A. (2018). Skill shift: Automation and the future of the workforce. *McKinsey Global Institute*, 1, 3–84.

Chavlis, S., & Poirazi, P. (2021). Drawing inspiration from biological dendrites to empower artificial neural networks. *Current Opinion in Neurobiology*, 70, 1–10. DOI: 10.1016/j.conb.2021.04.007 PMID: 34087540

Crawford, K., & Calo, R. (2016). There is a blind spot in AI research. *Nature*, 538(7625), 311–313. DOI: 10.1038/538311a PMID: 27762391

Cusumano, M. A., Gawer, A., & Yoffie, D. B. (2019). *The business of platforms: Strategy in the age of digital competition, innovation, and power* (Vol. 320). Harper Business.

de la Escosura, L. P., & Rodríguez-Caballero, C. V. (2022). War, pandemics, and modern economic growth in Europe. *Explorations in Economic History*, 86, 101467. DOI: 10.1016/j.eeh.2022.101467

de Soyres, F., & Gaillard, A. (2022). Global trade and GDP comovement. *Journal of Economic Dynamics & Control*, 138, 104353. DOI: 10.1016/j.jedc.2022.104353

Engseth, E. (2018). Cultural competency: A framework for equity, diversity, and inclusion in the archival profession in the United States. *The American Archivist*, 81(2), 460–482. DOI: 10.17723/0360-9081-81.2.460

Eubanks, V. (2018). *Automating Inequality: How High-Tech Tools Profile, Police, and Punish the Poor*. St. Martin's Press.

Eum, W., & Lee, J. D. (2022). The co-evolution of production and technological capabilities during industrial development. *Structural Change and Economic Dynamics*, 63, 454–469. DOI: 10.1016/j.strueco.2022.07.001

Fierro, L. E., Caiani, A., & Russo, A. (2022). Automation, job polarisation, and structural change. *Journal of Economic Behavior & Organization*, 200, 499–535. DOI: 10.1016/j.jebo.2022.05.025

Forti, V., Balde, C. P., Kuehr, R., & Bel, G. (2020). The Global E-waste Monitor 2020: Quantities, flows and the circular economy potential.

Gungor, V. C., Sahin, D., Kocak, T., Ergut, S., Buccella, C., Cecati, C., & Hancke, G. P. (2011). Smart grid technologies: Communication technologies and standards. *IEEE Transactions on Industrial Informatics*, 7(4), 529–539. DOI: 10.1109/TII.2011.2166794

Hampton, K., & Wellman, B. (2003). Neighboring in Netville: How the Internet supports community and social capital in a wired suburb. *City & Community*, 2(4), 277–311. DOI: 10.1046/j.1535-6841.2003.00057.x

Iturralde, K., Pan, W., Linner, T., & Bock, T. (2022). Automation and robotic technologies in the construction context: research experiences in prefabricated façade modules. In *Rethinking building skins* (pp. 475–493). Woodhead Publishing. DOI: 10.1016/B978-0-12-822477-9.00009-7

Jabbar, M. A., Shandilya, S. K., Kumar, A., & Shandilya, S. (2022). Applications of cognitive internet of medical things in modern healthcare. *Computers & Electrical Engineering*, 102, 108276. DOI: 10.1016/j.compeleceng.2022.108276 PMID: 35958351

Janssen, M., & Estevez, E. (2013). Lean government and platform-based governance—Doing more with less. *Government Information Quarterly*, 30, S1–S8. DOI: 10.1016/j.giq.2012.11.003

Khan, A., Rezaei, S., & Valaei, N. (2022). Social commerce advertising avoidance and shopping cart abandonment: A fs/QCA analysis of German consumers. *Journal of Retailing and Consumer Services*, 67, 102976. DOI: 10.1016/j.jretconser.2022.102976

Kline, R. R. (2001). Technological determinism.

Kumar, Y., Kaul, S., & Hu, Y. C. (2022). Machine learning for energy-resource allocation, workflow scheduling and live migration in cloud computing: State-of-the-art survey. *Sustainable Computing : Informatics and Systems*, 36, 100780. DOI: 10.1016/j.suscom.2022.100780

Lamba, R., & Subramanian, A. (2020). Dynamism with incommensurate development: The distinctive Indian model. *The Journal of Economic Perspectives*, 34(1), 3–30. DOI: 10.1257/jep.34.1.3

Lu, Y., Zheng, H., Chand, S., Xia, W., Liu, Z., Xu, X., Wang, L., Qin, Z., & Bao, J. (2022). Outlook on human-centric manufacturing towards Industry 5.0. *Journal of Manufacturing Systems*, 62, 612–627. DOI: 10.1016/j.jmsy.2022.02.001

Mergel, I., Edelmann, N., & Haug, N. (2019). Defining digital transformation: Results from expert interviews. *Government Information Quarterly*, 36(4), 101385. DOI: 10.1016/j.giq.2019.06.002

Mokyr, J., Vickers, C., & Ziebarth, N. L. (2015). The history of technological anxiety and the future of economic growth: Is this time different? *The Journal of Economic Perspectives*, 29(3), 31–50. DOI: 10.1257/jep.29.3.31

Mullin, A. E., Coe, I. R., Gooden, E. A., Tunde-Byass, M., & Wiley, R. E. (2021, November). Inclusion, diversity, equity, and accessibility: From organizational responsibility to leadership competency. []. Sage CA: Los Angeles, CA: SAGE Publications.]. *Healthcare Management Forum*, 34(6), 311–315. DOI: 10.1177/08404704211038232 PMID: 34535064

Munsamy, M., & Telukdarie, A. (2019, December). Digital HRM model for process optimization by adoption of industry 4.0 technologies. In 2019 IEEE international conference on industrial engineering and engineering management (IEEM) (pp. 374-378). IEEE.

Pirogova, O., & Makarevich, M. (2020). The formation of the enterprises human capital in the context of digitalization. In *E3S Web of Conferences* (Vol. 164, p. 09011). EDP Sciences. DOI: 10.1051/e3sconf/202016409011

Plotnikov, V., & Pirogova, O. (2018).Key Competencies as an Enterprise Value Management Tool." In 31 IBIMA Conference (Milan: IBIMA), 1716–21.

Rigger, E., Shea, K., & Stanković, T. (2022). Method for identification and integration of design automation tasks in industrial contexts. advanced engineering informatics, 52, 101558..

Stepputat, M., Beuss, F., Pfletscher, U., Sender, J., & Fluegge, W. (2021). Automated one-off production in woodworking by Part-to-Tool. *Procedia CIRP*, 104, 307–312. DOI: 10.1016/j.procir.2021.11.052

Sufi, A., & Taylor, A. M. (2022). Financial crises: A survey. Handbook of international economics, 6, 291-340.

Uzquiano, A., & Arlotta, P. (2022). Brain organoids: The quest to decipher human-specific features of brain development. *Current Opinion in Genetics & Development*, 75, 101955. DOI: 10.1016/j.gde.2022.101955 PMID: 35816938

Vachha, B. A., & Middlebrooks, E. H. (2022). Brain Functional Imaging Anatomy. *Neuroimaging Clinics of North America*, 32(3), 491–505. DOI: 10.1016/j.nic.2022.04.001 PMID: 35843658

Xin-gang, Z., & Jin, Z. (2022). Industrial restructuring, energy consumption and economic growth: Evidence from China. *Journal of Cleaner Production*, 335, 130242. DOI: 10.1016/j.jclepro.2021.130242

Zhang, H., Zhu, L., Dai, T., Zhang, L., Feng, X., Zhang, L., & Zhang, K. (2023). Smart object recommendation based on topic learning and joint features in the social internet of things. *Digital Communications and Networks*, 9(1), 22–32. DOI: 10.1016/j.dcan.2022.04.025

Zuboff, S. (2019). *The Age of Surveillance Capitalism: The Fight for a Human Future at the New Frontier of Power, edn.* PublicAffairs.

Chapter 14
Risk Assessment of Intellectual Captial of the University by Ranking Methods:
Digital Decisions and Expert Methods of Assessment

Anush G. Airapetova
St. Petersburg State University of Economics, Russia

Vladimir V. Korelin
St. Petersburg State University of Economics, Russia

Galiya R. Khakimova
St. Petersburg State University of Economics, Russia

ABSTRACT

Risk management is one of the main areas of the theory of managing complex social and economic systems, the need for the development of which is due to the complexity of the structure of enterprises as risk management objects and the high level of uncertainty in the external socio-economic and internal environment of enterprises. Thanks to the rapid development of software and hardware, today it is possible to use all the potential capabilities of digital technologies in risk assessment. Also, computer technology allows you to quickly process this information, keep it in a secure form. There are different management decision-making processes in conditions of uncertainty (mainly due to incomplete or inaccurate risk information). Some approaches do not allow the use of a well-developed device for simulating

DOI: 10.4018/979-8-3693-3423-2.ch014

dangerous situations, which affects the conclusions of the algorithmic and software-information software and reduces the validity of decisions on risk management of industrial enterprises. The article describes the risk assessment procedure for allocating risks to individual groups.

NOVELTY STATEMENT

This chapter presents a novel approach to assessing the intellectual capital of universities through the integration of ranking methods with digital decision-making tools and expert assessment techniques. It uniquely combines quantitative and qualitative methodologies to provide a comprehensive evaluation framework, addressing the complexities and nuances of intellectual capital in higher education institutions. The chapter offers new insights into how digital technologies can enhance the accuracy and reliability of intellectual capital assessments, contributing to more informed decision-making and strategic planning in universities. This innovative approach bridges the gap between traditional assessment methods and modern digital tools, setting a new standard for evaluating intellectual capital in academic settings.

1.INTRODUCTION

There are different management decision-making procedures in conditions of uncertainty (mainly due to incomplete or inaccurate risk information) (Airapetova, Korelin, Karabekova, et al. 2017). These include the theory of expert assessments, the theory of artificial intelligence, the theory of fuzzy (blurred) sets, and methods of fuzzy-logical inference (Kharlamova et al. 2020). With their help, it is possible to assess the degree of influence of risk factors on the activities of enterprises. It would also be useful to identify the most significant risk factors through a variety of linguistic variables, fuzzy values of factors, and dangerous variables. Here, it is necessary to bear in mind the subjectivity of some of these approaches (for example, expert methods) and the inability to draw a specific conclusion due to the large spread of data (statistical methods). Some approaches do not allow the use of a well-developed device for simulating dangerous situations (Brogi et al. 2018), which affects the conclusions of the algorithmic and software-information software and reduces the validity of decisions on enterprise risk management (Airapetova & Korelin 2019).

In terms of analysis and risk management software approaches, they are divided into phenomenological, deterministic, probabilistic and expert (Airapetova, Korelin, & Karabekova 2017). The method's disadvantage is its ability to miss rarely implemented but important sequences of events during the development of an accident, which is a complex and long-term process (Provotorov et al. 2019). The probabilistic method of risk analysis involves calculating the relative probabilities of process development and evaluating them by analyzing branched chains of events to select suitable models. Computational mathematical models are simplified, but this reduces the reliability of the estimates obtained, which is especially important for considering the risk of severe accidents. The probabilistic method of risk assessment ensures the relative reliability of the analysis results, but changes in the external environment should be insignificant. Therefore, the best option is a combination of probabilistic and expert methods. The expert method involves assessing risks by processing the opinions of experts – specialists in the area under study. In fact, this is a hybrid method, that is, an integrated approach for assessing and managing risks.

Despite the variety of approaches, they all include formal stages of analysis and assessment: the formation of a risk factor; the occurrence of a risk situation; occurrence of damage caused by risk (Kozlov et al. 2019). Risk of the intellectual capital of enterprises is the possibility of obtaining an unplanned result under the influence of both external and internal factors during the use of the intellectual capital of the enterprise (Gromova & Pupentsova 2020). Here is a summary table of methods for quantifying risks. Organizational, client and human capital are allocated as part of the IC (Tishkov et al. 2017), evaluation of each type takes into account their specificity, and when determining human capital as the totality of knowledge, competencies and connections of company employees, ways of influencing it and possibilities of its management are evaluated. At the same time, the specific characteristics of industries and forms of organization of entrepreneurial activity are taken into account (Evseeva et al. 2019). So, for example, Universities not only train high-profile specialists for enterprises of various industries, including high-tech enterprises, but also are scientific centers where methods are developed, scientific and economic contracts are executed, theories are put forward and proved, competitions are held, including at the international level.

By considering human resources as a flexible relationship between competencies, installations, and intelligent flexibility, it is possible to expand the components to be regulated and managed (Airapetova & Korelin 2019). So, in the field of competencies, special areas of knowledge are identified and analyzed, covering implicit aspects, mental abilities, the ability to create networks of personal contacts and use them (Okorokov et al. 2019). In modern conditions, universities develop methodological guidelines based on competencies that are formed on the basis of standards, and a set of documentary support is also compiled, which can be called an element of intellectual capital management (Tabolina et al. 2020). For its effective growth, they provide an expansion of the areas of competence, attitudes and flexibility of employees (Vetrova, Khakimova 2020). The need to manage the human capital of universities is due to objective circumstances (Kozlov et al. 2019). The absence of a market mechanism that would allow potential employers to influence the educational process at the university leads to problems in the employment of graduates, problems of non-compliance of professional graduates with the requirements of enterprises, especially high-tech enterprises (Vetrova et al. 2018). Also, the high load on teachers does not allow to fully use the intellectual potential of teaching staff to improve the educational process (Nurulin et al. 2019). Low graduate activity also reduces the ability to develop new technologies to improve the efficiency of intelligent capital management.

Difficulties are added to the inconsistency of the representations of the teaching staff, training algorithms and expectations of employers, the lack of direct communication between enterprises and universities leads to problems in employment. The intellectual capital of educational services institutions includes: knowledge and competencies of the administrative and managerial apparatus and faculty; content of educational programs; educational technologies; organizational and management structure of the institution and a set of mechanisms for information exchange and internal communication; staff development programs; the image of an educational institution; students; cooperative relations with enterprises and organizations that hire (or potentially hire) graduates of an educational institution, as well as with other educational institutions as consumers of its educational and methodological products; relations and experience of cooperation with organizations sending their employees for retraining, with state bodies ordering the development and implementation of educational programs for the training and retraining of managers and entrepreneurs; as well as with other groups directly or indirectly interested in the activities of the educational institution (Vetrova et al. 2017).

1.1 Theoretical Framework for Assessing Intellectual Capital in Universities

The meaning of intellectual capital (IC) in universities is complex and it refers to the different types of knowledge, abilities and links of an organization that have a direct impact on organizational performance (Frondizi et al., 2019). It is therefore important to comprehend and evaluate this specific intangible asset in furtherance of which this paper seeks to: Intellectual capital in universities can be broadly categorized into three key components: These resources include: Human capital, Structural capital and Relational capital (Iacoviello et al., 2019). Skills consist of all the knowledge, talents, and skills of university employees, teachers, researchers and administrators. Another structural form is the internal system of the institution, which in turn is the databases, the intellectual property it holds and the educational technologies used. This type of capital looks at the systems of relationship outside the university with key players such as other companies, ex-students, and government entities (Secundo et al., 2017). When it comes to assessing the above mentioned components, one has to use a complex approach which combines several methodological tools, all targeting the general nature of IC. Reasons for this become clear when one tries to make use of the traditional tools such as the financial instruments in measurement of IC where knowledge is involved it cannot be determined. Hence, the new approaches include both the quantities and the qualitative analysis as well as the expert estimations, the Fuzzy logic and the artificial intelligence (Frondizi et al., 2019). For example, expert methods make use of opinions or decisions of particular specialists in the area and offer qualitative assessments which are vital for indicating and measuring the peculiarities of the human and relational capital. Nevertheless, these methods may be bias and not consistent in their approach, this has increase the call for the use of statistical analysis /Machine learning and artificial intelligence models (Secundo, et al., 2016).

Human capital is highly significant in the case of universities as it defines their performance as educational establishments and their efficiency in providing quality education and coming up with new innovations. The competence of the faculty, their capacity to foster students and engage in research work are all factors highlighting human capital. Furthermore, the creation of human capital in other sectors of the economy is also attributed to universities since it churns out graduates who are able to meet the requirements of the various sectors in the job market (Leitner, 2004). Therefore, the evaluation of human capital in universities should not only focus on the current state of the faculty but also the future potential of the teacher staff and how successfully the university addresses issues of talent management acquirement, retention and development. Structural capital in universities is also vital, especially since higher education institutions have embraced the use of technology to support learning delivery and institutional processes (Ramirez et al. 2007). Structural capital represents the embedding of LMS, digital libraries and data analytics into the educational process as an example of the evolution process. Moreover, these technologies do not only augment the information and communication technology-based teaching and learning processes but also assist in the administrative procedures and organization of the processes that involve data collection to serve the purpose of decision making (Leitner, 2002). In addition, the structural capital includes the flow of patents and publication, as an outcome of the research intervention, undertaken by a university. Proper management and monetization of these assets can help to strengthen the positioning of the university and its financial position, which makes the IC assessment of this area mandatory (Frondizi et al., 2019).

External relational capital, on the other hand, points to the quality of connection that a university enjoys outside it. Universities are required to sustain and develop cooperative partnerships and networks with the industries, alumni and other Universities, governments and other funding sources for partnership, funding and programed accreditation purposes (Liu et al., 2023). These relationships are imperative for sharing of knowledge, nodal research collaboration and formation of avenue for students by providing internships and jobs. Evaluating relational capital consists of assessing the intensity and effects of the relationships within a firm, some possible criteria of which are the existing collaborations, the relations with external entities, and the experiences with partners (Paloma Sánchez et al., 2009). When it comes to the evaluation of the IC of universities, the theoretical background entails human, capital structural, and relational capital. All of these components are important subsystems of the institution and their performance affects the global sustainability of the organization. In the following, it is proposed that expert opinions, quantitative data, and technological platforms can be used to improve the understanding of knowledge-intensive assets and make sound decisions for universities' learning and research endeavors.

1.2 Digital Tools and Expert Methods in Intellectual Capital Assessment

The improvement of ICT has made it possible to conceive new approaches and methods in University IC valuation, these are accurate, fast and more extensive (Pokrovskaia et al., 2021). AI is already utilized in different aspects evaluating educational outcome and other methods such as big data analytics and machine learning algorithms provide universities with extra possibility to analysis a great amount of data and obtain the information formerly untouched (Lee & Wong, 2019). These technologies help to minimize the amount of time spent on carrying out non-essential routines hence providing experts with more time to attend to issues of assessment concerning the intellectual capital, issues of result analysis and decision making. For instance, AI can demographically classify human capital of the university based on an analysis of performance records, researches done, qualification, etc. In this way, by applying analysis on this data, it is possible to have a more accurate picture on the competencies of the faculty members and their potential (Liu et al., 2023). In addition, AI-based model scans help in forecasting potential future performance of the faculty and outcomes of students that would be relevant for planning and investment decisions. This predictive capability is most valuable in the context of the university's fixation on strengthening the collection of professionals as its primary source of intellectual capital.

Another effective approach that increases the evaluation of structural and relational capital is big data analytics. In universities, diverse data to do with operational activities, research and outside collaborations are produced and archived. Big data analytics enables institutions to analyze and find out information in this data that can assist in making decisions (Švarc et al., 2021). For instance, if university research collaborations are measured and the information revealed, then universities can determine the most fruitful relationships that need to be profited from or developed further. Likewise, big data can be employed to evaluate the productivity of technologies and other components of structural capital to create a coherent picture of different elements of the institution's intellectual capital (Liu et al., 2023). The professional approaches still come out as one of the essential elements of the assessment of the intellectual capital, especially in cases when qualitative evaluation is possible. These methods include gaining information form experts in the particular field of interest which may include academic administrators, faculty members and professionals in the sector (Izzo et al., 2022). Consultants 'opinions can be more credible in assessing aspects of the intellectual capital which are not easily measurable, such as the nature of the external interactions or the significance of the scholarly production. However, to avoid possible biases and contradictions of expert methods, their use should be backed up with objective digital tools, which would verify and adjust the assessments.

In this attitude, the use of digital tools along with the expert method provides a strong basis in the evaluation of intellectual capital in universities. Although digital technologies can generate the data necessary for lean and accurate assessments of various situations, methods developed by experts can supply the context and a clear understanding of the results obtained necessary to assess these situations adequately. Taken together, they let universities evaluate their IC stock in a more comprehensive manner and on the basis of both quantitative and qualitative criteria (Wudhikarn, 2017). Besides improving the validity and accuracy of the assessments it also brings about greater clarity in the functioning, thereby enabling the management of intellectual assets more effectively and the overall performance of the university can be improved. All in all, the combination of the digital instruments with the methods applied by experts is a major improvement in the assessment of the intellectual capital in universities (Lee & Wong, 2019). In this way, the advantages of using both paradigms can be realized: a stronger, more applicable estimation of the institution's intellectual assets can be obtained. In this context, it becomes possible for universities to arrive at strategic choices that are geared to improve the organizations' education and research foundations so as to thrive in a progressively intricate and technological advanced academicians' environment (Khoso et al., 2022).

2. LITERATURE REVIEW

The evaluation of IC in universities is a rather complex task and includes the assessment of various types of risks connected with personnel and performance of the organization. The intellectual capital of a university is the knowledge base and expertise inclusive of its officials and personnel, the competencies, experience, and networks it can leverage, and the relationships it can foster in ensuring that the university delivers on its mandate to the institution's added value (Serenko & Bontis, 2022). The current literature review examines the essential indices of risk factors within universities with a focus on personnel and organizational risk factors that put into account the whole performance of the university in regard to its personnel and IC sustainability (Khalili et al., 2017). Personnel risks are especially important in the academic area in which the faculty and staff comprises of intellectual human capital that is important to offer quality education and conduct research. Few factors are as closely tied to the academic status and the ability of the university to obtain students, teachers, and funds as the qualification of persons who work at the university (Chen & Chen, 2013). The common risks connected with personnel qualifications are inadequacy of academic staff professional background, difficulties in managing academic staff, and variety of risks linked with the turnover of key faculty members as the result of high workload, lack of professional growth opportunities or low motivation (Bakhru, 2019). Other certain drawbacks include reduction of the quality of education and research output since highly qualified personnel are likely to be replaced by relatively less qualified personnel this a factor that can severely affect the reputation and competitive position of any university.

Universities are faced with an additional challenge of staff retraining due to the ever changing dynamic of knowledge and technology therefore increasing the challenges of personnel risks management (Frondizi et al., 2019). Development of new technologies and new achievements in different areas of knowledge after preparing in university it is required containable in-service training to help staff of universities to maintain themselves up-to-date in their fields (Liu et al., 2023). However, this also brings into question issues to do with the deprecation of knowledge and ill-preparedness to train the workforce again and compromise on the standards of education and research in the institution. Furthermore, proliferation of existing faculty through inadequate staffing partially poses high job workloads that may increase turnover and reduce the satisfaction level of the faculty members (Khoso et al., 2022). Organizational risks are yet another relevant factor of the ICM in universities. These risks relate to the systems, procedures, and culture aspects by which the optimized exploitation of the ITC is ensured by an institution. Lack of efficiency in managing the human resources involves the improper use of the management tools, the wrong management procedures, and the ineffective management strategies result in suboptimal performance of the staff members and conduced to the low efficiency of the institution (Frondizi et al., 2019). For example, poor actualization of this center is evidenced by the poor management decisions leading to inefficiency in the use of resources, poor alignment between the strategic goals of the university and the academic strand, and poor coherence where there are two strands of decision making. I such inefficiencies affect strategic accomplishment of the university's strategic goals and objectives as well as increase vulnerability of the enterprise to risk occurrence (Veltri et al., 2014).

As one of the many risks within universities, the overbearing risk is a conflict of interest and abuse of authority. Misleader ship in an academic context may thus have demeaning repercussions especially where autonomy and academic freedom are accorded utmost importance; staff and student mistrust; uninformed stakeholders impugned the institutional authority. This results to a formation of a poison organizational culture that does not foster the creation of new ideas and cooperation and consequently does not tap the full potential of the institution's intellectual assets. Moreover, the absence of integration between the university management and faculties increases the risks; hence, the direction, intent, and action may not advance the university's mission and goals properly (Bakhru, 2019). Also the fact that many aspects of their personality- their ability to deal with people, their temper, and the degree of their extraversion are rather subjective and can hardly be measured on an objective scale pose a rather stern threat that cannot be measured adequately (Khalili et al., 2017). These characteristics can thereby impinge on the general organizational climate and the nature of work relations, such as teamwork and collaborative work that are central to the proper functioning of learning institutions. For instance, low motivation or rivalry between groups will contribute to low performances, and consequently affect the institution's Intellectual capital. The above risks bring out the need to take both professional and personal characteristics of the assessment of the intellectual capital in University.

2.1 Strategies for Mitigating Risks in University Intellectual Capital

To effectively manage the risks which stem from the management of intellectual capital in universities, there is therefore a need to put in place solutions that address personnel issues and organization issues (Rohani et al., 2015). Since IC is central to the goals and performance of HEIs, effective risk management strategies are crucial for thinking about IC in the long term and for actual practice. This subsection discusses some of the options that can be pursued by a university to ensure that the harms arising from inadequate personnel qualifications, poor management practices, and un-desirable personal attributes of employees are minimized; hence enhancing the university's intellectual capital (Khan & Ali, 2017). Among the basic measures that can be used to control personnel risks, the following can be distinguished: the use of professional development and training programs. Skills update and development are important in the campus as knowledge and technology keep on changing from time to time (Dalwai & Salehi, 2021). It is recommended that universities provide and/or arrange opportunities for the professional development for faculty and staff on an ongoing basis so as to maintain their academicians' cutting-edge expertise in the corresponding area (Pereira et al., 2024). This entails inculcation of new technical competencies, as well as expanding the existing leadership, communication, and teamwork proficiencies as a way of sustaining innovation in the institution. In this way, universities can avoid such risks as knowledge obsolescence, and lack of necessary qualifications that might affect the competence and motivation of the organization's employees (Frondizi et al., 2019).

The second conclusion is that, besides the aspects of professional development, the Universities should also enhance the management activity aimed at the reduction of organizational risks within the organization (Royal et al., 2014). It could be done with the help of introducing more progressive and adaptable approaches to management characteristic for modern academic settings (Khoso et al., 2022). For instance, universities can enhance management by decentralizing some decision making activities to involve the members of the academia hence enhancing the involvement of human resource. It not only enhances the decision making process but also increases the morale of the employees and hence the productivity decreases the chances of having dual agendas within the company (Jones, 2020). Furthermore, the openness of the communication channels that exist between the management and employees will help in passing on the strategic goals of the university and obtaining the support of all the members of the institution.

Another important tactic for reducing the risks linked with the personal characteristics of staff is the systematic employment and personnel policies stipulating the valuing of cultural adaption and team cohesiveness (Liu et al., 2023). Universities in the course of recruiting their employees needs to evaluate the applicants not merely for the technical skills they bring in, but especially for their compatibility to fit into the existing organizational culture, and their perceived capabilities to enhance the universities' existing pool of intellectual capital. These measures should cover the part of retention policies that consist in making the personnel more satisfied regarding possible discrimination on the base of sex, age, etc.; possible conflicts with private life and more generous opportunities for cooperation between the workers (Al-Omoush et al., 2022). As a result of a healthy working environment, issues of low motivation, conflicts, and other issues to do with personal attributes that may harm the productivity of the team as well as the general intellectual capital of the university will be handled comprehensively. Last but not the least; the use of digital tools and data analytics can go a long way in improving the risk management approaches. These technologies can be utilized in universities in order to track different aspects of an institution's intellectual capital, particularly in the cases of employees' productivity, organizational processes optimization, and efficacy of professional development initiatives (De Santis & Presti, 2018). With help of data-driven analytics, risks, which are in the process of emergence, can be recognized on time and specific actions targeting their mitigation can be taken to secure the idea that the stock of academic knowledge will stay viable no matter the circumstances outside of it (Mkumbuzi, 2016).

2.2 Determination of risk impact factors on university activity by individual risks

Risk factors for indicators are based on deviations of actual indicators from planned values according to the formula:

$$k_i = 1 - \frac{P_i^p}{P_i^p + \Delta P_i}$$

, (1)

where P_i^p - the planned value of the i-th indicator, ΔP_i – deviation of the actual value of the i-th key figure from the planned one. The deviation of the actual value of the key figure from the planned calculation is determined by the formula:

$$\Delta P_i = |P_i^f - P_i^p|$$

, (2)

where P_i^f is the actual value of the i-th indicator. The mathematical meaning of formula 2 is that the level of risk of each indicator is depending on the amount of deviation ΔP_i, which forms boundary values which generates boundary values k_i. In risk-free situations, deviations ΔP_i are minimal or tend to zero, which turns the fraction into one, and the risk level reduces to zero.

$$\begin{cases} \Delta P_i \to 0 \\ P_i^p + \Delta P_i \to P_i^p \\ \frac{P_i^p}{P_i^p + \Delta P_i} \to 1 \\ k_i \to 0 \end{cases} \quad (3)$$

If there are fluctuations, the fraction value tends to zero, and the risk level is 1:

$$\begin{cases} \Delta P_i \to \infty \\ P_i^p + \Delta P_i \to \infty \\ \dfrac{P_i^p}{P_i^p + \Delta P_i} \to 0 \\ k_i \to 1 \end{cases} \qquad (4)$$

Therefore, the value of the risk level of the *i*-th indicator lies in the range from zero to one: $0 < k_i < 1$. *Determination of risk impact ratios on university activity by risk groups.*

The risk impact ratios on the activity of the university for a particular group of calculations are considered as the average progressive ratio for individual risks. The set of indicators may vary depending on the significance and dynamics of each indicator. So, if the indicator constantly goes to zero to include it in the group does not make sense. Indicators below 0.05 are also not included in the estimate. The calculation of the coefficient is made according to the following formula:

$$K_i^p = \sqrt[n]{\prod k_i^p}$$

, (5)

where K_i^p – risk ratio of an individual group, k_i^p – risk level of a single indicator, n – number of key figures in the group. Determination of the aggregate risk impact factor on the activity of the university. The total risk ratio is determined after the races of the main types of risks formed under the influence of factors of the internal and external environment of the enterprise, its teaching, financial, organizational activities, and direction of activity. This display reflects the general level of risk that is taken into account in the main points of management decisions: for example, when launching a new master's program, changing teaching technology, concluding commercial contracts.

The integral risk factor is calculated using the formula:

$$K_{int} = \sum v_i / K_i^n$$

, (6)

where K_{int} – aggregate risk ratio of the IC University, v_i – weight of risk group, K_i^n – risk group ratio. The integral risk factor is calculated using the formula: The weight of the risk group is calculated by experts (Delphi method, survey variant, etc.). It is mandatory to check the representativeness of the sample, the consistency of the opinion of experts, the calculation of the concordance coefficient. The positivity of the eigenvalues is the determining factor for establishing the stability property of parabolic evolutionary systems with distributed parameters on the graph.

2.3 Study of solution stability

The author compiled an evaluation model for calculating risk indicators based on Microsoft Excel, an excerpt from which is shown in table 2. According to the methodology proposed by the author, it is further necessary to calculate the aggregate risk of a group nature: production, financial, organizational and environmental. At the same time, those indicators which risk ratios turned out to be zero were excluded from the calculations. The calculation of the industrial risk factor was carried out according to the formula (5).

$$K^p_{p.p.} = \sqrt[4]{\prod k^p_{p1}, k^p_{p2}, k^p_{p3}, k^p_{p4}}$$

, (7)

where

$$k^p_{p1} = \sqrt[2]{k^p_{mto} * k^p_{y.o}}$$

, (8)

k^p_{p1} – risk ratio of lack of educational process support;

Table 1. The result of calculations of risk indicators for the average university for 2015 to 2020

Risk	Indicator	Code	Probability (p)	Degree of threat (s)	Deviation ΔP_i	k_i
Risk of insufficient provision of the educational process k_{p1}	Insufficient MTO coefficient	Ki	0.5	0.6	4.67	0.13
	Capital productivity	Cp			0.21	0.67
Risk of non-competitive services k_{p3}	Development	Vp	0.2	0.4	64.93	0.04
	Scope of implementation	Op			1339.67	0.09

continued on following page

Table 1. Continued

Risk	Indicator	Code	Probability (p)	Degree of threat (s)	Deviation ΔP_i	k_i
Insolvency risk k_{fl}	Absolute liquidity ratio	Kaliq	0.2	0.4	0.07	0.24
	Total coverage factor	Ktcov			0.06	0.05
Risk of financial volatility k_{f2}	Autonomy factor	Ka	0.2	0.3	0.08	0.09
	Financial sustainability ratio	Kfs			0.18	0.01
	Debt Cover Ratio	K_{DC}			0.01	0.15
	Exceeding the number of hours for extracurricular work on the norm, %	Nvperunit	0.3	0.4	0.05	0.13

$$k_{p2}^p = \sqrt[3]{k_{p.m}^p * k_{z.o}^p * k_{i.o}^p}$$

, (9)

k_{p2}^p – risk ratio of decrease in teaching activity;

$$k_{p3}^p = \sqrt[2]{k_v^p * k_{z.o}^p}$$

, (10)

k_{p3}^p – risk ratio of non-competitive service; k_{p4}^p, k_{p5}^p, k_{p6}^p – additional risks according to the specifics of the university, respectively. Based on the evaluation methodology proposed by the author, an estimated model of the university's risk ratios can be compiled in the Microsoft Excel program and the components of all types of risk can be calculated by year, average values, standard deviations and aggregate indicators by risk groups can be found. Thus, it is possible to develop an IC risk assessment model for their management zones, which will allow us to distribute risks and verify the validity of the model. We can highlight the most critical risks, which will allow only for them to use preventive measures that can reduce the negative consequences of their offense or prevent their offense in general.

3. DISCUSSION

The section of introduction, and the literature review of this chapter also point towards the nature of managing and acquiring IC in the university. As a form of capital, human, structural and relational capital has been identified to determine the success of universities and their competence to deliver education and execute research, as well as their ability to sustain their competitiveness. However, the assessment and management of IC are hard to embark and even more difficult to sustain mainly due to a number of challenges arising from the dynamic technological environment enhanced by ever changing educational needs and pressures from the outside world. To a large extent, the literature review reveals the significance of identifying risk probabilities relating to personnel and organizational factors. It is also important for universities which quality of education and research activities depends on qualified faculty and motivated staff and their associated personnel risks. The review also identifies other key organizational risks such as poor work flow that is in relation to management of the organization or conflict of interest which is a blow to the university's capacity to capitalize on the intellectual capital.

From the literature review, one of the important considerations derived is that there is a requirement for a holistic and coordinated approach towards risk management for universities. All the mentioned methodologies such as the usage of changes and odds, the application of mathematical models and expert assessments contribute to approaches in risks quantification and control. These approaches enable universities to develop a structured procedure for defining potential risks, estimating the effects of different risk factors, and designing approaches to managing these risks. By adopting such models, which are described in the above formulas, it is easier to have less bias and more realistic assessment of IC risks that are crucial for decision making. Furthermore, the discussion shows that risk management is an issue that is under development in relation to the academic environment. The trend towards greater use of digital tools and technologies in universities also poses new possibilities for using these innovations to develop IC management. The globalization of risk assessment, with the application of data analytics, AI, and other related digital technologies holds the prospect of enhancing the reliability of such evaluations. However, it also poses some risks, like the required qualified staff to operate these technologies and technology relatively rapid cycle of becoming obsolete.

In summary, it is important to stress that the University IC management does not imply exclusion of the traditional risk assessment tools, and at the same time is closely connected with the digital technologies use. As it reveals both personnel and organizational risks and as the universities learn how to mitigate these risks with reference to the changes that occur in the external environment, universities will be in a position to protect and improve their intellectual capital thereby making them sustainable in the long-run. The information described in this chapter forms the basis for building more effective and more sustainable strategies in managing IC in HE institutions.

4. CONCLUSION

In this chapter, an attempt has been made to delineate the status of IC measurement and management and complexities in universities, and to stress the objective to consider personnel and organizational risks in the IC measurement and management process. Human capital, the structure that makes up a university, and the relationships which are the primary wealth of a university are known as intellectual capital, which is useful to sustain the university competitive advantage and its successful future. The chapter also systematically analyzed personnel risks in connection with IC, for example inadequate qualifications and inefficient management, as well as organizational risks that include bad management and possible conflict of interest. The techniques described in this chapter, especially those of mathematical modeling and professionals' evaluation, provide solid means of measuring and controlling these risks. These approaches give a framework through which the effects of various risk factors in the running of the university can be assessed hence advancing in informed decisions being made. The use of digital inures and technologies also contributes to the improvement of these risk assessments to better capability and efficiency of the assessment. Finally, it is clear that the practice of managing intellectual capital in universities need to search for an optimal solution, preserving the use of traditional tools for evaluating risks and taking into account digital tools. By being alert and identifying risks threatening the institutions as well as the universities own internal risks, one can protect the valuable capital that is the intellectual capital: sustainability and quality of the universities and the entire academic sector. Thus, it is highlighted that continuous evolution and change in managing IC in universities is imperative because of the growing globalization and competitiveness of education system.

REFERENCES

Airapetova, A. G., & Korelin, V. V. (2018, November). The Problem with Modern Secondary and Higher Education Systems in Russia–Free to Choose or the Problem with the Public Education System. In *2018 XVII Russian Scientific and Practical Conference on Planning and Teaching Engineering Staff for the Industrial and Economic Complex of the Region (PTES)* (pp. 162-164). IEEE. https://doi.org/DOI: 10.1109/PTES.2018.8604266

Airapetova, A. G., Korelin, V. V., & Karabekova, A. A. (2017, November). Specifics of human resource policies at various life stages of an enterprise. In 2017 IEEE VI Forum Strategic Partnership of Universities and Enterprises of Hi-Tech Branches (Science. Education. Innovations)(SPUE) (pp. 169-171). IEEE. https://doi.org/DOI: 10.1109/IVForum.2017.8246082

Al-Omoush, K. S., Palacios-Marqués, D., & Ulrich, K. (2022). The impact of intellectual capital on supply chain agility and collaborative knowledge creation in responding to unprecedented pandemic crises. *Technological Forecasting and Social Change*, 178, 121603.

Babkin, A. V., Kuzmina, S. N., Oplesnina, A. V., & Kozlov, A. V. (2019). Selection of Tools of Automation of Business Processes of a Manufacturing Enterprise. 2019 International Conference "Quality Management, Transport and Information Security, Information Technologies" (IT&QM&IS), 226–229. https://doi.org/DOI: 10.1109/ITQMIS.2019.8928302

Bakhru, K. M. (2019). Importance of intellectual capital in ranking of business school of India. *International Journal of Business and Globalisation*, 22(2), 258–278.

Brogi, A., Forti, S., Ibrahim, A., Kim, S. S., Gupta, H., Vahid Dastjerdi, A., Ghosh, S. K., Buyya, R., D'Angelo, G., Ferretti, S., & Ghini, V., AbdElhalim, E., Obayya, M., Kishk, S., Chiariotti, F., Condoluci, M., Mahmoodi, T., Zanella, A., Anglano, C., ... Guazzone, M. (2018). Survey High-Performance Modelling and Simulation for Selected Results of the COST Action IC1406 cHiPSet. *Future Generation Computer Systems*, 29(1).

Chen, I. S., & Chen, J. K. (2013). Present and future: A trend forecasting and ranking of university types for innovative development from an intellectual capital perspective. *Quality & Quantity*, 47, 335–352.

Dalwai, T., & Salehi, M. (2021). Business strategy, intellectual capital, firm performance, and bankruptcy risk: Evidence from Oman's non-financial sector companies. *Asian Review of Accounting*, 29(3), 474–504.

De Santis, F., & Presti, C. (2018). The relationship between intellectual capital and big data: A review. *Meditari Accountancy Research*, 26(3), 361–380.

Evseeva, S., Kalchenko, O., Plis, K., & Evseeva, O. (2019). The role of information and communication technologies as a part of business intelligence in improving the wealth of nations. *IOP Conference Series. Materials Science and Engineering*, 618(1), 012080. DOI: 10.1088/1757-899X/618/1/012080

Frondizi, R., Fantauzzi, C., Colasanti, N., & Fiorani, G. (2019). The evaluation of universities' third mission and intellectual capital: Theoretical analysis and application to Italy. *Sustainability*, 11(12), 3455.

Frondizi, R., Fantauzzi, C., Colasanti, N., & Fiorani, G. (2019). The evaluation of universities' third mission and intellectual capital: Theoretical analysis and application to Italy. *Sustainability*, 11(12), 3455.

Gromova, E. A., & Pupentsova, S. V. (2020). Simulation modelling as a method of risk analysis in real estate valuation. *IOP Conference Series. Materials Science and Engineering*, 898(1), 012048. DOI: 10.1088/1757-899X/898/1/012048

Iacoviello, G., Bruno, E., & Cappiello, A. (2019). A theoretical framework for managing intellectual capital in higher education. *International Journal of Educational Management*, 33(5), 919–938.

Izzo, M. F., Fasan, M., & Tiscini, R. (2022). The role of digital transformation in enabling continuous accounting and the effects on intellectual capital: The case of Oracle. *Meditari Accountancy Research*, 30(4), 1007–1026.

Jones, L. A. (2020). *Reputation risk and potential profitability: Best practices to predict and mitigate risk through amalgamated factors*. Capitol Technology University.

Khalili, Y., Fakhari, H., Basti, E. M. K., & Aghajani, H. (2017). Intellectual capital indicators ranking in the universities of Iran using Delphi fuzzy technique. *RISK GOVERNANCE & CONTROL: Financial markets and institutions*, 147.

Khan, S. N., & Ali, E. I. E. (2017). The moderating role of intellectual capital between enterprise risk management and firm performance: A conceptual review. *American Journal of Social Sciences and Humanities*, 2(1), 9–15.

Kharlamova, T., Kharlamov, A., & Gavrilova, R. (2020). The development of the Russian economy under the influence of the Fourth Industrial Revolution and the use of the potential of the Arctic. IOP Conference Series: Materials Science and Engineering, 940(1), 0–7. https://doi.org/DOI: 10.1088/1757-899X/940/1/012113

Khoso, A. K., Darazi, M. A., Mahesar, K. A., Memon, M. A., & Nawaz, F. (2022). The impact of ESL teachers' emotional intelligence on ESL Students academic engagement, reading and writing proficiency: Mediating role of ESL students motivation. *Int. J. Early Childhood Spec. Educ*, 14, 3267–3280.

Kozlov, A., Kankovskaya, A., Teslya, A., & Zharov, V. (2019). Comparative study of socio-economic barriers to development of digital competences during formation of human capital in Russian Arctic. *IOP Conference Series. Earth and Environmental Science*, 302(1), 012125. DOI: 10.1088/1755-1315/302/1/012125

Lavrov, N., Druzhinin, A., & Alekseeva, N. (2020). Description and Analysis of Economic Efficiency of the Real Estate Model Transformed in the Framework of Digitalization. *IOP Conference Series. Materials Science and Engineering*, 940(1), 012042. DOI: 10.1088/1757-899X/940/1/012042

Lee, C. S., & Wong, K. Y. (2019). Advances in intellectual capital performance measurement: A state-of-the-art review. *The Bottom Line (New York, N.Y.)*, 32(2), 118–134.

Leitner, K. H. (2002). Intellectual Capital Reporting for Universities: Conceptual background and application within the reorganisation of Austrian universities. *The Value of Intangibles. Autonomous University of Madrid Ministry of Economy*, (November), 25–26.

Leitner, K. H. (2004). Intellectual capital reporting for universities: Conceptual background and application for Austrian universities. *Research Evaluation*, 13(2), 129–140.

Liu, H., Zhu, Q., Khoso, W. M., & Khoso, A. K. (2023). Spatial pattern and the development of green finance trends in China. *Renewable Energy*, 211, 370–378.

Mkumbuzi, W. P. (2016). Influence of intellectual capital investment, risk, industry membership and corporate governance mechanisms on the voluntary disclosure of intellectual capital by UK listed companies. *Asian Social Science*, 12(1), 42.

Nurulin, Y., Skvortsova, I., Tukkel, I., & Torkkeli, M. (2019). Role of Knowledge in Management of Innovation. *Resources*, 8(2), 87. DOI: 10.3390/resources8020087

Okorokov, R., Timofeeva, A., & Kharlamova, T. (2019). Building intellectual capital of specialists in the context of digital transformation of the Russian economy. *IOP Conference Series. Materials Science and Engineering*, 497(1), 012015. DOI: 10.1088/1757-899X/497/1/012015

Paloma Sánchez, M., Elena, S., & Castrillo, R. (2009). Intellectual capital dynamics in universities: A reporting model. *Journal of Intellectual Capital*, 10(2), 307–324.

Pereira, V., Jayawardena, N. S., Sindhwani, R., Behl, A., & Laker, B. (2024). Using firm-level intellectual capital to achieve strategic sustainability: Examination of phenomenon of business failure in terms of the critical events. *Journal of Intellectual Capital*.

Podvalny, S. L., Podvalny, E. S., & Provotorov, V. V. (2017). The Controllability of Parabolic Systems with Delay and Distributed Parameters on the Graph. Procedia Computer Science, 103(October 2016), 324–330. https://doi.org/DOI: 10.1016/j.procs.2017.01.115

Pokrovskaia, N. N., Korableva, O. N., Cappelli, L., & Fedorov, D. A. (2021). Digital regulation of intellectual capital for open innovation: Industries' expert assessments of tacit knowledge for controlling and networking outcome. *Future Internet*, 13(2), 44.

Provotorov, V. V., Sergeev, S. M., & Part, A. A. (2019). Solvability of hyperbolic systems with distributed parameters on the graph in the weak formulation. *Vestnik of Saint Petersburg University Applied Mathematics Computer Science Control Processes*, 15(1). Advance online publication. DOI: 10.21638/11702/spbu10.2019.108

Ramirez, Y., Lorduy, C., & Rojas, J. A. (2007). Intellectual capital management in Spanish universities. *Journal of Intellectual Capital*, 8(4), 732–748.

Rohani, A., Keshavarz, E., & Keshavarz, A. (2015). Prioritising (ranking) of indexes for measuring intellectual capital using FAHP and fuzzy TOPSIS techniques. *International Journal of Industrial and Systems Engineering*, 21(3), 356–376.

Royal, C., Evans, J., & Windsor, S. S. (2014). The missing strategic link–human capital knowledge, and risk in the finance industry–two mini case studies. *Venture Capital*, 16(3), 189–206.

Secundo, G., Dumay, J., Schiuma, G., & Passiante, G. (2016). Managing intellectual capital through a collective intelligence approach: An integrated framework for universities. *Journal of Intellectual Capital*, 17(2), 298–319.

Secundo, G., Perez, S. E., Martinaitis, Ž., & Leitner, K. H. (2017). An Intellectual Capital framework to measure universities' third mission activities. *Technological Forecasting and Social Change*, 123, 229–239.

Serenko, A., & Bontis, N. (2022). Global ranking of knowledge management and intellectual capital academic journals: A 2021 update. *Journal of Knowledge Management*, 26(1), 126–145.

Švarc, J., Lažnjak, J., & Dabić, M. (2021). The role of national intellectual capital in the digital transformation of EU countries. Another digital divide? *Journal of Intellectual Capital*, 22(4), 768–791.

Tabolina, A. V., Olennikova, M. V., Tikhonov, D. V., Kozlovskii, P., Baranova, T. A., & Gulk, E. B. (2020). Project Activities in Technical Institutes as a Mean of Preparing Students for Life and Professional Self-determination. In Advances in Intelligent Systems and Computing: Vol. 1134 AISC (pp. 800–807). Springer. https://doi.org/DOI: 10.1007/978-3-030-40274-7_77

Tishkov, P. I., Khakimova, G. R., Diasamidze, M. A., & Minchenko, L. V. (2017). Organization and methodological basis for company's risk monitoring and control. *2017 XX IEEE International Conference on Soft Computing and Measurements (SCM)*, 799–800. https://doi.org/DOI: 10.1109/SCM.2017.7970728

Veltri, S., Mastroleo, G., & Schaffhauser-Linzatti, M. (2014). Measuring intellectual capital in the university sector using a fuzzy logic expert system. *Knowledge Management Research and Practice*, 12(2), 175–192.

Vetrova, E. N., Khakimova, G. R., Gladysheva, I. V., & Lapochkina, L. V. (2018). Development Strategy of University Interaction with Employers. *2018 XVII Russian Scientific and Practical Conference on Planning and Teaching Engineering Staff for the Industrial and Economic Complex of the Region (PTES)*, 245–248. https://doi.org/DOI: 10.1109/PTES.2018.8604249

Vetrova, E. N., Khakimova, G. R., Tihomirov, N. N., & Diasamidze, M. A. (2017). Structurization of intellectual capital risks in the conditions of labor market integration and globalization. *2017 XX IEEE International Conference on Soft Computing and Measurements (SCM)*, 801–803. https://doi.org/DOI: 10.1109/SCM.2017.7970729

Wudhikarn, R. (2017, March). Determining key performance indicators of intellectual capital in logistics business using Delphi method. In *2017 International Conference on Digital Arts, Media and Technology (ICDAMT)* (pp. 164-169). IEEE.

Chapter 15
Integrated Management System Implementation Prospects at the Aerospace Complex Enterprises:
Enterprise's Digital-Integrated Management System

Ekaterina M. Messineva
Moscow Aviation Institute (National Research University), Russia

Alexander G. Fetisov
Moscow Aviation Institute (National Research University), Russia

ABSTRACT

A comprehensive study of the current situation related to certification on ISO functional management systems (ISO 9001, ISO 14001 and ISO 45001) was carried out to assess the prospects for the integrated management systems creation and implementation in aerospace industry. This study analyzes complex data on the certificates number issued for aerospace enterprises in the world leading aerospace countries during the period 2009-2019. Paper demonstrates the current trends in the implementation and certification of quality, environmental, occupational health, and safety management systems by enterprises in this countries. During the investigation, both global trends and the domestic Russian trends were identified. Then Pearson correlation coefficients were calculated between the time series of the number of ISO 9001 and ISO 14001 certificates. Based on the obtained results, a basic scheme

DOI: 10.4018/979-8-3693-3423-2.ch015

was proposed for the development of a digital integrated management system for enterprises in the aerospace industry, taking into account the specifics of the Russia.

1. INTRODUCTION

The efficiency and competitiveness of Russia's aerospace industry depend heavily on the adoption of modern regulatory frameworks. A critical factor in this is the development of Integrated Management Systems (IMS), which utilize digital technologies to streamline processes and ensure compliance with international standards (Kozlov et al., 2021). An IMS integrates key management systems such as the Quality Management System (QMS) ISO 9001:2015, the Environmental Management System (EMS) ISO 14001:2015, and the Occupational Health and Safety Management System ISO 45001:2018, which are essential for the aerospace sector (Domingues et al., 2017). However, many aerospace companies in Russia continue to implement these systems separately, hindering the full realization of their potential. The ISO 9000 series of standards, introduced in 1987, established the foundation for modern quality management practices across various industries.

The aerospace industry, in particular, has adopted industry-specific standards such as AS/EN/JISQ 9100, which adapt ISO 9001 for the specific safety and technological needs of aerospace manufacturing and services (Soare & Militaru, 2018). These standards are critical for maintaining compliance with global leaders such as Boeing and Airbus, which require suppliers to hold certifications under AS/EN/JISQ 9100 (Tomic et al., 2012).

Nevertheless, due to the extended usage of these standards, the emergence of a single IMS has been slow in Russia. The use of integrated systems in the assessment processes could also help eradicate issues such as process duplication and enable conformity to several standards (Talapatra et al., 2019; Ahmad et al., 2019). This is especially true where the integration of the systems is digital as it allows for better tracking of documents, constant supervision, and prompt modifications where necessary to remain competitive in a technologically sensitive industry such as aerospace (Kovrigin & Vasiliev, 2020). The purpose of this chapter is to identify the prospects for IMS implementation in Russia's aerospace industry. This comprises exploring the certification patterns of ISO 9001 and ISO 14001 in the selected aerospace countries, and, in particular, defining successfully implemented practices to develop an optimize IMS model for Russian companies. These management systems provide aerospace organizations with ways to enhance performance, pursue compliance and also address domestic and global markets.

1.1 Importance of Integrated Management Systems in Aerospace

The aerospace industry is not only highly diversified but also one of the most stringent in its regulation in the international market (Yan et al., 2022). This complexity stems from the issues of safety, environmental, and quality that need to be adhered to together with the need for efficiency and cost control. Among the strategies for handling these issues is the IMS where key management systems that include QMS, EMS, and OHSMS are combined in one approach (Domingues et al., 2017). This approach not only eliminates the issue of duplicated through eliminating the point of effort but also guarantees compliance with set international standard across all functionalities within the organization. In the aerospace industry particularly, where product quality and safety is central to any manufacturing firm's goal, the process of implementing QMS normally anchored on the ISO 9001 is usually the first stage of IMS evolution. ISO 9001 guarantees that the procedure is well-followed in line with the customer's and the regulatory body's requirements (Hamid et al., 2019). However, the real value of an IMS is how the QMS interacts with other management systems, for example: ISO 14001 for environmental management and ISO 45001 for Occupational Health & Safety. Such integration enables aerospace firms to monitors their operational impacts on the environment, promote the health and safety of their employees and maintain credible and quality systems of production concurrently (Bernardo et al., 2012).

The implementation of IMS is also essential for meeting international aerospace-specific standards like AS 9100 which is established on ISO 9001, but has extra criteria on safety as well as technology accuracy (Soare & Militaru, 2018). The establishment of AS 9100 is mandatory for aerospace firms for them to work with leaders in the industry like Boeing, Airbus and Lockheed Martin. AS 9100 when implemented in conjunction with an IMS makes it easier to operate and deal with multiple audits thereby improving the manner in which organizations handle complex regulatory needs (Talapatra et al., 2019). Furthermore, implementation of IMS can bring positive changes to the company in terms of coordination and internal communications among different organizational departments especially in the aerospace industry which requires high levels of specialization. As the quality, environmental and safety management systems are integrated under one system, it leads to the overall improvement of the management systems hence making the companies to respond easily to the changing market requirements and; compliance with regulatory requirements (Sampaio et al., 2012). This is especially so in aerospace that faces the need to adopt new technologies such as sustainable aviation fuel and new safety measures that have to be integrated within the existing management systems with ease (Kozlov et al., 2021). In conclusion, the integration of QMS, EMS, and

OHSMS into a single IMS framework is not only a best practice but also a necessity for aerospace companies aiming to remain competitive in a global market. By ensuring compliance with multiple standards, improving operational efficiency, and fostering a culture of continuous improvement, IMS provides aerospace enterprises with a robust platform for sustainable growth and innovation.

1.2 Digital Transformation and Its Role in IMS Implementation

Digital transformation is an important factor that can has a positive impact on the improvement of the IMS in the aerospace industries concerning the aspects of efficiency and effectiveness. The application of IT solutions within IMS provides the organizations with an opportunity for the enhancement of data collection, efficiency of business processes, and quality of decision making. IT enhancements including cloud computing, processing of big data, and AI may help in minimizing errors when executing repetitive tasks while enhancing the efficiency of the management systems (Vasiliev et al., 2020). In the case of aerospace companies, where accuracy and risks are significant, digital transformation offers the framework to oversee and control intricate procedures in real-time with conformance to these standards which include ISO 9001 and AS 9100 (AKIMOV & TIKHONOV, 2023)

Among the opportunities to be gained from digitalizing IMS, the most important is developing an efficient and multifunctional System of Document and Information Management (Glazner, 2006). Aerospace firms can effectively address actual information and documentation demands of the quality, environmental, and safety management systems and other essential records made available of the cloud-based tools accessible in real time to the various functional divisions and locations (Kobzev et al., 2020). These conditions make it easy for the various teams to work together to attain corporate goals and objectives while at the same time ensuring that they adhere to international standards as well as increasing organizational effectiveness. Further, digital technologies' utilization in IMS implementation addresses the Plan-Do-Check-Act (PDCA) cycle that is Augmenting continuous improvement in management systems. Automated system can also automate the observation of performance indicators, issue alarms in case of changes in the indicators and suggest options to handle the changes (Bi et al., 2014). However, this capability does not only provide reinforcement of measures concerning ISO 45001 and ISO 14001, but also minimizes the paperwork on keeping multiple management systems (Olaru et al., 2014).

It also leads to effective management of risks and since the aerospace companies are operating within highly regulated industries, this is an added advantage. For instance, the tools such as predictive analytics and machine learning algorithms in the operations can reveal various risks in operations hence allowing organizations to address these risks before they cause significant problems (Aliyev & Shahverdiyeva, 2023). This predictive capability is especially important in the aerospace industry that, for instance, if a quality or safety issue arises could lead to a disaster (Soare & Militaru, 2018). Moreover, with the digitization of IMS aerospace enterprises can meet the increased demand of sustainability and offer tools that are required to control and decrease such negative effects. For instance, data analytics is useful in the monitoring of emissions, resource consumption, and waste production in real-time would assist organizations in protection of environment to meet ISO 14001 requirements (Bernardo et al., 2012). This inclusion of sustainability metrics in IMS assists not only in the compliance with the legal requirements but also enables the emphasis on corporate responsibility for business sustainability in the long run (Kozlov, Pavlova, & Królas, 2021).

2. LITERATURE REVIEW

The aerospace industry operates in an environment that demands high precision, safety, and adherence to international standards. As a result, Integrated Management Systems (IMS) have become critical to managing quality, environmental sustainability, and occupational safety efficiently. This section reviews the existing literature on IMS in aerospace, focusing on its development, implementation challenges, and the growing role of digital transformation. The review also examines how IMS can be optimized for operational efficiency and compliance with international standards through digital tools.

2.1 Evolution and Challenges of IMS in Aerospace

Integrated Management Systems (IMS) is a concept that has developed over the recent decades, when managers realized that having several management systems were counterproductive. First, each of the top business strategies including Quality Management Systems (QMS), Environmental Management Systems (EMS), and Occupational Health and Safety Management Systems (OHSMS) were adopted separately. But, it was seen that many organizations started having problems related to duplicity of work, redundancy and extra burden of administrative work for having these different systems (Sheng, 2019). To address these issues, the concept to amalgamate these systems into a single system was proposed and hence IMS was evolved.

In the aerospace industry, IMS has been implemented in a bid to match international standards like the internationally recognized standards for quality management-ISO 9001, environmental management-ISO 14001 and occupational health and safety management system-ISO 45001. Furthermore, the aerospace organizations also follow the other sector standards like AS/EN/JISQ 9100 that involves a set of safety and technology requirements related to aerospace industry only (Hubbard, 2015). These standards have emerged as a result of growing sophistication of the aerospace operations and to accommodate for enhanced risk management and compliance with the regulators.

The literature also reveals various issues that can hinder the adoption of IMS in aerospace organizations. There are two objections, one of which is the integration process. When such systems have been developed under different departments and have different architectures and goals, it is often a challenge for many companies to integrate them into a common system, as has been pointed out by Bernardo et al. (2012). Another major challenge is culture and resistance to change within organizations. Often, people are resistant to change, especially if they are using particular systems in the organization for several years as employees and managers do. It is most apparent in sectors like aerospace that involve considerable risks of product quality and safety, which makes organizations cautious (Amrani & Ducq, 2020). Additional challenges are incurred because of the necessity to develop the mechanism further and adapt it to the challenging shifts in the regulatory requirements. IMS frameworks should have the ability to change along the particular period because new regulations are developed and at the same time there should be conformity throughout the organization. This is because, on one hand, there is a need to establish standardized processes for manufacturing different aerospace products, while on the other hand, there is need to be flexible, given that the companies operate within complex environments (Al-Momani et al., 2020). However, different management systems need to be integrated where different management systems have different requirements. For instance, whereas the ISO 9001 deals with customer and product, therefore quality assurance, the ISO 14001 is about environment, whereas the ISO 45001 is about safety. Sustaining multiple and often conflicting objectives within one IMS is not an easy task and therefore needs strategic planning and management (Bernardo et al., 2012).

Despite these challenges, there are significant benefits to implementing an IMS in aerospace. Research shows that companies with an IMS can reduce costs by eliminating redundant processes, improve efficiency through streamlined operations, and enhance compliance with international standards (Olaru et al., 2014). Furthermore, IMS can foster a culture of continuous improvement, which is essential for maintaining competitiveness in the rapidly evolving aerospace industry (Liu et al., 2023).

2.2 The Role of Digital Transformation in IMS Implementation

High precision industry such as aerospace, it was revealed that, digital transformation is a vital key in the success of the IMS implementation. IMS has significantly transformed the working of organizations by incorporating the digital technologies like big data analytics, cloud computing, and artificial intelligence (AI). In quality, environmental and safety performance indicators, digital tools make it possible to have real-time tracking of the performance and achievement of organizational goals and any discrepancies can be tackled before getting out of hand (Vasiliev et al., 2020). In the context of IMS, the use of digital resources can facilitate company operations and enhance decision-making based on accessible and timely information. Another benefit of digital business for IMS is the replacement of many manual processes, thus minimizing the potential for errors to occur, and making employees more productive and efficient. For instance, the AI system can help assess records of production processes and identify weak points, which may lead to some threats (Khoso et al., 2023). Likewise, cloud-based platforms are instrumental in delivering aerospace industries with a central hub for documentation that is compliant with international standards. This largely reduces the need for paperwork and guarantees that all information is availed to the stakeholders in the organization (Kobzev et al., 2020).

Big data analysis also contributes towards the effectiveness of IMS by allowing the firms to have a 'look' into their Operational data to predict trends. For example, predictive analytics can also help identify possible quality problem areas, or safety concerns, so that one can avoid those issues before are major risks occur (Oche et al., 2021). This is especially so in the aerospace industry where issues of quality and safety if not well dealt with can result to negative high impact effects (Soare & Militaru, 2018). Digital transformation also enhances enhanced risk management since organizations obtain the tools required to regularly measure and evaluate risks faced within organizations.

Furthermore, the aspects of digital transformation are largely aligned with the Plan-Do-Check-Act (PDCA) concept which is one of the most well-known management techniques dealing with the continuous improvement of various systems. Whereas digital tools are capable of recording and continuously monitoring the performance levels and then preparing integrated reports on performance anomalies with recommendations on the necessary corrective actions to be taken (Badiru et al., 2018). This helps aerospace companies to create a culture of continuous improvement aiming at responding to new needs in the market or new requirements of the regulations that rule the industry. Furthermore, many analysts have pointed out that, through digital processes of change aerospace manufacturing companies can re-allocate their resources in a more effective way and, thereby, enhance their performances in terms of the level of wastage to the environment in the light of ISO 14001 standards (Butt, 2020).

The literature nevertheless indicates some various difficulties of digital transformation concerning IMS implementation. There is a recurring issue of high cost of adopting digital technologies which hampers the efforts of the smaller organizations (Hamid et al., 2019). Also, we get into concern the issue of qualified staff to support and operate these digital systems. For instance, the aerospace industry has been admitted to be struggling with a shortage of skilled human resources that are capable of practicing IMS digitally as well as maintaining it (Vasiliev et al., 2020). In addition, the implementation of digital tools to management systems may expose the latter to certain challenges that make it mandatory for managers to spend ample amount of time to integrate these tools with the existing systems (Kobzev et al., 2020).

3. METHODOLOGY

3.1 Quantitative Research Design

This study employs a quantitative research design to assess the level of Integrated Management Systems (IMS) implementation and to evaluate the role of digital transformation in enhancing operational efficiency within aerospace enterprises. The quantitative method is suitable for this research as it allows for objective measurement and analysis of numerical data regarding the extent of IMS implementation, the use of digital tools, and the performance outcomes associated with these processes. The focus on measurable outcomes such as compliance rates, cost reductions, and improvements in efficiency ensures that the study's findings can be generalized across the aerospace industry. By relying solely on quantitative data, the research aims to provide clear, data-driven insights into how IMS and digital transformation contribute to the performance of aerospace companies.

3.2 Data Collection

Data for this study will be collected using a structured survey designed to capture the current state of IMS implementation in aerospace companies, as well as the extent to which digital technologies such as big data analytics, artificial intelligence (AI), and cloud computing have been integrated into these systems. The survey will consist of a series of closed-ended questions that ask respondents to rate the degree of IMS adoption within their organizations, as well as the specific digital tools that are used to enhance management systems. Respondents will also be asked to assess the operational benefits derived from IMS and digital integration, including improvements in regulatory compliance, cost efficiency, and safety performance. To ensure comprehensive data collection, the survey will be distributed electronically to a random sample of aerospace companies, targeting managers, quality control personnel, and digital transformation teams. The use of electronic surveys will facilitate broader participation and ensure that data is collected efficiently.

3.3 Variables and Measures

This study will focus on several key variables to measure the impact of IMS implementation and digital transformation on operational outcomes. The independent variable in this research will be the degree of IMS implementation, which will be measured by the number of management systems integrated within each organization. This includes Quality Management Systems (QMS), Environmental Management Systems (EMS), and Occupational Health and Safety Management Systems (OHSMS). The dependent variables will consist of operational efficiency, which will be measured in terms of cost savings, error reduction, and compliance with international standards. Another dependent variable will be digital integration, which will assess the extent to which digital technologies such as AI and cloud-based platforms have been incorporated into the IMS framework. Finally, challenges associated with IMS implementation, such as costs, technical difficulties, and workforce training, will be measured by respondents' assessments of the most significant barriers to successful integration.

3.4 Data Analysis

Once the survey data has been collected, it will be analyzed using a range of statistical techniques to ensure that the results provide meaningful insights into the research questions. Descriptive statistics will be used to summarize the data, highlighting the extent of IMS implementation and digital technology adoption across the sample. The analysis will also explore how many aerospace companies have fully or partially adopted IMS, and what digital tools are most commonly used to support these systems. Correlation analysis will then be conducted to explore the relationship between IMS implementation and operational efficiency, allowing the study to determine whether higher levels of integration are associated with improved performance outcomes. Regression analysis will be employed to further examine the impact of digital transformation on operational efficiency, cost reductions, and regulatory compliance. Additionally, the challenges associated with IMS and digital integration will be analyzed to identify the most common obstacles faced by organizations during the implementation process.

3.5 Validity and Reliability

To ensure the validity of the survey instrument, the questions will be based on existing frameworks and literature related to IMS and digital transformation in aerospace enterprises. Industry experts will be consulted during the development phase of the survey to ensure that the questions accurately reflect the real-world challenges and benefits associated with IMS implementation. Reliability will be assessed using Cronbach's alpha to test the consistency of responses across the survey items. Prior to full data collection, the survey will be pilot tested with a small group of aerospace companies to ensure that the questions are clear and the instrument is effective in capturing the necessary data. Feedback from the pilot test will be used to refine the survey before it is distributed to the full sample.

4. RESULTS

According to Aero Dynamic Advisory (URL: https://aerodynamicadvisory.com, accessed on 2021/08/10), in 2017, the combined aerospace industry in these countries was 731 billion USD, representing 87% of the global aerospace industry. At that time, according to this indicator, the Russian Federation was in 6th place (see Table 1).

Table 1. Aerospace industry volume, billions of dollars, in the world's leading aerospace countries for 2017

Nº	Country	Industry volume
1	USA	408.4
2	France	69.0
3	China	61.2
4	UK	48.8
5	Germany	46.2
6	Russia	27.1
7	Canada	24.0
8	Japan	21.0
9	Spain	14.4
10	India	11.0

According to ISO data, for the period from 2009 to 2019, Russian enterprises belonging to 39 sectors of the economy received 168737 ISO 9001 certificates (92.3% of the total number of ISO certificates issued in the Russian Federation),

17837 ISO 14001 certificates (9.7%) and 178 ISO certificates 45001. From the given data, it is clear that in the Russian Federation, much more attention is paid to QMS certification than to other management systems (EMS and OHSAS) certification. It should be noted that the maximum number of received certificates fell in 2009 and 2010, the following local peak was observed in 2013, and after 2015 the activity of Russian enterprises in general in obtaining ISO certificates decreased significantly (see Figure 1). Foreign policy reasons can explain this situation. Figure 2 shows the dynamics of the ISO 9001 and ISO 14001 certificates relative share received by aerospace enterprises of the Russian Federation during the analyzed period. It shows that in 2009–2010, the ISO 14001 certificates proportion in relation to the total number of these certificates received in the country category was relatively high (about 5%). Then, it dropped sharply and became approximately the same as the share of ISO 9001 certificates. It should be noted that since 2015, the relative share of any ISO certifications received by aerospace enterprises in Russia has declined significantly, so much so that certification in this industry has declined more than in the whole economy.

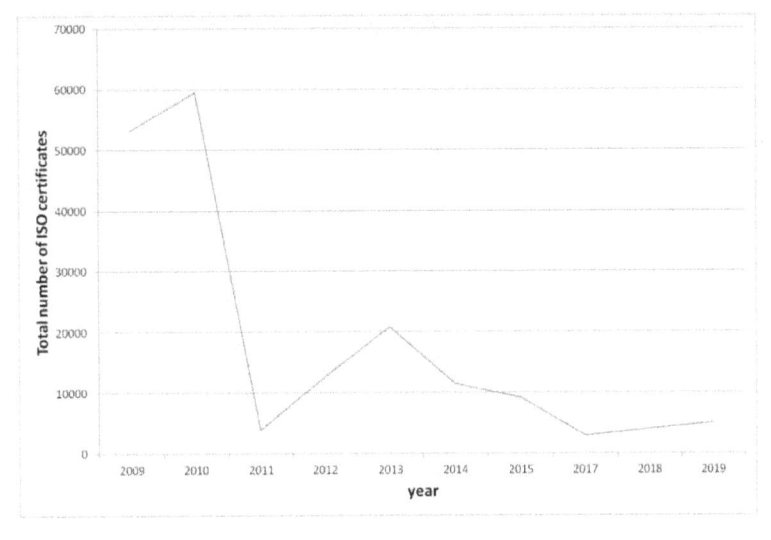

Figure 1. The total dynamics of obtaining ISO certificates in Russia during 2009–2019 period

Figure 2. The share (%) of ISO 9001 and ISO 14001 certificates received by aerospace enterprises during 2009–2019 of the total number of corresponding certificates received by Russian enterprises and organizations

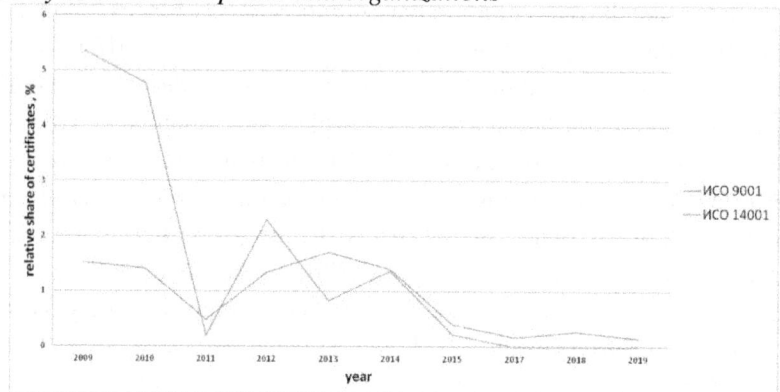

Table 2(a). ISO data on the number of ISO 9001 certificates issued in the world's leading aerospace countries

Country	2009	2010	2011	2012
USA	211	73	212	129
France	1	17	248	135
China	82	74	142	83
UK	81	56	24	102
Germany	89	97	86	143
Russia	791	820	16	158
Canada	17	13	10	8
Japan	182	182	177	187
Spain	54	78	73	85
India	39	28	46	45
Average	154.7	143.8	103.4	107.5

Table 2(b). ISO data on the number of ISO 9001 certificates issued in the world's leading aerospace countries

2013	2014	2015	2017	2018	2019
279	306	314	129	76	121
272	252	231	180	4	8
70	79	59	115	111	128
49	99	63	72	23	28
68	73	74	52	27	69
272	145	33	4	9	6
17	17	18	8	3	2
192	210	222	238	239	70
59	54	4	8	51	70
39	46	53	73	88	14
131.7	128.1	107.1	87.9	63.1	51.6

Table 3(a). ISO data on the number of ISO 14001 certificates issued in the world leading aerospace countries

Country	2009	2010	2011	2012
USA	11	8	9	17
France	1	7	54	30
China	28	31	43	10
UK	18	10	7	14
Germany	52	14	14	24
Russia	75	87	1	22
Canada	9	0	0	1
Japan	31	28	28	31
Spain	13	13	11	17
India	6	2	1	4
Average	24.4	20	16.8	17

Table 3(b). ISO data on the number of ISO 14001 certificates issued in the world leading aerospace countries

2013	2014	2015	2017	2018	2019
22	18	18	23	18	38
41	37	36	30	0	0
41	42	45	56	55	70
14	12	9	6	3	4
11	9	9	6	3	11
41	16	2	0	0	0
0	0	0	3	1	59
32	34	31	36	34	35
12	9	5	7	15	15
2	4	5	4	2	8
21.6	18.1	16	17.1	13.1	24

Figures 3 and 4 show the aerospace enterprises dynamics of QMS (ISO 9001) and EMS (ISO 14001) certification in the Russian Federation in comparison with the world average. It shows that in 2009-2010, the total quantity of certificates issued for the studied functional management systems in the Russian Federation in the aerospace industry was significantly higher than the world average (almost 3.57 times in 2009 and 4.2 times). However, in 2011, the quantity of received certificates dropped sharply. Later, in 2012–2014, the certification rate in the Russian aerospace industry, both in environmental management systems and in quality management systems, stayed approximately at the average global level. Since 2015, it has sharply decreased. Thus, the total number of certificates received by aerospace enterprises in the Russian Federation in 2019 was 14.5 times lower than the world average.

It is worth noting that the number of ISO 9001 certificates issued to organizations related to aviation and space during the analyzed period gradually decreased all over the world. Since 2014, this trend has been observed for all leading aerospace countries except Japan. Perhaps this trend is due to the limited number of such enterprises and the fact that new market participants in this industry appear relatively rarely. On the contrary, between 2009 and 2019, the number of ISO 14001 certificates did not decrease, and for a country like China, one can even note a trend towards them increasing.

Figure 3. The ISO 9001 certificates were obtained in the Russian Federation aerospace during the period from 2009 to 2019 in comparison with the world average values

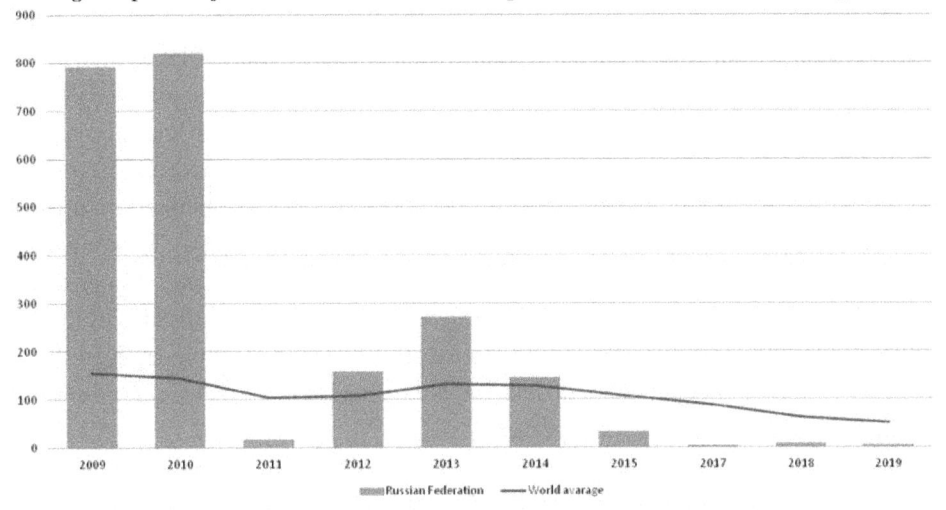

Figure 4. The ISO 14001 (B) certificates number of obtained in the Russian Federation aerospace during the period from 2009 to 2019 in comparison with the world average values

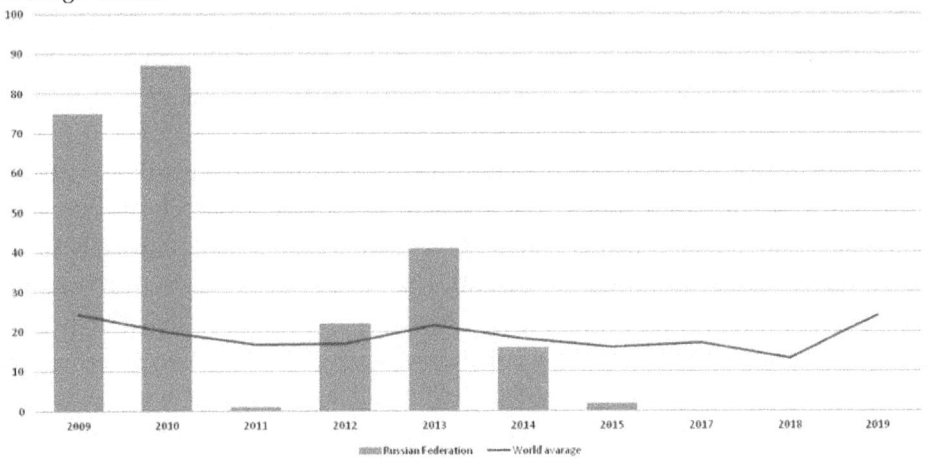

5. DISCUSSION

The dynamics of ISO certification within the Russian aerospace industry from 2009 to 2019 provide important insights into the implementation of Integrated Management Systems (IMS), particularly with regard to Quality Management Systems (QMS) and Environmental Management Systems (EMS). As demonstrated by the data, Russian aerospace enterprises have shown a strong emphasis on QMS certification, with a notably higher number of ISO 9001 certificates obtained compared to other types of certifications, such as ISO 14001 and ISO 45001. This suggests that, for Russian aerospace companies, quality management remains a top priority, likely due to the stringent safety and performance standards required within this highly specialized industry. The lower number of EMS certifications indicates that while environmental management is acknowledged, it does not receive the same level of attention as quality management.

Another factor to consider is the variation in ISO certifications which reached their high in 2009 and 2010 but declined in the subsequent years. In this period, Russian aerospace enterprises were registering many more certificates of compliance with the ISO 9001 standard than averages in the world, indicating an initial trend towards compliance with international standards of quality. But, this momentum was not continued and, in fact, the number of certifications dropped significantly in the next years. By the year 2015, the certification activity was significantly below the worldwide average in the Russian Federation, which evidences its outlets or reduced desire/ability to acquire new credentials. There could be many reasons for this: changes in the foreign policy, economic sanctions, and the general political climate that could have made it either impossible for Russian enterprises to seek international certifications or that they preferred not to do it.

This analysis also reveals that the Russian aerospace industry pays little attention to the ISO 14001 certifications. The percentages of ISO 14001 certifications in 2009–2010 was relatively higher at about 5 percent but this reduced drastically and is now closely situated to the ISO 9001 percent. This means that although environmental management was an early consideration, it was not a significant concern as time went by. Thus, the Russian aerospace industry, along with many other industries in the world, could be more concerned with performance and safety requirements than the environmental management systems. The above trend, however, is in a different direction with most of the world's countries including China that has increasingly presented higher certifications to ISO 14001 standard. On the other hand, the Russian aerospace industry is yet to adapt with such trends since there are possibly some underlying reasons such as absence or scarcity of regulatory compliance pressure or absence of competitive force to compel the Russians to adopt green technologies.

6. CONCLUSION

Examining the patterns of ISO certification in the Russian aerospace industry are as follows: During the period between 2009 and 2019, the Russian aerospace industry was characterized by the adoption of the IMS. Although the first years depicted relatively increased interest in quality management with many organizations achieving ISO 9001 then, the overall IMS activity in terms of ISO 14001 and 45001 has been low. The trend analysis shows that the number of certifications after 2010 has greatly reduce and this political decisions, economic sanctions, domestic regulation has played a major role in reducing the industry reliance on international standards. Despite the fact that the Russian aerospace sector continues to attach importance to quality management demonstrated by the number of actual ISO 9001, the development of environmental and OHS standards within IMS scope remains rather limited. This is in contrast to the global trends whereby other leading aerospace nations including the Chinese are still reinforcing environmental management through integration of ISO 14001 certifications. The freezing of certification activity, some of which are in the field of environment and occupational health and safety, may indicate that Russian aerospace companies are likely to have issues in relation to international standards of sustainability and responsible management. As the competition within aerospace apparels intensify including with global industries Russian companies must begin to rebalance their focus on the IMS system and consider the extension of the scope/aim of IMS beyond a mere vehicle for effective quality management.

REFERENCES

Ahmad, M., & Beddu, S. (2019). State-of-the-art compendium of macro and micro energies. *Advances in Science and Technology Research Journal.*, 13(1), 88–109. DOI: 10.12913/22998624/103425

Akimov, A., & Tikhonov, A. (2023). Implementation of Digital Technologies in Personnel Management System of Enterprises of Rocket and Space Industry. *Journal of Theoretical and Applied Information Technology*, 101(5), 1761–1770.

Al-Momani, H., Al Meanazel, O. T., Kwaldeh, E., Alaween, A., Khasaleh, A., & Qamar, A. (2020). The efficiency of using a tailored inventory management system in the military aviation industry. *Heliyon*, 6(7), e04424. DOI: 10.1016/j.heliyon.2020.e04424 PMID: 32695911

Aliyev, A. G., & Shahverdiyeva, R. O. (2023). Development of a conceptual model of effective management of innovative enterprises based on digital twin technologies. [IJIEEB]. *Int. J. Inf. Eng. Electron. Bus.*, 15(4), 34–47. DOI: 10.5815/ijieeb.2023.04.04

Amrani, A., & Ducq, Y. (2020). Lean practices implementation in aerospace based on sector characteristics: Methodology and case study. *Production Planning and Control*, 31(16), 1313–1335. DOI: 10.1080/09537287.2019.1706197

Badiru, A. B., Ibidapo-Obe, O., & Ayeni, B. J. (2018). *Manufacturing and enterprise: An integrated systems approach*. CRC Press. DOI: 10.1201/9780429055928

Bernardo, M., Casadesús, M., Karapetrovic, S., & Heras, I. (2012). Do integration difficulties influence management system integration levels? *Journal of Cleaner Production*, 21(1), 23–33. DOI: 10.1016/j.jclepro.2011.09.008

Bernardo, M., Simon, A., Tarí, J. J., & Molina-Azorín, J. F. (2009). Integrated management systems: Development and testing of a theoretical model. *Journal of Cleaner Production*, 17(5), 742–750. DOI: 10.1016/j.jclepro.2008.11.003

Bernardo, M., Simon, A., Tarí, J. J., & Molina-Azorín, J. F. (2012). Benefits of integrated management systems: The views of managers. *The TQM Journal*, 24(5), 386–402. DOI: 10.1108/17542731211261550

Bi, Z., Da Xu, L., & Wang, C. (2014). Internet of things for enterprise systems of modern manufacturing. *IEEE Transactions on Industrial Informatics*, 10(2), 1537–1546. DOI: 10.1109/TII.2014.2300338

Butt, J. (2020). A conceptual framework to support digital transformation in manufacturing using an integrated business process management approach. *Designs*, 4(3), 17. DOI: 10.3390/designs4030017

Domingues, J. P., Sampaio, P., & Arezes, P. M. (2017). Integrated management systems assessment: A maturity model proposal. *Journal of Cleaner Production*, 142(1), 145–158. DOI: 10.1016/j.jclepro.2016.07.100

Glazner, C. G. (2006). Enterprise integration strategies across virtual extended enterprise networks: a case study of the F-35 Joint Strike Fighter Program enterprise (Doctoral dissertation, Massachusetts Institute of Technology).

Hamid, A. R. A., Yusof, S. M., Rahman, S. A., & Idris, M. A. (2019). The implementation of ISO 9001:2015 in the aerospace sector: Issues and challenges. *International Journal of Productivity and Performance Management*, 68(3), 504–523. DOI: 10.1108/IJPPM-09-2017-0218

Hubbard, P. D. (2015). Fault management via dynamic reconfiguration for integrated modular avionics (Doctoral dissertation, Loughborough University).

Karapetrovic, S., & Casadesús, M. (2009). Implementing environmental management systems in Spanish universities. *The TQM Journal*, 21(5), 507–519. DOI: 10.1108/17542730910983396

Khoso, A. K., Darazi, M. A., Mahesar, K. A., Memon, M. A., & Nawaz, F. (2022). The impact of ESL teachers' emotional intelligence on ESL Students academic engagement, reading and writing proficiency: Mediating role of ESL students motivation. *Int. J. Early Childhood Spec. Educ*, 14, 3267–3280.

Khoso, A. K., Khurram, S., & Chachar, Z. A. (2024). Exploring the Effects of Embeddedness-Emanation Feminist Identity on Language Learning Anxiety: A Case Study of Female English as A Foreign Language (EFL) Learners in Higher Education Institutions of Karachi. *International Journal of Contemporary Issues in Social Sciences*, 3(1), 1277–1290.

Kobzev, A. S., Smirnova, E. N., & Fedorov, V. I. (2020). The implementation of integrated management systems in aerospace organizations using digital technologies. *Russian Engineering Research*, 40(1), 67–72. DOI: 10.3103/S1068798X20010136

Kovrigin, E. A., & Vasiliev, V. A. (2020, September). Barriers in the integration of modern digital technologies in the system of quality management of enterprises of the aerospace industry. In 2020 International Conference Quality Management, Transport and Information Security, Information Technologies (IT&QM&IS) (pp. 331-335). IEEE. DOI: 10.1109/ITQMIS51053.2020.9322960

Kozlov, A., Pavlova, E., & Królas, A. (2021). Development of integrated management systems in the aerospace industry: A Russian perspective. *International Journal of Engineering Research & Technology (Ahmedabad)*, 10(2), 110–117. DOI: 10.17577/IJERTV10IS020067

Liu, H., Zhu, Q., Khoso, W. M., & Khoso, A. K. (2023). Spatial pattern and the development of green finance trends in China. *Renewable Energy*, 211, 370–378. DOI: 10.1016/j.renene.2023.05.014

Oche, P. A., Ewa, G. A., & Ibekwe, N. (2021). Applications and challenges of artificial intelligence in space missions. *IEEE Access : Practical Innovations, Open Solutions*, 12, 44481–44509. DOI: 10.1109/ACCESS.2021.3132500

Olaru, M., Maier, D., Nicoara, D., & Maier, A. (2014). Establishing the basis for development of an organization by adopting the integrated management systems: Comparative study of various models and concepts. *Procedia: Social and Behavioral Sciences*, 109, 693–697. DOI: 10.1016/j.sbspro.2013.12.531

Sampaio, P., Saraiva, P., & Domingues, P. (2012). Management systems: Integration or addition? *International Journal of Quality & Reliability Management*, 29(4), 402–424. DOI: 10.1108/02656711211224857

Sheng, R. (2019). *Systems engineering for aerospace: A practical approach*. Academic Press.

Soare, A., & Militaru, C. (2018). Quality management systems in the aerospace industry: An approach based on AS/EN/JISQ 9100 standards. *Management of Sustainable Development*, 10(1), 39–44. DOI: 10.1515/msd-2018-0010

Talapatra, S., Uddin, M. K., & Rahman, M. H. (2019). Development of an implementation framework for integrated management systems. *The TQM Journal*, 31(3), 472–489. DOI: 10.1108/TQM-08-2018-0107

Tomic, S., Spasojevic-Brkic, V., & Klarin, M. (2012). The importance of AS/EN/JISQ 9100 quality standards in the aerospace industry. *Quality and Reliability Engineering International*, 28(4), 459–467. DOI: 10.1002/qre.1234

Vasiliev, D., Kuznetsov, I., Ivanov, V., & Smirnov, A. (2020). Digital transformation in aerospace: The role of integrated management systems. *Journal of Aerospace Engineering*, 233(1), 105–114. DOI: 10.1177/0954410020922057

Yan, M. R., Hong, L. Y., & Warren, K. (2022). Integrated knowledge visualization and the enterprise digital twin system for supporting strategic management decision. *Management Decision*, 60(4), 1095–1115. DOI: 10.1108/MD-02-2021-0182

Chapter 16
Top Trends of the Transport-and-Logistic Activity Development in the Conditions of Digitalization:
Identification of the Main Trends in Logistics

Elena Yu. Vasilyeva
Moscow State University of Civil Engineering, Russia

ABSTRACT

The factors, which influenced the activity of the transport-and-logistic companies most significantly, are analysed in the article. The top trends, peculiar to transport-and-logistic activity in modern conditions, are pointed out. According to the results of the analysis that was carried out, nowadays digitalization is the main subject in logistics. In particular, paperless registration of cargo transportation, robotization during freight processing, the use of unmanned vehicles for transportation, and the introduction of obligatory marking for separate types of commodities are revealed. The author finds out the logical communication between the development of electronic commerce and the adaptation of logistics for it, and concludes, that only the complex logistics will be able to conform to modern requirements. At the same, some factors which slow down the digitalization of the transport-and-logistic sphere in Russia are revealed. The results of the analysis and the conclusions, drawn by the author, can be useful for the further development of the transport-and-logistic complex.

DOI: 10.4018/979-8-3693-3423-2.ch016

NOVELTY STATEMENT

The novelty of this book chapter lies in its comprehensive exploration of the transformative impact of digitalization on the transport and logistics sector. By identifying and analyzing key trends such as the integration of artificial intelligence, blockchain technology, and automation, this chapter provides a forward-looking perspective on how these advancements are reshaping logistical operations. It highlights the critical role of data-driven decision-making and innovative digital tools in enhancing efficiency, reducing costs, and fostering sustainability, offering valuable insights for both practitioners and researchers.

1. INTRODUCTION

In the last few years most of the drivers that shaped the transport and logistics activities have been intimately related to the COVID-19 pandemic. They are border closure, shortage of capacities in available transport, limitations on movement of cargoes, changes in supply chain, increased e-commerce, decrease in warehouse excess capacity and exponential growth in the digital space. Finally, the changes that occurred in the world in the year 2020 presented a serious threat to transport-and-logistic industry in general, and to the specific segments that make it up in particular. Due to the above changes, logistics providers and carriers had to transform quickly and replace unsustainable future resource use with unsustainable methods. The pandemic showed the shift of the markets depending on the changing epidemiological situation in various industries. In some sectors, for example building materials and metallurgy international logistics services were reduced by 40–50% during the peak time. But variables from other industries for instance beverages and consumer goods increased by 20- 30%. In this context, it should be pointed out that occasional drop in demand in a certain sector is inconsequential for the viabilit of logistics service providers. At the same time, the condition of an industry in the long run (the demand for logistics services) which means that even temporary decreases are possible but the industry still remains attractive. For instance, the COVID-19 pandemic affected decrease in mechanical engineering, chemical industry and building materials industry but all those industries has high capacity and they demand the logistics services. While the companies like the pharmaceutical industry and woodworking which experienced a rise in delivery requests during the pandemic crises, they are still limited in the market capacity of their products. Looking at the behavior of different industries in the periods of crises and in view of the overall transport capacity some sectors can be classified as the sectors with high logistic potential – for example the sector of beverage (juice, water) consumables, auto motor, pulp and paper, chemical goods, foods, mechanics (Container Intermodal Technology, 2021).

After the target industries that can provide the maximum potential for logistics organizations have been prioritized, it is necessary to assess the trends observed within the framework of logistics in the contemporary environment (Rudakova, Panshin, & Vlasov, 2021). The most significant feature of the contemporary market of logistics is that it is truly global and increasingly digitalized. In 2021, using the information of the RBC Market Research, it was stated that 58% of Russian transport companies started using various tools of automation to manage employees, plans and finances, warehouse and fuel consumption in 2020. Additionally, the study by the Institute for Statistical Studies and Economics of Knowledge, Higher School of Economics (2021) identified that the demand for the advanced digital technologies in the transport and logistics was 89%. 4 billion rubles in 2020 and will increase to 626 billion rubles in 2025. From 9 to 14 billion rubles by 2030, with a growth rate of 21% per year on average. Thus, digitalization is expected to enhance the overall industry performance by 20% by 2030 (CNews, 2021). The accelerated digitalization of logistics is largely a response to the pandemic (Kharitonova, Grokhotova, & Bogdanova, 2021). Recognizing its benefits, leading logistics companies have automated more than 60% of their operations. However, the Russian logistics sector still lags significantly behind global leaders in this regard, necessitating further research on the issue.

1.1 Impact of the COVID-19 Pandemic on the Transport and Logistics Industry

The COVID-19 impacted the global transport and logistics industry particularly with the disruption of the global supply chain and brought significant change with it. Some of the most frantic impacts include border closures and limitations on the movement of goods and people which in turn increased the time and cost involved in moving products especially those which required cross-border transportation (Notteboom, Pallis, & Rodrigue, 2021). Following the implementation of measures that surrounded the virus, the loading of multiple sectors such as automotive, construction, and the manufacturing sector depressed significantly. Global trade of goods was stated to have reduced between 13 – 32% in 2020 by the World trade organization (2020) with transport and logistics providers largely affected. Additionally, the sudden shift towards e-commerce during the pandemic placed further strain on logistics networks. With many consumers turning to online shopping, logistics companies faced unprecedented demand for last-mile delivery services, requiring them to scale operations rapidly (Ivanov, 2021). The surge in e-commerce activity was particularly evident in the U.S., where online retail sales grew by 44%

in 2020, forcing logistics providers to adapt to changing consumer behaviors (U.S. Census Bureau, 2021).

Yet, the pandemic also contributed towards the process of digitalization, especially in the field of logistics. The changes forced organizations to adopt mindful supply chain management strategies and leverage tools such as automation, tracking, and data analysis to address disruptions (Deloitte, 2020). For instance, getting admission to the route optimization systems and predictive analytics based on the al system enabled firms to handle the intricate features of global supply networks (Alicke et al., 2020). These measures eased some of the effects of the pandemic on the industry and are expected to persist in the future more especially after the virus. In conclusion, while the pandemic brought significant challenges to the transport and logistics industry, it also triggered rapid innovation and digital transformation, setting the stage for a more resilient and technology-driven future.

1.2 The Role of Digitalization in Post-Pandemic Logistics

The COVID-19 pandemic fast-tracked the disruption of the traditional chain of logistics; there is a need for a strong digital system to power the chain. Pressure have been piled on supply chain in the global networks thereby making organizations to seek enhanced digital technologies, complexity, operations efficiency and risk minimization. This new shift was made possible by the integration of automation as well as artificial intelligence (AI) in the responses of logistics providers to changing demand (Gong et al., 2021). Digitalization helped not only to optimize supply chains in terms of their performance but also to increase their robustness, allowing companies to plan potentially disruptive events and adjust routes, if necessary. Live data analysis was also very crucial applying in the pandemic situation so that businesses can make the right decisions. A recent report by Accenture (2020) revealed that organizations that integrated data analytics to their logistics networks achieved improved supply chain visibility and was thus able to manage inventory status and delivery lead times from delaying significantly. Moreover, the adoption of cloud-based logistics management systems enhanced cooperation between various shippers in the supply chain logistics by increasing the level of openness (Ivanov, 2021).

It also brought the question of contactless technology in logistics to the foreground, as SI's supply chain and distribution division discovered. It became very important to ensure that contactless delivery was implemented, use of digital means in making payments, and electronic proof of delivery implementation (McKinsey & Company, 2021). Besides managing health risks, these technologies also advanced the customers' experiences through eliminating manual work and papers. Prospects for the development of the Logistics industry in the following years are likely to be informed by the continuous enhancement of the digital frontier. In the future, block chain technology coupled with artificial intelligence and the Internet of Things are expected to transform the supply chain through processes that will increase supply chain visibility, security, and efficacy, according to Sharma et al. (2020). The integration of such technologies in the logistics industry has been further accelerated during the pandemic and created a basis for development of a more intelligent, adaptive and digital supply chain.

1.3 Problem Statement

Numerous researchers have explored the challenges associated with the implementation of digital technologies in the transport and logistics industry. Scholars such as Zaboyev and Zhuravleva (2014), Korovyakovsky and Kupriyanovsky (2018), and Panychev et al. (2020) have contributed significantly to the understanding of these issues. However, much of the attention in both Russian and international research has been focused on isolated experiments or case studies (SteadieSeifi et al., 2014). This has left a gap in the literature regarding a comprehensive, systematic analysis of the key trends and broader obstacles in the digitalization of transport and logistics during the COVID-19 pandemic. The primary aim of this research is to identify the main trends in the development of transport and logistics activities during the pandemic, with a focus on the ongoing digitalization across various sectors of the economy.

Furthermore, this study seeks to uncover the obstacles preventing faster and more effective digitalization within the logistics sector, which is critical for maintaining global supply chain resilience. This research utilizes a wide range of sources, including analytical materials from the Logirus company, forecasts by Avtodor Group specialists, expert polls from RBC Market Research, reports by the Higher School of Economics, and reviews from the CNews internet portal. Additionally, key government projects such as "Smart City" and "Safe and High-Quality Roads" were analyzed. The study also reviewed the prototype of the state information system developed by the Federal State Unitary Enterprise ZashchitaInfoTrans. The methods applied in this research include generalization and systematization of scientific and statistical data, comparative analysis, synthesis, and a systems approach. Additionally, economic analysis and statistical theory were employed to process the data effectively.

2. LITERATURE REVIEW

This section explores the key literature surrounding the digital transformation of transport and logistics, focusing on the challenges and opportunities that have emerged in response to the COVID-19 pandemic. It also examines the main trends influencing the industry, with an emphasis on the impact of digitalization on operational efficiency and supply chain resilience.

2.1 Digitalization in Transport and Logistics

Technology has become a strong factor that is affecting the conventional transport and logistics business by making their operations more efficient, transparent, and timely in the global supply chain. The adoption of innovations like AI, IoT, big data, and blockchain has caused disruption in how logistics companies oversee the movement of goods, storage, and client relations. these technologies allow the kinetics of flow acquisition and its analysis, which, in turn, increases the efficiency of decision-making, productivity, and customer satisfaction (Ivanov and Dolgui, 2020). Perhaps the most obvious and important reason for the digitalization of the logistics sector is the concerns for transparency and timely tracking information. About IoT for example, it assists logistics firms to monitor the status of shipments in near-real time, monitor the functioning of vehicles, and even plan routes based on current traffic conditions.

This level of transparency does not only enhance the satisfaction of the customer, but it also aids organizations in realizing the problems within their processes (Alicke, Barriball, & Trautwein, 2020). For example, the sensors installed in trucks can collect information about fuel used, which will help companies decrease emissions and adhere to the legislation on emissions. Also, the integration of artificial intelligence and machine learning in logistics has been revolutionary. These technologies enable logistics providers to anticipate changes in demand patterns, manage inventory more efficiently, and some routine functions like optimizing delivery paths, and planning timetables (Deloitte, 2020). This trend of automating has led to increased efficiency in delivery since more tasks are handled by the machines, thus leaving out little room for mistakes when doing things like stocking in the warehouse, moving stocks around, etc. Another of the components of digitalization in logistics is the use of blockchain technologies as a reliable register of transactions and document authorship. Using the example of data security and fraud minimization, it is possible to state that blockchain increases confidence in supply chain participants (Sharma et al., 2020).

2.2 Key Challenges in Implementing Digital Technologies

Many opportunities exist to leverage digitalization across the transport and logistics sector However, several major barriers exist. The risk of investment and other associated expenses are always a challenge, costs linked to implementing digital technologies including AI, IoT, and blockchain are tough to digest, especially for SMEs. Zaboyev and Zhuravleva (2014) pointed out that to obtain digital technologies, software, and specific equipment, the initial costs are too expensive for SMEs. This tends to create a divide between large, organizations that are in a position to incorporate efficient digital systems and other firms which are left to deal with the challenge.

Apart from financial issues, there is a shortcoming in relevant digital competencies or at least in adequate digital qualifications among the workforce, which is clearly defined as one of the most significant challenges in the context of digitization. According to Korovyakovsky and Kupriyanovsky (2018), it is common to find today's companies in the logistics sector to lack technical competencies of employees capable of managing and operating the involved digital systems. This situation is particularly so in the developing countries where skills and education in the use of digital technologies might not be easily accessed. Inability to create a digitally literate workforce exposes a firm to huge problems when it comes to exploiting the use of advanced technologies such as artificial intelligence and IoT to enhance operational effectiveness and efficiency. Another major problem is that of cyber security, especially with employees, clients or even third party vendors. The rise in the use of digital technology brings risks that affect transport and logistics firms in instances whereby their data is hacked, ransomware attacks, or system outages. Gong, Mitchell, and Krishnamurthy also pointed out that logistics firm must be deeply concerned with cybersecurity as strategic plans and other important information must be safeguarded. However, many Small and Medium-sized enterprises cannot afford to adopt wide-spread protection measures and are generally liable to cyber threats. Same as mentioned above in the area of supply chain, according to the European Union Agency for Cybersecurity (ENISA) (2020), cyberattacks in this sector surged by 66% in 2020 demonstrating the escalation of the threat in the context of digitalization.

Moreover, applying computing technologies in systems and operations also pose operational issues. Today's reinforcements, which are commonly applied within the logistics industry, may be incompatible with modern digital tools, and, therefore, necessitate significant expenses for modernization or replacement. The adoption of new technologies involves challenges in integrating the innovations with the existing logistics structures that may result in emergence of high costs, delays and interruptions (Deloitte, 2020). Some of the organizations may avoid implementing digital technologies because they feel that there are risks and steep barriers associated with change from legacy systems. Finally, the regulatory and legal concerns are also the factors that hinder digital transformation of logistics. Data protection laws are different in various countries and regions globally on the use of personal information, the conduct of digital transactions and on cross-border transfers of data which poses a challenge to the implementation of technology solutions including the block chain and IoT (Sharma et al., 2020). It is thus clear that managing these regulatory challenges demand ample legal know-how and this further compounds the costs and difficulties in digitalization.

3. METHODOLOGY

The methodology of this study was designed to examine the trends and challenges in the transport and logistics sector, particularly focusing on the impact of digitalization in the context of the COVID-19 pandemic. A combination of quantitative and qualitative methods was employed to ensure a comprehensive analysis of the industry's transition towards digital solutions, as well as to explore the barriers to and benefits of implementing these technologies in Russia's logistics sector.

3.1 Data Collection

Data for this study was collected from a variety of sources to ensure a comprehensive analysis. Official statistics from governmental and industry-specific reports, such as those from the Ministry of Transport of the Russian Federation and the Federal State Unitary Enterprise ZashchitaInfoTrans, were examined to understand the extent of digitalization within the logistics sector. Additionally, industry websites like CNews and Commerzbank provided data on electronic document flow, the use of unmanned vehicles, and advancements in robotic technologies. A range of thematic reviews and expert reports from consultancy firms were also analyzed to gain insights into emerging trends, including the shift towards electronic consignment notes and waybills, the development of logistics ecosystems, and the expansion of e-commerce during the pandemic. Key infrastructure and digitalization initiatives in Russia, such as the "Smart City" and "Safe and Qualitative Highways" projects, were also reviewed to assess the influence of government policies in promoting digital transformation. Experimental projects, including the paperless cargo transportation system and the Unmanned Logistics Corridors project, were incorporated to provide a deeper understanding of ongoing digitalization efforts in the logistics industry.

3.2 Experimentation

To better understand the application of digital technologies in logistics, the study evaluated the ongoing experiment in Russia involving the paperless registration of cargo transportation. During this experiment, various companies participated in the testing of electronic transportation documents, including consignment notes and waybills. Data was collected on the number of electronic documents issued, the companies involved, and the processes adopted. The prototype of the state information system developed by ZashchitaInfoTrans served as a focal point of the analysis, with particular attention given to the exchange of electronic data between businesses and government authorities. A statistical analysis of the volume and distribution of electronic consignment notes and waybills was conducted (as seen in Figures 1 and 2). This allowed for an assessment of the effectiveness and efficiency of electronic document flow within the transport sector. The project included participation from 17 companies and involved more than 525 electronic documents, providing real-time insights into the operational efficiencies gained from digitalization.

3.3 Qualitative Analysis

A qualitative approach was employed to assess the barriers to the mass adoption of digital technologies, particularly electronic document flow, within the logistics sector. Interviews and consultations were held with industry experts, representatives from government bodies, and participants in the paperless cargo transportation experiment. This provided insight into issues such as high implementation costs, the lack of standardization in document formats, and the challenges related to system integration and cybersecurity. Additionally, case studies of the implementation of digital logistics systems, such as the BaseTracK Logistics SF technology and autonomous vehicle testing in Russia, were used to assess the benefits of digitalization. Factors such as time savings, reduced fuel consumption, and improved cargo tracking were measured and analyzed based on data from companies participating in these digitalization efforts.

3.4 Comparative Analysis

Comparative analysis was employed to evaluate the differences in the adoption of digital technologies across various regions of Russia. By comparing the digitalization processes in major logistic hubs, such as Moscow and St. Petersburg, to those in smaller regions, such as the Tver and Tyumen regions, the study identified geographical discrepancies in the pace of digital transformation. This was further supported by data on the development of warehouse infrastructure and the growth of e-commerce logistics in the regions.

3.5 Analytical Tools

The study utilized several analytical tools to process and interpret the collected data effectively. Descriptive statistics were employed to describe and summarize key patterns within the data, particularly focusing on the distribution of electronic transportation documents among operators and carriers. This method also helped quantify the efficiency gains reported by companies that had adopted digital technologies, offering insight into how digitalization is impacting operational performance across the logistics sector. Comparative analysis was another essential tool used to assess the differences in adoption rates of digital technologies between various regions and companies. By comparing digitalization efforts in major logistics hubs with those in smaller regions, the study was able to identify geographic disparities and understand the broader economic impact of implementing digital solutions. This analysis also helped to highlight the variation in outcomes between large corporations and smaller, resource-constrained enterprises, providing a nuanced understanding of how digital transformation is progressing at different levels of the logistics industry. In addition to quantitative approaches, qualitative content analysis was applied to analyze interviews, thematic reviews, and expert reports. This method was particularly useful in exploring the barriers to digitalization, such as high costs, lack of infrastructure, and regulatory challenges. By examining the narratives provided by industry experts and participants in the logistics sector, the study was able to capture the complex social and policy dynamics that influence the adoption of digital technologies. This approach also offered valuable insights into the role of government regulations and initiatives in shaping the digital future of the logistics industry.

4. RESULTS

The analysis of the collected data, using descriptive statistics, comparative analysis, and qualitative content analysis, revealed several significant trends and insights regarding the digital transformation of the transport and logistics industry in Russia. These findings are categorized based on the specific areas of electronic document flow adoption, regional differences in digitalization, and the barriers and facilitators to further technological adoption.

4.1 Adoption of Electronic Document Flow

The study of the descriptive statistics helped to establish that electronic document flow has not been advanced equally throughout the T&L sector with regard to consignment notes and waybills. As shown in the figures 1 and 2 electronic transportation documents are much distributed among the operators as well as the carriers. In particular, the large companies of ICL-KPO VS and Korus Consulting used the electronic consignment notes more frequently by issuing 160 and 69 documents respectively during the experiment. It was also observed that the adoption rates of new technologies were slower with the LCCs and independent carriers mainly due to the constraints of resources vis-à-vis automation. Such gaps raise questions as to whether, although the prospects of digitalization are acknowledged, its application is highly biased towards the large and more resource-endowed players.

In addition, the study revealed that the companies that took part in the experiment were able to enhance efficiency by reducing paperwork, minimizing errors and shortening decision-making cycle hence noting a cut of 5-10% on overall operations cost. Such efficiency improvements were observed mainly where companies used automated systems for fuel control and monitoring, route scheduling and cargo tracking. Therefore, the implementation of these digital solutions proved the ability of the sector to generate huge economic returns once conducted in large scale.

Table 1. Distribution of electronic documents by operators and carriers

Operator/Carrier Name	Number of Electronic Consignment Notes	Number of Electronic Waybills
ICL-KPO VS	160	69
Korus Consulting	69	45
Editsoft	46	2
Others	42	9

The descriptive analysis shows significant variability in the adoption of electronic document flow among logistics operators and carriers. Larger companies, such as ICL-KPO VS and Korus Consulting, issued the majority of electronic consignment notes, accounting for 160 and 69 documents respectively. In contrast, smaller operators contributed fewer electronic notes, reflecting the uneven adoption of digital technologies across the sector (see Figure 1).

Figure 1. Distribution of the total quantity of electronic consignment notes, among the operators

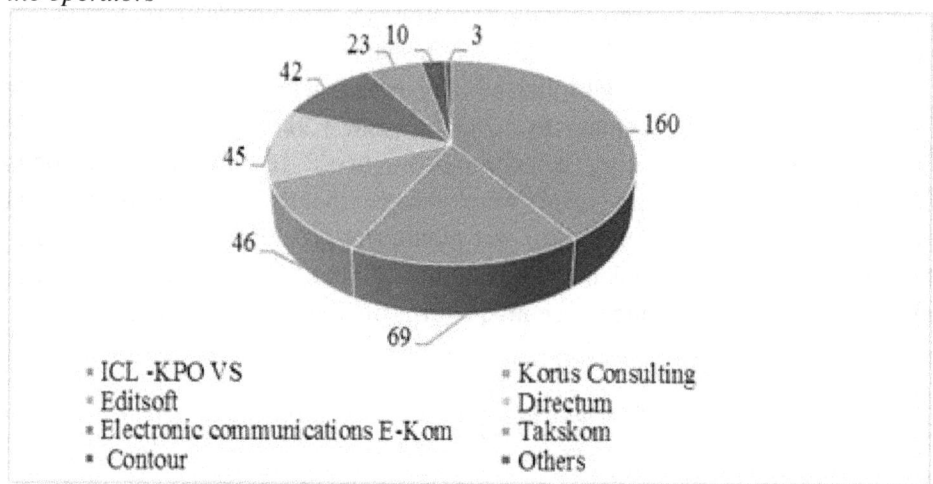

- ICL-KPO VS
- Editsoft
- Electronic communications E-Kom
- Contour
- Korus Consulting
- Directum
- Takskom
- Others

Similarly, the distribution of electronic waybills also varied, with Resource Trans and Auto PEK handling the highest numbers (69 and 45, respectively), while smaller carriers contributed minimally to the total waybill count (see Figure 2). This highlights that larger carriers are more likely to adopt electronic documentation, potentially due to better resources for digital infrastructure.

Figure 2. Distribution of the total quantity of electronic waybills, among the carriers

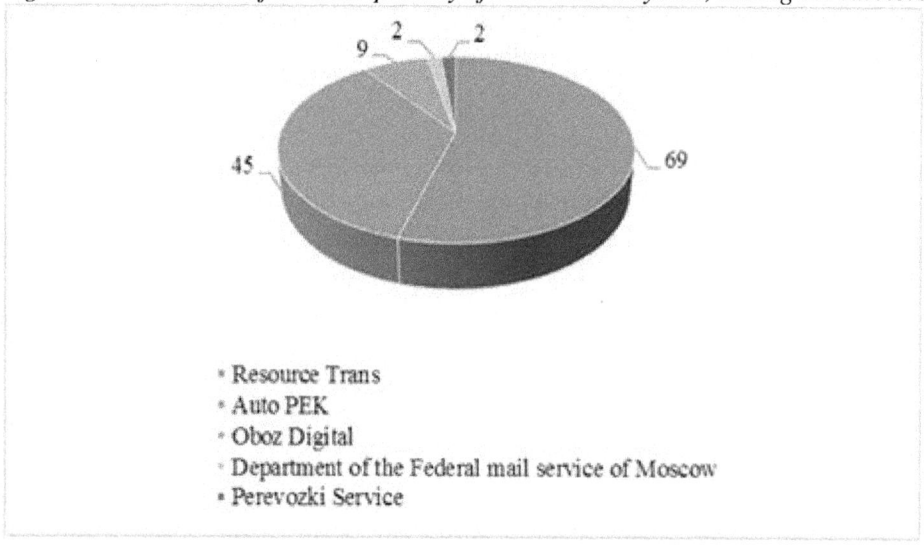

- Resource Trans
- Auto PEK
- Oboz Digital
- Department of the Federal mail service of Moscow
- Perevozki Service

4.2 Regional Differences in Digitalization

Cross-sectional analysis provided evidence of the variation among the regions with regards to the use and integration of digital technologies in the logistics sector. The logistics hubs in the Russian Federation including Moscow, St. Petersburg and Krasnodar region was found to adopt a high level of digitalization with more than forty percent of logistics companies participating in programs including the Unmanned Logistics Corridors. Meanwhile, some of the least developed regions inclining Tver and Tyumen regions demonstrated lower rates of usage due to downs in the availability of infrastructure as well as lower investments in the diffusion of digital technologies.

Moreover, the provinces that receive higher governmental support and infrastructure investment like 'Smart City', 'Safe and Qualitative Highways' also achieved higher progress of digitalization. Such regions were inclined to engage in such experiments as the paperless cargo transportation system that has illustrated that public policy and private investment are critical to the development of digital systems. The results of the comparison suggest that more specific initiatives should be implemented in order to remove the gap between large and developed metropolitan areas and remote, often less-developed regions.

Table 2. Regional comparison of digitalization adoption in logistics

Region	Percentage of Companies Adopting Digital Solutions	Participation in Government Initiatives
Moscow	65%	Yes
St. Petersburg	60%	Yes
Tver Region	30%	No
Tyumen Region	25%	No

4.3 Barriers and Facilitators to Digitalization

Consequently, the qualitative content analysis pointed to the following challenges affecting the adoption of digital technologies in the logistics sector. A significant challenge mentioned was the high implementation costs with emphasis on the situation where SMEs had to embark on implementation. Some of the leading challenges mentioned by many small businesses include the high cost of infrastructure required to support digital solutions like the AI-based algorithms and tools for predictive analytical models, RPA, and advanced fuel-monitoring systems. Moreover, the shortage of qualified workforce with specific knowledge of the technology also became an issue that was evident most in the rural and less developed areas. However, the analysis also identified several enablers of digitalization. Businesses that had earlier adopted digital solutions to their supply chains and operations were able to adapt to closures and other such disruptions and their functions did not get severely affected. Besides, other policies and regulation such as the ones that enable the use of electronic consignment notes and waybills were deemed vital in the transformation towards paperless documents. Seaports also contributed to the digitalization of the sector through initiatives such as experimentation, as exemplified by the Unmanned Logistics Corridors program.

Table 3. Key barriers to digitalization in logistics

Barrier	Percentage of Companies Affected	Key Challenges Identified
High Implementation Costs	55%	Initial infrastructure investment
Lack of Skilled Workforce	40%	Shortage of technical expertise
Cybersecurity Concerns	35%	Lack of robust cybersecurity measures
Integration with Legacy Systems	45%	Difficulty in integrating with old systems

4.4 Impact of Digital Solutions on Efficiency and Cost Reduction

For further understanding, the descriptive statistics and the qualitative content analysis revealed that the companies which adopted the digital systems registered efficiency gains. A 40 000-ton reduction of paper usage year to year helped the company achieve environmental and cost saving objectives. The companies realized up to a 10% fuel saving per trip attributable to real-time optimization route and a 2% reduction in overall transportation generic costs, cargo monitoring, and real time tracking. Furthermore, the firms that incorporated robotic systems and automated cargo handling procedures mentioned a raised warehouse production by 10 percent and reduced down time by 20 percent. These improvements were credited to the use of Information Technology systems that improved the synchronization of the warehouse and transport logistics, and also, virtual decision of various aspects of the operations of the cargo.

Table 4. Efficiency gains and cost reductions from digitalization

Efficiency Metric	Average Improvement (%)	Cost Savings (%)
Fuel Consumption	10%	10%
Warehouse Productivity	10%	N/A
Overall Transportation Cost	2%	2%
Paper Document Reduction	40,000 tons annually	N/A

5. DISCUSSION

This chapter examines the state of digital transformation in the Russian transport and logistics industry, with reference to the COVID-19 crisis. The findings suggest that noteworthy changes occurred in the improvement of electronic document flow and automation, yet the level of transformation is rather unbalanced between aforementioned operators/regions. These implications are important for practitioners and policymakers in an attempt to increase digital integration within the logistics industry. Perhaps the most significant discovery is the fact that digital uptake in logistics operators and carriers is still not even. ICL KPO VS and Korus Consulting are among the larger players and are faster in deploying electronic consignment notes and waybills while the slow process is evident with the small players and independent players. This is in line with previous literature showing that larger organizations are more likely to have the financial and technological capabilities to support digital solutions (Zaboyev & Zhuravleva, 2014). While larger organizations are likely to better address these financial challenges, since they generate rather than spend large amounts of money, smaller organizations face problems with initial investment in digitalization which often comprises costs of personnel training, hardware acquisition, and new software integration (Korovyakovsky & Kupriyanovsky, 2018). This points to a trend of increasing polarization of digital benefits where only computerized, well-resourced companies are able to fully capitalize on the efficiency and cost-reducing prospects that digitalization offers.

The varying level of digital literacy across regions also contribute heavily to this problem. The analysis showed that logistic centers like Moscow and St. Petersburg are ahead of other regions in digital penetration. This is consistent with the pattern presented in the literature on regional economic development where governmental and infrastructural support enhances the readiness to adopt new technologies (Alicke et al., 2020). Thus, such regions' participation in the state initiatives such as 'Smart City' and 'Safe and Qualitative Highways' grants those territories infrastructure and funding requirement for Digital transformation. On the other hand, the regions which receive little governmental funding and have more limited access to economic capital tend to have lower rates of adoption, thus increasing the gap between the developed and the developing world.

The views on barriers to digitalization were also represented as a major finding of the study. Several challenges that came to the lime light were high costs, Skilled human resource restraint, and cyber security risks. The cost involved in the use or adoption of the digital technologies especially to the SMEs, hinders such organizations to match with powerful and technologically superior organizations. However, the skills in managing and maintaining these technologies are scarce making the adoption of these technologies even harder. The implication of this research finding is in line with Gong, Mitchell and Krishnamurthy (2021) assertion that there is a dearth of talents that companies in logistics industry, especially in developing countries, can hire to foster digital disruption. Furthermore, a problem of cyber security was observed since logistics companies depend much on digital systems thus exposing them to cyber criminals. Other studies have also highlighted this as a key issue of concern especially as digitization intensifies thus calling for enhanced cybersecurity measures (Sharma et al., 2020). Still, it is evident that there are considerable advantages in going digital. It has been found out that organizations which incorporate digital technologies have registered significant improvements in productivity, decreased fuel usage, accelerated decisions, and generally lower expenses. These findings alert with technology literature especially in the use of intelligent technologies such as AI and IoT with the potential of increasing the efficiency of operations and decreasing the effect of human mistakes on logistics performance (Ivanov & Dolgui, 2020). For example, adoption of new technologies in rout optimization and fuel control reduced fuel expenses by 10% per flight due to implemented automated technology. In addition, the employment of electronic document flow has minimized the usage of paper documents which in turn reduced the impact on the environment as well as time issues; thus, the prospects of digitalization in logistics for sustainable development would be boosted.

These findings have implications for the scenario of logistics industry in the future. To make the experiments more efficient, there is a necessity to allocate the support more evenly across the regions and the types of the companies, while the focused support of the small operators and less developed regions is still required. This could include supporting SMEs with incentives such as subsidies in order to encourage the use of digital services or more investment in the region's digital systems to make digital changeover more effective across the board. Furthermore, efforts to shape the skills deficiencies through education and training intervention is very crucial in preparing the workforce for management and sustained support of these emerging technologies. Furthermore, there is the need to embrace cybersecurity since there is a rise in the adoption of technology in organizations to prevent loss of valuable information and the honesty of the logistics business.

6. CONCLUSION

The use of new technologies can become key strategies for modernizing the transport and logistics industry in Russia and making further improvements to its performance, product delivery costs, and environmental impact. The results of this research show both the main successes and the key issues concerning the implementation of various digital tools, including electronic document flow and automation. The bigger companies and logistics centers of urban areas are the early adopters of these technologies while the SMEs and rural areas face financial, structural, and skills constraints. The study also discovered that those companies, which have embraced the process of digitalization, have also received the following benefits, which include the following; They have been in a position to cut down their expenses, gain a better understanding of their decision-making processes and also minimized the use of papers and fuel. However, the increased inequality in the distribution of digital technology and disparities between big and small enterprises and between the developed and less developed areas suggest the need for differential efforts.

Thus, the adoption of digital technologies in the logistics sector should be facilitated by the communication policymakers in terms of financial subsidies especially for SMEs and regions with low development. Furthermore, the development of the digital infrastructure and workforce skills availability is also essential for lifting the logistics company's productivity potential and adopting the advanced technologies. Other issues that should not go unattended are the security issues since there is a high propensity for cybercrimes due to increased reliance on digital platforms.

REFERENCES

Accenture. (2020). [*How companies are accelerating digital transformation in supply chain management*. Accenture Strategy. Retrieved from https://www.accenture.com]. *COVID*, 19, •••.

Alicke, K., Barriball, E., & Trautwein, V. (2020). Supply-chain recovery in coronavirus times—plan for now and the future. McKinsey & Company. Retrieved from https://www.mckinsey.com/business-functions/operations/our-insights/supply-chain-recovery-in-coronavirus-times-plan-for-now-and-the-future

Container Intermodal Technology. (2021, January 25). Transport and logistics news review [In Russian]. Retrieved September 29, 2021, from https://citekb.ru/feed/cit/2021/01/25/obzor-smi-novosti-transporta-i-logistiki-vypusk-02-65-11012021-25012021

Deloitte. (2020). COVID-19: Managing supply chain risk and disruption. Deloitte Insights. Retrieved from https://www2.deloitte.com/global/en/pages/risk/articles/covid-19-managing-supply-chain-risk-and-disruption.html

European Union Agency for Cybersecurity (ENISA). (2020). Threat landscape for supply chain attacks. ENISA Report. Retrieved from https://www.enisa.europa.eu/publications/threat-landscape-for-supply-chain-attacks

Gong, Y., Mitchell, P., & Krishnamurthy, A. (2021). The impact of digitalization on the logistics industry: Opportunities and challenges. *Journal of Business Logistics*, 42(1), 121–135. DOI: 10.1111/jbl.12264

Higher School of Economics. (2021). Market research of the market of transport-and-logistic services in Russia in 2016-2020, forecast till 2025 [In Russian]. Retrieved September 29, 2021, from https://marketing.rbc.ru/research/40161/

Ivanov, D. (2021). Digital supply chain and resilience in the context of COVID-19: A dynamic capability approach. *Supply Chain Management Review*, 27(1), 38–49. DOI: 10.1108/SCMR-06-2020-0235

Ivanov, D., & Dolgui, A. (2020). A digital supply chain twin for managing the disruption risks and resilience in the era of Industry 4.0. *Production Planning and Control*, 32(9), 775–788. DOI: 10.1080/09537287.2020.1768450

Kharitonova, T., Grokhotova, M., & Bogdanova, O. (2021). Accelerated digitalization in logistics: A forced response to the pandemic. *Digital Logistics Journal*, 7(2), 45–58.

Korovyakovsky, V. P., & Kupriyanovsky, O. N. (2018). Digitalization in logistics: Challenges and opportunities. *Journal of Logistics and Supply Chain Management*, 6(4), 233–245. DOI: 10.1016/j.jlscm.2018.06.004

McKinsey & Company. (2021). *The future of logistics: Digitalization and automation post-COVID-19*. Retrieved from https://www.mckinsey.com/industries/travel-transport-and-logistics/our-insights/the-future-of-logistics-post-covid-19

Notteboom, T., Pallis, A. A., & Rodrigue, J. P. (2021). Disruptions and resilience in global container shipping and ports: A critical review. *Maritime Economics & Logistics*, 23(2), 179–210. DOI: 10.1057/s41278-020-00180-5

Panychev, A. Y., Sigova, M. V., & Sokolov, G. V. (2020). The impact of digital transformation on logistics. *Rivista Internazionale di Economia dei Trasporti*, 47(3), 123–134. DOI: 10.1080/00207543.2020.1818273

RBC Market Research. (2021). Market research of the transport and logistics services market in Russia (2016-2020) with a forecast until 2025 [In Russian]. Retrieved September 29, 2021, from https://marketing.rbc.ru/research/40161/

Rudakova, E., Panshin, I., & Vlasov, A. (2021). Evaluating trends in modern logistics: Post-pandemic challenges and opportunities. *Transport and Logistics Journal*, 15(3), 22–35.

Sharma, R., Kamble, S., Gunasekaran, A., & Kumar, V. (2020). A systematic review of blockchain technology adoption in supply chains. *Journal of Cleaner Production*, 262, 121–344. DOI: 10.1016/j.jclepro.2020.121344

SteadieSeifi, M., Dellaert, N. P., Nuijten, W., Van Woensel, T., & Raoufi, R. SteadieSeifi. (2014). Multimodal freight transportation planning: A literature review. *European Journal of Operational Research*, 233(1), 1–15. DOI: 10.1016/j.ejor.2013.06.055

U.S. Census Bureau. (2021). Quarterly retail e-commerce sales. U.S. Department of Commerce. Retrieved from https://www.census.gov/

Zaboyev, A. I., & Zhuravleva, E. K. (2014). Challenges of digital technology integration in logistics. *Russian Transport Review*, 12(2), 45–56.

Compilation of References

Abbes, N., Sejri, N., Xu, J., & Cheikhrouhou, M. (2022). New lean six sigma readiness assessment model using fuzzy logic: Case study within clothing industry. *Alexandria Engineering Journal*, 61(11), 9079–9094. DOI: 10.1016/j.aej.2022.02.047

Abdullaev, A., & Bekmurodova, S. (2023). Transformation of the regulation of commercial banks in the conditions of the development of the digital economy of Uzbekistan. Scientific Collection. *InterConf*, (142), 93–102.

Abid, N., Marchesani, F., Ceci, F., Masciarelli, F., & Ahmad, F. (2022). Cities trajectories in the digital era: Exploring the impact of technological advancement and institutional quality on environmental and social sustainability. *Journal of Cleaner Production*, 377, 134378. DOI: 10.1016/j.jclepro.2022.134378

Abushova, E., Burova, E., Suloeva, S., & Shcheglova, A. (2017). Complex approach to selecting priority lines of business by an enterprise. In *Proceedings of the 2017 6th International Conference on Reliability, Infocom Technologies and Optimization (Trends and Future Directions)(ICRITO)* (pp. 565-569). IEEE. DOI: DOI: 10.1109/ICRITO.2017.8342491

Accenture. (2020). [*How companies are accelerating digital transformation in supply chain management*. Accenture Strategy. Retrieved from https://www.accenture.com]. *COVID*, 19, •••.

Acs, Z. J., Stam, E., Audretsch, D. B., & O'Connor, A. (2017). The lineages of the entrepreneurial ecosystem approach. *Small Business Economics*, 49(1), 1–10. DOI: 10.1007/s11187-017-9864-8

Adachi, Y. (2010). *Building Big Business in Russia: The Impact of Informal Corporate Governance Practices* (1st ed.). Routledge., DOI: 10.4324/9780203856000

Adel, A. (2022). Future of industry 5.0 in society: Human-centric solutions, challenges and prospective research areas. *Journal of Cloud Computing (Heidelberg, Germany)*, 11(1), 40. DOI: 10.1186/s13677-022-00314-5 PMID: 36101900

Adner, R. (2017). Ecosystem as structure: An actionable construct for strategy. *Journal of Management*, 43(1), 39–58. DOI: 10.1177/0149206316678451

Adner, R., & Kapoor, R. (2010). Value creation in innovation ecosystems: How the structure of technological interdependence affects firm performance in new technology generations. *Strategic Management Journal*, 31(3), 306–333. DOI: 10.1002/smj.821

Agarkov, S., Motina, T., & Matviishin, D. (2018, August). The environmental impact caused by developing energy resources in the Arctic region. In *IOP Conference Series: Earth and Environmental Science* (Vol. 180, No. 1, p. 012007). IOP Publishing. **DOI** DOI: 10.1088/1755-1315/180/1/012007

Ahmad, M., & Beddu, S. (2019). State-of-the-art compendium of macro and micro energies. *Advances in Science and Technology Research Journal.*, 13(1), 88–109. DOI: 10.12913/22998624/103425

Ahmad, T., Zhang, D., Huang, C., Zhang, H., Dai, N., Song, Y., & Chen, H. (2021). Artificial intelligence in sustainable energy industry: Status Quo, challenges and opportunities. *Journal of Cleaner Production*, 289, 125834. DOI: 10.1016/j.jclepro.2021.125834

Ahmed, I. E. S., & Sabah, A. (2021). The determinants of capital structure of the GCC oil and gas companies. *International Journal of Energy Economics and Policy*, 11(2), 30–39. DOI: 10.32479/ijeep.10570

Ahsan, M., Cornelis, E. F. I., & Baker, A. (2018). Understanding backers' interactions with crowdfunding campaigns: Co-Innovators or consumers? *Journal of Research in Marketing and Entrepreneurship*, 20(2), 252–272. DOI: 10.1108/JRME-12-2016-0053

Aibar-Guzmán, B., García-Sánchez, I. M., Aibar-Guzmán, C., & Hussain, N. (2022). Sustainable product innovation in agri-food industry: Do ownership structure and capital structure matter? *Journal of Innovation & Knowledge*, 7(1), 1–10. DOI: 10.1016/j.jik.2021.100160

Airapetova, A. G., Korelin, V. V., & Karabekova, A. A. (2017, November). Specifics of human resource policies at various life stages of an enterprise. In 2017 IEEE VI Forum Strategic Partnership of Universities and Enterprises of Hi-Tech Branches (Science. Education. Innovations)(SPUE) (pp. 169-171). IEEE. https://doi.org/DOI: 10.1109/IVForum.2017.8246082

Airapetova, A. G., & Korelin, V. V. (2018, November). The Problem with Modern Secondary and Higher Education Systems in Russia–Free to Choose or the Problem with the Public Education System. In *2018 XVII Russian Scientific and Practical Conference on Planning and Teaching Engineering Staff for the Industrial and Economic Complex of the Region (PTES)* (pp. 162-164). IEEE. https://doi.org/DOI: 10.1109/PTES.2018.8604266

Akhmetshin, E., Ilyina, I., Kulibanova, V., & Teor, T. (2020). The Methods of Identification and Analysis of Key Indicators Affecting the Place Reputation in the Modern Information and Communication Space. *IOP Conference Series. Materials Science and Engineering*, 940(October), 012100. DOI: 10.1088/1757-899X/940/1/012100

Akhtar, S., Tian, H., Alsedrah, I. T., Anwar, A., & Bashir, S. (2024). Green mining in China: Fintech's contribution to enhancing innovation performance aimed at sustainable and digital transformation in the mining sector. *Resources Policy*, 92, 104968. DOI: 10.1016/j.resourpol.2024.104968

Akimov, A., & Tikhonov, A. (2023). Implementation of Digital Technologies in Personnel Management System of Enterprises of Rocket and Space Industry. *Journal of Theoretical and Applied Information Technology*, 101(5), 1761–1770.

Alghamdi, B., Potter, L. E., & Drew, S. (2021). Validation of architectural requirements for tackling cloud computing barriers: Cloud provider perspective. *Procedia Computer Science*, 181, 477–486. DOI: 10.1016/j.procs.2021.01.193

Alicke, K., Barriball, E., & Trautwein, V. (2020). Supply-chain recovery in coronavirus times—plan for now and the future. McKinsey & Company. Retrieved from https://www.mckinsey.com/business-functions/operations/our-insights/supply-chain-recovery-in-coronavirus-times-plan-for-now-and-the-future

Alieksieienko, T. F., Kryshtanovych, S., Noskova, M., Burdun, V., & Semenenko, A. (2022). The use of modern digital technologies for the development of the educational environment in the system for ensuring the sustainable development of the region. *International Journal of Sustainable Development and Planning*, 8(17), 2427–2434. DOI: 10.18280/ijsdp.170810

Alimoradi Jaghdari, S., Mehrabanpour, M. R., & Najafi Moghadam, A. (2020). Determining the effective factors on financing the optimal capital structure in oil and gas companies. *Petroleum Business Review*, 4(3), 1–20. DOI: 10.22050/pbr.2020.255708.1133

Ali, T., Marc, B., Omar, B., Soulaimane, K., & Larbi, S. (2021). Exploring Destination's Negative e-Reputation Using Aspect Based Sentiment Analysis Approach: Case of Marrakech Destination on TripAdvisor. *Tourism Management Perspectives*, 40(October), 100892. DOI: 10.1016/j.tmp.2021.100892

Aliyev, A. G., & Shahverdiyeva, R. O. (2023). Development of a conceptual model of effective management of innovative enterprises based on digital twin technologies. [IJIEEB]. *Int. J. Inf. Eng. Electron. Bus.*, 15(4), 34–47. DOI: 10.5815/ijieeb.2023.04.04

Allison, T. H., Davis, B. C., Webb, J. W., & Short, J. C. (2017). Persuasion in crowdfunding: An elaboration likelihood model of crowdfunding performance. *Journal of Business Venturing*, 32(6), 707–725. DOI: 10.1016/j.jbusvent.2017.09.002

Al-Momani, H., Al Meanazel, O. T., Kwaldeh, E., Alaween, A., Khasaleh, A., & Qamar, A. (2020). The efficiency of using a tailored inventory management system in the military aviation industry. *Heliyon*, 6(7), e04424. DOI: 10.1016/j.heliyon.2020.e04424 PMID: 32695911

Al-Omoush, K. S., Palacios-Marqués, D., & Ulrich, K. (2022). The impact of intellectual capital on supply chain agility and collaborative knowledge creation in responding to unprecedented pandemic crises. *Technological Forecasting and Social Change*, 178, 121603.

Al-Shaiba, A. S., Al-Ghamdi, S. G., & Koc, M. (2019). Comparative review and analysis of organizational (in) efficiency indicators in Qatar. *Sustainability (Basel)*, 11(23), 6566. DOI: 10.3390/su11236566

Alvarez-Herault, M. C., Gouin, V., Chardin-Segui, T., Malot, A., Coignard, J., Raison, B., & Coulet, J. (2023). *Distribution system planning: Evolution of methodologies and digital tools for energy transition* (1st ed.). Wiley-ISTE., DOI: 10.1002/9781394209477

Amrani, A., & Ducq, Y. (2020). Lean practices implementation in aerospace based on sector characteristics: Methodology and case study. *Production Planning and Control*, 31(16), 1313–1335. DOI: 10.1080/09537287.2019.1706197

Anaba, D. C., Kess-Momoh, A. J., & Ayodeji, S. A. (2024). Digital transformation in oil and gas production: Enhancing efficiency and reducing costs. *International Journal of Management & Entrepreneurship Research*, 6(7), 2153–2161. DOI: 10.51594/ijmer.v6i7.1263

Andreas, W., Holmesland, M., & Efrat, K. (2019). It is not all about money: Obtaining additional benefits through equity crowdfunding. *The Journal of Entrepreneurship*, 28(2), 270–294. DOI: 10.1177/0971355719851899

Aneslagon, D. M. C., & Lim, B., L. B., Tomongha, M. A. S., R. Legaspi, J. E., P. Limbaga, A. J., Casayas, J. S., & Talaboc, D. Q. (. (2024). Assessing the Nexus between Social Responsibility, Environmental Initiatives, and Profitability: A Sustainable Finance Perspective of the Universal Banks in the Philippines. *International Journal of Management Thinking*, 2(1), 38–51. DOI: 10.56868/ijmt.v2i1.40

Anouar, A., Touhafi, A., & Tahiri, A. (2017). Towards an SOA architectural model for AAL-PaaS design and implimentation challenges. *International Journal of Advanced Computer Science and Applications*, 8(7). Advance online publication. DOI: 10.14569/IJACSA.2017.080708

Ardebili, A. A., Longo, A., & Ficarella, A. (2021, October). Digital twin (DT) in smart energy systems-systematic literature review of dt as a growing solution for energy internet of the things (EIoT). In *E3S Web of Conferences, 76th Italian National Congress ATI (ATI 2021),* Vol. 312(2), (pp. 1-18). https://doi.org/DOI: 10.1051/e3sconf/202131209002

Ardolino, M., Rapaccini, M., Saccani, N., Gaiardelli, P., Crespi, G., & Ruggeri, C. (2018). The role of digital technologies for the service transformation of industrial companies. *International Journal of Production Research*, 56(6), 2116–2132. DOI: 10.1080/00207543.2017.1324224

Arenkov, I., Tsenzharik, M., & Vetrova, M. (2019, September). Digital technologies in supply chain management. In *International Conference on Digital Technologies in Logistics and Infrastructure (ICDTLI 2019)* (pp. 448-453). Atlantis Press. https://doi.org/DOI: 10.2991/icdtli-19.2019.78

Arinze, C. A., Ajala, O. A., Okoye, C. C., Ofodile, O. C., & Daraojimba, A. I. (2024). Evaluating the integration of advanced IT solutions for emission reduction in the oil and gas sector. *Engineering Science & Technology Journal*, 5(3), 639–652. DOI: 10.51594/estj.v5i3.862

Ashrafi, M., Magnan, G. M., Adams, M., & Walker, T. R. (2020). Understanding the conceptual evolutionary path and theoretical underpinnings of corporate social responsibility and corporate sustainability. *Sustainability (Basel)*, 12(3), 760. DOI: 10.3390/su12030760

Autor, D. H. (2015). Why are there still so many jobs? The history and future of workplace automation. *The Journal of Economic Perspectives*, 29(3), 3–30. DOI: 10.1257/jep.29.3.3

Azarenko, N., Kazakov, O., Kulagina, N., & Rodionov, D. (2020, September). The model of human capital development with innovative characteristics in digital economy. [). IOP Publishing.]. *IOP Conference Series. Materials Science and Engineering*, 940(1), 012032. DOI: 10.1088/1757-899X/940/1/012032

Babkin, A. V., Kuzmina, S. N., Oplesnina, A. V., & Kozlov, A. V. (2019). Selection of Tools of Automation of Business Processes of a Manufacturing Enterprise. 2019 International Conference "Quality Management, Transport and Information Security, Information Technologies" (IT&QM&IS), 226–229. https://doi.org/DOI: 10.1109/ITQMIS.2019.8928302

Babkin, A., Glukhov, V., Shkarupeta, E., Kharitonova, N., & Barabaner, H. (2021). Methodology for assessing industrial ecosystem maturity in the framework of digital technology implementation. *International Journal of Technology*, 12(7), 1397–1406. DOI: 10.14716/ijtech.v12i7.5390

Badiru, A. B., Ibidapo-Obe, O., & Ayeni, B. J. (2018). *Manufacturing and enterprise: An integrated systems approach*. CRC Press. DOI: 10.1201/9780429055928

Bakhru, K. M. (2019). Importance of intellectual capital in ranking of business school of India. *International Journal of Business and Globalisation*, 22(2), 258–278.

Balog-Way, D., McComas, K., & Besley, J. (2020). The evolving field of risk communication. *Risk Analysis*, 40(S1), 2240–2262. DOI: 10.1111/risa.13615 PMID: 33084114

Bambulyak, A., & Frantzen, B. (2009). Oil transport from the Russian part of the Barents Region. *Status; Riksorgan för Sveriges Lungsjuka*, (January), 107.

Bañales, S. (2020). The enabling impact of digital technologies on distributed energy resources integration. *Journal of Renewable and Sustainable Energy*, 12(4), 045301. DOI: 10.1063/5.0009282

Barnewold, L., & Lottermoser, B. G. (2020). Identification of digital technologies and digitalisation trends in the mining industry. *International Journal of Mining Science and Technology*, 30(6), 747–757. DOI: 10.1016/j.ijmst.2020.07.003

Bartnitzki, T. (2017). Mining 4.0: Importance of Industry 4.0 for the raw materials sector. *Artificial Intelligence*, 2(1), 25–31.

Battistella, C., Colucci, K., De Toni, A. F., & Nonino, F. (2013). Methodology of business ecosystems network analysis: A case study in Telecom Italia Future Centre. *Technological Forecasting and Social Change*, 80(6), 1194–1210. DOI: 10.1016/j.techfore.2012.11.002

Bauer, J. M. (2010). Regulation, public policy, and investment in communications infrastructure. *Telecommunications Policy*, 34(1-2), 65–79. DOI: 10.1016/j.telpol.2009.11.011

Belaïd, F., Al-Sarihi, A., & Al-Mestneer, R. (2023). Balancing climate mitigation and energy security goals amid converging global energy crises: The role of green investments. *Renewable Energy*, 205, 534–542. DOI: 10.1016/j.renene.2023.01.083

Bell, F. (2016). Looking beyond Place Branding: The Emergence of Place Reputation. *Journal of Place Management and Development*, 9(3), 247–254. DOI: 10.1108/JPMD-08-2016-0055

Belmonte-Ureña, L. J., Plaza-Úbeda, J. A., Vazquez-Brust, D., & Yakovleva, N. (2021). Circular economy, degrowth and green growth as pathways for research on sustainable development goals: A global analysis and future agenda. *Ecological Economics*, 185, 107050. DOI: 10.1016/j.ecolecon.2021.107050

Bennatan, A., Choukroun, Y., & Kisilev, P. (2018, June). Deep learning for decoding of linear codes-a syndrome-based approach. In *IEEE International Symposium on Information Theory (ISIT)* (pp. 1595-1599). IEEE. https://doi.org/DOI: 10.1109/ISIT.2018.8437530

Berkvens, L., Roets, A., Haesevoets, T., & Verschuere, B. (2024). Local civil society organizations' appreciation of different local policy decision-making instruments. *Local Government Studies*, 50(3), 545–572. DOI: 10.1080/03003930.2023.2227592

Bernardo, M., Casadesús, M., Karapetrovic, S., & Heras, I. (2012). Do integration difficulties influence management system integration levels? *Journal of Cleaner Production*, 21(1), 23–33. DOI: 10.1016/j.jclepro.2011.09.008

Bernardo, M., Simon, A., Tarí, J. J., & Molina-Azorín, J. F. (2009). Integrated management systems: Development and testing of a theoretical model. *Journal of Cleaner Production*, 17(5), 742–750. DOI: 10.1016/j.jclepro.2008.11.003

Bernardo, M., Simon, A., Tarí, J. J., & Molina-Azorín, J. F. (2012). Benefits of integrated management systems: The views of managers. *The TQM Journal*, 24(5), 386–402. DOI: 10.1108/17542731211261550

Bertayeva, K., Panaedova, G., Natocheeva, N., & Belyanchikova, T. (2019). Industry 4.0 in the mining industry: Global trends and innovative development. In *E3S Web of conferences* (Vol. 135, p. 04026). EDP Sciences.

Beyers, F., & Heinrichs, H. (2020). Global partnerships for a textile transformation? A systematic literature review on inter- and transnational collaborative governance of the textile and clothing industry. *Journal of Cleaner Production*, 261, 121131. DOI: 10.1016/j.jclepro.2020.121131

Bianco, I., Ilin, I., & Iliinsky, A. (2021). Digital technology risk reduction mechanisms to enhance ecological and human safety in the northern sea route for oil and gas companies. In *E3S Web of Conferences* (Vol. 258, p. 06047). EDP Sciences. DOI: 10.1051/e3sconf/202125806047

Bithas, G., Kutsikos, K., Warr, A., & Sakas, D. (2018). Managing transformation within service systems networks: A system viability approach. *Systems Research and Behavioral Science*, 35(4), 469–484. DOI: 10.1002/sres.2543

Bi, Z., Da Xu, L., & Wang, C. (2014). Internet of things for enterprise systems of modern manufacturing. *IEEE Transactions on Industrial Informatics*, 10(2), 1537–1546. DOI: 10.1109/TII.2014.2300338

Blanco, J., García, A., & Morenas, J. D. L. (2018). Design and implementation of a wireless sensor and actuator network to support the intelligent control of efficient energy usage. *Sensors (Basel)*, 18(6), 1892. DOI: 10.3390/s18061892 PMID: 29890737

Bocken, N. M., De Pauw, I., Bakker, C., & Van Der Grinten, B. (2016). Product design and business model strategies for a circular economy. *Journal of Industrial and Production Engineering*, 33(5), 308–320. DOI: 10.1080/21681015.2016.1172124

Bogan, V. L. (2012). Capital structure and sustainability: An empirical study of microfinance institutions. *The Review of Economics and Statistics*, 94(4), 1045–1058. DOI: 10.1162/REST_a_00223

Boiko, J. M., & Eromenko, A. I. (2014). Improvements encoding energy benefit in protected telecommunication data transmission channels. *Communications*, 2(1), 7–14. DOI: 10.11648/j.com.20140201.12

Bondarev, A., & Krysiak, F. C. (2021). Economic development and the structure of cross-technology interactions. *European Economic Review*, 132, 103628. DOI: 10.1016/j.euroecorev.2020.103628

Borgia, E. (2014). The Internet of Things vision: Key features, applications and open issues. *Computer Communications*, 54(1), 1–31. DOI: 10.1016/j.comcom.2014.09.008

Borodina, M., Idrisov, H., Kapustina, D., Zhildikbayeva, A., Fedorov, A., Denisova, D., Gerasimova, E., & Solovyanenko, N. (2023). State Regulation of Digital Technologies for Sustainable Development and Territorial Planning. *International Journal of Sustainable Development and Planning*, 18(5), 1615–1624. DOI: 10.18280/ijsdp.180533

Boyd, B. K., Bergh, D. D., & Ketchen, D. J.Jr. (2010). Reconsidering the Reputation—Performance Relationship: A Resource-Based View. *Journal of Management*, 36(3), 588–609. DOI: 10.1177/0149206308328507

Boyer, M. M., & Filion, D. (2007). Common and fundamental factors in stock returns of Canadian oil and gas companies. *Energy Economics*, 29(3), 428–453. DOI: 10.1016/j.eneco.2005.12.003

Boy, G. A. (2017). A human-centered design approach. In *The handbook of human-machine interaction* (pp. 1–20). CRC Press. DOI: 10.1201/9781315557380-1

Bozhuk, S. G., Maslova, T. D., Pletneva, N. A., & Evdokimov, K. V. (2019). Improvement of the Consumers' Satisfaction Research Technology in the Digital Environment. *IOP Conference Series. Materials Science and Engineering*, 666(1), 012055. DOI: 10.1088/1757-899X/666/1/012055

Braun, E., Eshuis, J., Klijn, E. H., & Zenker, S. (2018). Improving Place Reputation: Do an Open Place Brand Process and an Identity-Image Match Pay Off? *Cities (London, England)*, 80, 22–28. DOI: 10.1016/j.cities.2017.06.010

Brezet, H., & van Hemel, C. (Eds.). (1997). *Ecodesign: A promising approach to sustainable production and consumption* (1st ed.). U.N.E.P.

Brogi, A., Forti, S., Ibrahim, A., Kim, S. S., Gupta, H., Vahid Dastjerdi, A., Ghosh, S. K., Buyya, R., D'Angelo, G., Ferretti, S., & Ghini, V., AbdElhalim, E., Obayya, M., Kishk, S., Chiariotti, F., Condoluci, M., Mahmoodi, T., Zanella, A., Anglano, C., ... Guazzone, M. (2018). Survey High-Performance Modelling and Simulation for Selected Results of the COST Action IC1406 cHiPSet. *Future Generation Computer Systems*, 29(1).

Brynjolfsson, E. (2014). The second machine age: Work, progress, and prosperity in a time of brilliant technologies.

Buck, T. (2003). Modern Russian corporate governance: Convergent forces or product of Russia's history? *Journal of World Business*, 38(4), 299–313. DOI: 10.1016/j.jwb.2003.08.017

Bughin, J., Hazan, E., Lund, S., Dahlström, P., Wiesinger, A., & Subramaniam, A. (2018). Skill shift: Automation and the future of the workforce. *McKinsey Global Institute*, 1, 3–84.

Bukhonova, S., & Yablonskaya, A. (2022). Study of the Level of Digitalization of the Banking Sector of Russia in the Context of the Pandemic and the Development of Transport Technologies. In *International School on Neural Networks, Initiated by IIASS and EMFCSC* (pp. 703–710). Springer International Publishing.

Butticè, V., Colombo, M. G., & Wright, M. (2017). Serial crowdfunding, social capital, and project success. *Entrepreneurship Theory and Practice*, 41(2), 183–207. DOI: 10.1111/etap.12271

Butticè, V., & Noonan, D. (2020). Active backers, product commercialisation and product quality after a crowdfunding campaign: A comparison between first-time and repeated entrepreneurs. *International Small Business Journal*, 38(2), 111–134. DOI: 10.1177/0266242619883984

Butt, J. (2020). A conceptual framework to support digital transformation in manufacturing using an integrated business process management approach. *Designs*, 4(3), 17. DOI: 10.3390/designs4030017

Cai, D., Hu, J., Jiang, H., Ai, F., & Bai, T. (2023). Research on the relationship between defense technology innovation and high-quality economic development: Gray correlation analysis based on panel data. *MDE. Managerial and Decision Economics*, 44(7), 3867–3877. DOI: 10.1002/mde.3925

Cai, W., Polzin, F., & Stam, E. (2021). Crowdfunding and social capital: A systematic review using a dynamic perspective. *Technological Forecasting and Social Change*, 162(1), 120412. DOI: 10.1016/j.techfore.2020.120412

Cao, L., Gu, M., Jin, D., & Wang, C. (2023). Geopolitical risk and economic security: Exploring natural resources extraction from BRICS region. *Resources Policy*, 85, 103800. DOI: 10.1016/j.resourpol.2023.103800

Carayannis, E. G., Ilinova, A., & Cherepovitsyn, A. (2021). The future of energy and the case of the arctic offshore: The role of strategic management. *Journal of Marine Science and Engineering*, 9(2), 134. DOI: 10.3390/jmse9020134

Cavallo, A., Ghezzi, A., & Sanasi, S. (2021). Assessing entrepreneurial ecosystems through a strategic value network approach: Evidence from the San Francisco area. *Journal of Small Business and Enterprise Development*, 28(2), 261–276. DOI: 10.1108/JSBED-05-2019-0148

Chanysheva, A., & Ilinova, A. (2021). The future of Russian arctic oil and gas projects: Problems of assessing the prospects. *Journal of Marine Science and Engineering*, 9(5), 528. DOI: 10.3390/jmse9050528

Chavlis, S., & Poirazi, P. (2021). Drawing inspiration from biological dendrites to empower artificial neural networks. *Current Opinion in Neurobiology*, 70, 1–10. DOI: 10.1016/j.conb.2021.04.007 PMID: 34087540

Che, H., Erdem, T., & Öncü, T. S. (2015). Consumer learning and evolution of consumer brand preferences. *Quantitative Marketing and Economics*, 13(3), 173–202. DOI: 10.1007/s11129-015-9158-x

Chen, W. M., Wang, S. Y., & Wu, X. L. (2022). Concept refinement, factor symbiosis, and innovation activity efficiency analysis of innovation ecosystem. In E. G. Nepomuceno (Ed.), Mathematical Problems in Engineering, Vol. 2022(4), 1-5. DOI: 10.1155/2022/1942026

Chen, D., Heyer, S., Ibbotson, S., Salonitis, K., Steingrímsson, J. G., & Thiede, S. (2015). Direct digital manufacturing: Definition, evolution, and sustainability implications. *Journal of Cleaner Production*, 107, 615–625. DOI: 10.1016/j.jclepro.2015.05.009

Chen, I. S., & Chen, J. K. (2013). Present and future: A trend forecasting and ranking of university types for innovative development from an intellectual capital perspective. *Quality & Quantity*, 47, 335–352.

Cherepovitsyn, A., Rutenko, E., & Solovyova, V. (2021). Sustainable development of oil and gas resources: A system of environmental, socio-economic, and innovation indicators. *Journal of Marine Science and Engineering*, 9(11), 1307. DOI: 10.3390/jmse9111307

Chernikov, A. D., Eremin, N. A., Stolyarov, V. E., Sboev, A. G., Semenova-Chashchina, O. K., & Fitsner, L. K. (2020). Application of Artificial Intelligence Methods for Identifying and Predicting Complications in the Construction of Oil and Gas Wells: Problems and Solutions. *Georesursy*, 22(3), 87–96. DOI: 10.18599/grs.2020.3.87-96

Chistobaev, A., & Kulakovskiy, E. (2020, September). Digital technologies in ensuring local government in the Russian Federation. [). IOP Publishing.]. *IOP Conference Series. Materials Science and Engineering*, 940(1), 012039. DOI: 10.1088/1757-899X/940/1/012039

Choros, K. A., Job, A. T., Edgar, M. L., Austin, K. J., & McAree, P. R. (2022). Can Hyperspectral Imaging and Neural Network Classification Be Used for Ore Grade Discrimination at the Point of Excavation? *Sensors (Basel)*, 22(7), 2687. DOI: 10.3390/s22072687 PMID: 35408301

Chu, L. K., Ghosh, S., Doğan, B., Nguyen, N. H., & Shahbaz, M. (2023). Energy security as new determinant of renewable energy: The role of economic complexity in top energy users. *Energy*, 263, 125799. DOI: 10.1016/j.energy.2022.125799

Clarysse, B., Wright, M., Bruneel, J., & Mahajan, A. (2014). Creating value in ecosystems: Crossing the chasm between knowledge and business ecosystems. *Research Policy*, 43(7), 1164–1176. DOI: 10.1016/j.respol.2014.04.014

Classen, M., Blum, C., Osterrieder, P., & Friedli, T. (2019, September). Everything as a service? Introducing the St. Gallen IGaaS management model. In J. Meierhofer, S.S. West (Ed.), *Proceedings of the 2nd Smart Services Summit* (pp. 61-65). Data Innovation Alliance.

Colombo, M. G., Franzoni, C., & Rossi-Lamastra, C. (2015). Internal social capital and the attraction of early contributions in crowdfunding. *Entrepreneurship Theory and Practice*, 39(1), 75–100. DOI: 10.1111/etap.12118

Compagnoni, M., & Stadler, M. (2021). *Growth in a circular economy* (No. 145). [Working Papers, University of Tübingen]. UT Campus Repository. DOI: 10.15496/publikation-56495

Container Intermodal Technology. (2021, January 25). Transport and logistics news review [In Russian]. Retrieved September 29, 2021, from https://citekb.ru/feed/cit/2021/01/25/obzor-smi-novosti-transporta-i-logistiki-vypusk-02-65-11012021-25012021

Cordes, E. E., Jones, D. O., Schlacher, T. A., Amon, D. J., Bernardino, A. F., Brooke, S., Carney, R., DeLeo, D. M., Dunlop, K. M., Escobar-Briones, E. G., Gates, A. R., Génio, L., Gobin, J., Henry, L.-A., Herrera, S., Hoyt, S., Joye, M., Kark, S., Mestre, N. C., & Witte, U. (2016). Environmental impacts of the deep-water oil and gas industry: A review to guide management strategies. *Frontiers in Environmental Science*, 4, 58. DOI: 10.3389/fenvs.2016.00058

Crawford, K., & Calo, R. (2016). There is a blind spot in AI research. *Nature*, 538(7625), 311–313. DOI: 10.1038/538311a PMID: 27762391

Crosetto, P., & Regner, T. (2018). It's never too late: Funding dynamics and self pledges in reward-based crowdfunding. *Research Policy*, 47(8), 1463–1477. DOI: 10.1016/j.respol.2018.04.020

Crozier, M. (2007). Recursive governance: Contemporary political communication and public policy. *Political Communication*, 24(1), 1–18. DOI: 10.1080/10584600601128382

Curtis, J., & Kaufman, S. (2020). "It's not what you see but what you hear…": Understanding environment protection officers' responsive decision making. *Journal of Environmental Management*, 262, 110336. DOI: 10.1016/j.jenvman.2020.110336 PMID: 32250813

Cusumano, M. A., Gawer, A., & Yoffie, D. B. (2019). *The business of platforms: Strategy in the age of digital competition, innovation, and power* (Vol. 320). Harper Business.

da Silva, M. M. (2020). *Power and gas asset management regulation, planning and operation of digital energy systems* [LNEN Series]. Springer., DOI: 10.1007/978-3-030-36200-3

Dalwai, T., & Salehi, M. (2021). Business strategy, intellectual capital, firm performance, and bankruptcy risk: Evidence from Oman's non-financial sector companies. *Asian Review of Accounting*, 29(3), 474–504.

Daneeva, Y., Glebova, A., Daneev, O., & Zvonova, E. (2020, August). Digital transformation of oil and gas companies: energy transition. In *Russian Conference on Digital Economy and Knowledge Management (RuDEcK 2020)* (pp. 199-205). Atlantis Press. DOI: 10.2991/aebmr.k.200730.037

Danylyshyn, B., Onyshchenko, S., & Maslii, O. (2019). Socio-economic security: modern approach to ensuring the socio-economic development of the region. *Науковий журнал «Економіка і регіон»*, (4 (75)), 6-13.

Darazi, M. A., Khoso, A. K., & Mahesar, K. A. (2023). INVESTIGATING THE EFFECTS OF ESL TEACHERS' FEEDBACK ON ESL UNDERGRADUATE STUDENTS' LEVEL OF MOTIVATION, ACADEMIC PERFORMANCE, AND SATISFACTION: MEDIATING ROLE OF STUDENTS' MOTIVATION. *Pakistan Journal of Educational Research*, 6(2).

de la Escosura, L. P., & Rodríguez-Caballero, C. V. (2022). War, pandemics, and modern economic growth in Europe. *Explorations in Economic History*, 86, 101467. DOI: 10.1016/j.eeh.2022.101467

De Luca, F., Iaia, L., Mehmood, A., & Vrontis, D. (2022). Can social media improve stakeholder engagement and communication of Sustainable Development Goals? A cross-country analysis. *Technological Forecasting and Social Change*, 177(1), 121525. DOI: 10.1016/j.techfore.2022.121525

De Santis, F., & Presti, C. (2018). The relationship between intellectual capital and big data: A review. *Meditari Accountancy Research*, 26(3), 361–380.

de Soyres, F., & Gaillard, A. (2022). Global trade and GDP comovement. *Journal of Economic Dynamics & Control*, 138, 104353. DOI: 10.1016/j.jedc.2022.104353

Defra. (2013). *Waste Prevention Programme for England* [Policy Paper, Department for Environment, Food & Rural Affairs]. GOV.UK.

Deloitte. (2020). COVID-19: Managing supply chain risk and disruption. Deloitte Insights. Retrieved from https://www2.deloitte.com/global/en/pages/risk/articles/covid-19-managing-supply-chain-risk-and-disruption.html

Dikaputra, R., Sulung, L. A. K., & Kot, S. (2019). Analysis of success factors of reward-based crowdfunding campaigns using multi-theory approach in ASEAN-5 countries. *Social Sciences (Basel, Switzerland)*, 8(10), 293. DOI: 10.3390/socsci8100293

Dlouhá, J., Barton, A., Janoušková, S., & Dlouhý, J. (2013). Social learning indicators in sustainability-oriented regional learning networks. *Journal of Cleaner Production*, 49, 64–73. DOI: 10.1016/j.jclepro.2012.07.023

Dlouhá, J., Henderson, L., Kapitulčinová, D., & Mader, C. (2018). Sustainability-oriented higher education networks: Characteristics and achievements in the context of the UN DESD. *Journal of Cleaner Production*, 172, 4263–4276. DOI: 10.1016/j.jclepro.2017.07.239

Dmitrieva, D., Cherepovitsyna, A., Stroykov, G., & Solovyova, V. (2021). Strategic sustainability of offshore arctic oil and gas projects: Definition, principles, and conceptual framework. *Journal of Marine Science and Engineering*, 10(1), 23. DOI: 10.3390/jmse10010023

Dmitrievsky, A. N., Eremin, N. A., Filippova, D. S., & Safarova, E. A. (2020). Digital oil and gas complex of Russia. Georesursy= Georesources, Special issue, 32-35.

Doan, N., Hashemi, S. A., Ercan, F., Tonnellier, T., & Gross, W. J. (2019, October). Neural dynamic successive cancellation flip decoding of polar codes. In *2019 IEEE International Workshop on Signal Processing Systems (SiPS)* (pp. 272-277). IEEE. https://doi.org/DOI: 10.1109/SiPS47522.2019.9020513

Doğan, B., Shahbaz, M., Bashir, M. F., Abbas, S., & Ghosh, S. (2023). Formulating energy security strategies for a sustainable environment: Evidence from the newly industrialized economies. *Renewable & Sustainable Energy Reviews*, 184, 113551. DOI: 10.1016/j.rser.2023.113551

Dolnicar, S., & Grün, B. (2013). Validly Measuring Destination Image in Survey Studies. *Journal of Travel Research*, 52(1), 3–14. DOI: 10.1177/0047287512457267

Domingues, J. P., Sampaio, P., & Arezes, P. M. (2017). Integrated management systems assessment: A maturity model proposal. *Journal of Cleaner Production*, 142(1), 145–158. DOI: 10.1016/j.jclepro.2016.07.100

Domingues, M. S., Baptista, A. L., & Diogo, M. T. (2017). Engineering complex systems applied to risk management in the mining industry. *International Journal of Mining Science and Technology*, 27(4), 611–616. DOI: 10.1016/j.ijmst.2017.05.007

Dondossola, G., Garrone, F., & Szanto, J. (2011, July). *Cyber risk assessment of power control systems—A metrics weighed by attack experiments. In 2011 IEEE Power and Energy Society General Meeting*. IEEE., DOI: 10.1109/PES.2011.6039589

Dong, P., Zhang, H., Li, G. Y., Gaspar, I. S., & NaderiAlizadeh, N. (2019). Deep CNN-based channel estimation for mmwave massive MIMO systems. *IEEE Journal of Selected Topics in Signal Processing*, 13(5), 989–1000. DOI: 10.1109/JSTSP.2019.2925975

Dragičević, Z., & Bošnjak, S. (2019). Digital transformation in the mining enterprise: The empirical study. *Mining and Metallurgy Engineering Bor*, (1-2), 73–90. DOI: 10.5937/mmeb1902073D

Druzhinin, A., & Alekseeva, N. (2020). Efficiency assessment and research of the internal risks in the innovation-active industrial cluster. In *SPBPU IDE '20: Proceedings of the 2nd International Scientific Conference on Innovations in Digital Economy* (pp. 1-7). ACM Digital Library. DOI: DOI: 10.1145/3444465.3444501

Dubolazov, V., Simakova, Z., Leicht, O., & Shchelkonogov, A. (2021). The impact of digitalization on production structures and management in industrial enterprises and complexes. In Schaumburg, H., Korablev, V., & Ungvari, L. (Eds.), *Technological transformation: A new role for humans, machines, and management: TT-2020* (pp. 39–47). Springer., DOI: 10.1007/978-3-030-64430-7_4

Dunlop, C. A., Ongaro, E., & Baker, K. (2020). Researching COVID-19: A research agenda for public policy and administration scholars. *Public Policy and Administration*, 35(4), 365–383. DOI: 10.1177/0952076720939631

Eder, L. V., Filimonova, I. V., Provornaya, I. V., Komarova, A. V., & Nikitenko, S. M. (2017). New directions for sustainable development of oil and gas industry of Russia: Innovative strategies, regional smart specializations, public-private partnership. *International Multidisciplinary Scientific GeoConference: SGEM, 17*(1.5), 365-372.

Efrat, K., & Gilboa, S. (2020). Relationship approach to crowdfunding: How creators and supporters interaction enhances projects' success. *Electronic Markets*, 30(4), 899–911. DOI: 10.1007/s12525-019-00391-6

Efrat, K., Gilboa, S., & Sherman, A. (2020). The role of supporter engagement in enhancing crowdfunding success. *Baltic Journal of Management*, 15(2), 199–213. DOI: 10.1108/BJM-09-2018-0337

Egorov, D. P. (2021). A CLASSIFICATION OF OIL AND GAS-BEARING TERRITORIES AND A QUALITATIVE ASSESSMENT OF THE RUSSIAN OIL AND GAS INDUSTRY. International Research Journal, 2021(5107).

Eldor, N., & Mamlakat, K. (2024). ISSUES OF INCREASING INVESTMENT ATTRACTIVENESS IN THE DEVELOPMENT OF THE COUNTRY'S ECONOMY. *EUROPEAN JOURNAL OF INNOVATION IN NONFORMAL EDUCATION*, 4(3), 474–478.

Elnaiem, A., Mohamed-Ahmed, O., Zumla, A., Mecaskey, J., Charron, N., Abakar, M. F., Raji, T., Bahalim, A., Manikam, L., Risk, O., Okereke, E., Squires, N., Nkengasong, J., Rüegg, S. R., Abdel Hamid, M. M., Osman, A. Y., Kapata, N., Alders, R., Heymann, D. L., & Dar, O. (2023). Global and regional governance of One Health and implications for global health security. *Lancet*, 401(10377), 688–704. DOI: 10.1016/S0140-6736(22)01597-5 PMID: 36682375

Engseth, E. (2018). Cultural competency: A framework for equity, diversity, and inclusion in the archival profession in the United States. *The American Archivist*, 81(2), 460–482. DOI: 10.17723/0360-9081-81.2.460

Ershova, I., Obukhova, A., & Belyaeva, O. (2020). Implementation of innovative digital technologies in the world. *Economic Annals-XXI/Ekonomičnij Časopis-XXI*, 186.

Esiri, A. E., Babayeju, O. A., & Ekemezie, I. O. (2024). Implementing sustainable practices in oil and gas operations to minimize environmental footprint.

Eubanks, V. (2018). *Automating Inequality: How High-Tech Tools Profile, Police, and Punish the Poor*. St. Martin's Press.

Eum, W., & Lee, J. D. (2022). The co-evolution of production and technological capabilities during industrial development. *Structural Change and Economic Dynamics*, 63, 454–469. DOI: 10.1016/j.strueco.2022.07.001

European Union Agency for Cybersecurity (ENISA). (2020). Threat landscape for supply chain attacks. ENISA Report. Retrieved from https://www.enisa.europa.eu/publications/threat-landscape-for-supply-chain-attacks

Evseeva, S., Kalchenko, O., Plis, K., & Evseeva, O. (2019). The role of information and communication technologies as a part of business intelligence in improving the wealth of nations. *IOP Conference Series. Materials Science and Engineering*, 618(1), 012080. DOI: 10.1088/1757-899X/618/1/012080

Ezu, G. (2020). Effect of capital structure on financial performance of oil and gas companies quoted on the Nigerian Stock Exchange. *International Journal of Business and Management*, 8(4), 293–305. DOI: 10.24940/theijbm/2020/v8/i4/BM2004-046

Fadeev, A. M., & Tsukerman, V. A. (2020, April). Assessment of the prospects of developing the western arctic shelf fields on the basis of evaluation of the technical-economic potential of fields. In *IOP Conference Series: Earth and Environmental Science* (Vol. 459, No. 6, p. 062024). IOP Publishing. **DOI** DOI: 10.1088/1755-1315/459/6/062024

Fadeev, A., Kalyazina, S., Levina, A., & Dubgorn, A. (2021). Requirements for transport support of offshore production in the Arctic zone. *Transportation Research Procedia*, 54, 883–889. DOI: 10.1016/j.trpro.2021.02.143

Fang, X., Misra, S., Xue, G., & Yang, D. (2011). Smart grid—The new and improved power grid: A survey. *IEEE Communications Surveys and Tutorials*, 14(4), 944–980. DOI: 10.1109/SURV.2011.101911.00087

Farhoud, M., Shah, S., Stenholm, P., Kibler, E., Renko, M., & Terjesen, S. (2021). Social enterprise crowdfunding in an acute crisis. *Journal of Business Venturing Insights*, 15(2), e00211. DOI: 10.1016/j.jbvi.2020.e00211

Farzaneh, H., Malehmirchegini, L., Bejan, A., Afolabi, T., Mulumba, A., & Daka, P. P. (2021). Artificial intelligence evolution in smart buildings for energy efficiency. *Applied Sciences (Basel, Switzerland)*, 11(2), 763. DOI: 10.3390/app11020763

Fayziev, S. (2024). Complaens Risks of Using It Technologies in Banks in the Period of Digitization. Asian Journal of Technology & Management Research (AJTMR) ISSN, 2249(0892).

Fei, L., Shuang, M., & Xiaolin, L. (2023). Changing multi-scale spatiotemporal patterns in food security risk in China. *Journal of Cleaner Production*, 384, 135618. DOI: 10.1016/j.jclepro.2022.135618

Feng, Y., Chen, H. H., & Tang, J. (2018). The impacts of social responsibility and ownership structure on sustainable financial development of China's energy industry. *Sustainability (Basel)*, 10(2), 1–15. DOI: 10.3390/su10020301

Feng, Y., Lee, C. C., & Peng, D. (2023). Does regional integration improve economic resilience? Evidence from urban agglomerations in China. *Sustainable Cities and Society*, 88, 104273. DOI: 10.1016/j.scs.2022.104273

Ferlander, S. (2003). The internet, social capital and local community.

Fernandez-Vidal, F., Gonzalez, R., Gasco, J., & Llopis, J. (2022). Digitalization and corporate transformation: The case of European oil & gas firms. *Technological Forecasting and Social Change*, 174, 121293. DOI: 10.1016/j.techfore.2021.121293

Feroz, A. K., Zo, H., & Chiravuri, A. (2021). Digital transformation and environmental sustainability: A review and research agenda. *Sustainability (Basel)*, 13(3), 1530. DOI: 10.3390/su13031530

Fierro, L. E., Caiani, A., & Russo, A. (2022). Automation, job polarisation, and structural change. *Journal of Economic Behavior & Organization*, 200, 499–535. DOI: 10.1016/j.jebo.2022.05.025

Filimonova, I., Komarova, A., Nemov, V., & Provornaya, I. (2020). Sustainable development of Russian energy sector: Hydrocarbons of Eastern Siberia. *International Multidisciplinary Scientific GeoConference: SGEM*, 20(1.2), 777-783.

Filimonova, I. V., Komarova, A. V., Provornaya, I. V., Dzyuba, Y. A., & Link, A. E. (2020). Efficiency of oil companies in Russia in the context of energy and sustainable development. *Energy Reports*, 6, 498–504. DOI: 10.1016/j.egyr.2020.09.027

Filimonova, I. V., Komarova, A. V., Provornaya, I. V., & Mishenin, M. V. (2020). Changes in the structure of the oil raw material base as a factor determining federal budget revenues. *Mining Journal*, (4), 30–36. DOI: 10.17580/gzh.2020.04.06

Firoiu, D., Ionescu, G. H., Băndoi, A., Florea, N. M., & Jianu, E. (2019). Achieving sustainable development goals (SDG): Implementation of the 2030 agenda in Romania. *Sustainability (Basel)*, 11(7), 2156. DOI: 10.3390/su11072156

Fischer, J. R., González, S. A., Herran, M. A., Judewicz, M. G., & Carrica, D. O. (2013). Calculation-delay tolerant predictive current controller for three-phase inverters. *IEEE Transactions on Industrial Informatics*, 10(1), 233–242. DOI: 10.1109/TII.2013.2276104

Foroudi, P., Gupta, S., Kitchen, P., Foroudi, M. M., & Nguyen, B. (2016). A Framework of Place Branding, Place Image, and Place Reputation. *Qualitative Market Research*, 19(2), 241–264. DOI: 10.1108/QMR-02-2016-0020

Forti, V., Balde, C. P., Kuehr, R., & Bel, G. (2020). The Global E-waste Monitor 2020: Quantities, flows and the circular economy potential.

Francesca, B., Manganiello, M., & Ricci, O. (2021). Is equity crowdfunding the land of promise for female entrepreneurship? *PuntOorg International Journal*, 6(1), 12–36. DOI: 10.19245/25.05.pij.6.1.3

Freet, D., Agrawal, R., John, S., & Walker, J. J. (2015, October). Cloud forensics challenges from a service model standpoint: IaaS, PaaS and SaaS. In *Proceedings of the 7th International Conference on Management of Computational and Collective intElligence in Digital EcoSystems* (pp. 148-155). ACM Digital Library. https://doi.org/DOI: 10.1145/2857218.2857253

Frenken, K., & Schor, J. (2017). Putting the sharing economy into perspective. *Environmental Innovation and Societal Transitions*, 23, 3–10. DOI: 10.1016/j.eist.2017.01.003

Frick, S. A., & Rodríguez-Pose, A. (2023). What draws investment to special economic zones? Lessons from developing countries. *Regional Studies*, 57(11), 2136–2147. DOI: 10.1080/00343404.2023.2185218

Frondizi, R., Fantauzzi, C., Colasanti, N., & Fiorani, G. (2019). The evaluation of universities' third mission and intellectual capital: Theoretical analysis and application to Italy. *Sustainability*, 11(12), 3455.

Frosch, R. A., & Gallopoulos, N. E. (1989). Strategies for manufacturing. *Scientific American*, 261(3), 144–153. DOI: 10.1038/scientificamerican0989-144

Galazova, S. S. (2023). Digital banking ecosystems: Comparative analysis and competition regulation in Russia. [переводная версия]. *Journal of New Economy*, 24(4), 82–106. DOI: 10.29141/2658-5081-2023-24-4-5

Gamidullaeva, L., Tolstykh, T., Bystrov, A., Radaykin, A., & Shmeleva, N. (2021). Cross-sectoral digital platform as a tool for innovation ecosystem development. *Sustainability (Basel)*, 13(21), 11686. DOI: 10.3390/su132111686

Gandini, A., & Gandini, A. (2016). Reputation, the Social Capital of a Digital Society. The Reputation Economy: Understanding Knowledge Work in Digital Society, 27-43. DOI: 10.1057/978-1-137-56107-7_3

Gao, S., Hakanen, E., Töytäri, P., & Rajala, R. (2019). Digital transformation in asset-intensive businesses: Lessons learned from the metals and mining industry.

García, D., Bautista, J., & Esther, D. Q. P. (2016). The Complex Link of City Reputation and City Performance. Results for FsQCA Analysis. *Journal of Business Research*, 69(8), 2830–2839. DOI: 10.1016/j.jbusres.2015.12.052

Garcia-Gomez, S., Jimenez-Ganan, M., Taher, Y., Momm, C., Junker, F., Biro, J., Menychtas, A., Andrikopoulos, V., & Strauch, S. (2012). Challenges for the comprehensive management of cloud services in a PaaS framework. *Scalable Computing: Practice and Experience*, 13(3), 201–214. DOI: 10.1109/ISPA.2012.72

Gautier, D. L., Bird, K. J., Charpentier, R. R., Grantz, A., Houseknecht, D. W., Klett, T. R., Moore, T. E., Pitman, J. K., Schenk, C. J., Schuenemeyer, J. H., Sørensen, K., Tennyson, M. E., Valin, Z. C., & Wandrey, C. J. (2009). Assessment of undiscovered oil and gas in the Arctic. *Science*, 324(5931), 1175–1179. DOI: 10.1126/science.1169467 PMID: 19478178

Gawer, A., & Cusumano, M. A. (2014). Industry platforms and ecosystem innovation. *Journal of Product Innovation Management*, 31(3), 417–433. DOI: 10.1111/jpim.12105

Genga, Y., Oyerinde, O. O., & Versfeld, J. (2021). Iterative soft-input soft-output bit-level Reed-Solomon decoder based on information set decoding. *SAIEE Africa Research Journal*, 112(2), 52–65. DOI: 10.23919/SAIEE.2021.9432893

Gesvindr, D., & Buhnova, B. (2016). Performance challenges, current bad practices, and hints in PaaS cloud application design. *Performance Evaluation Review*, 43(4), 3–12. DOI: 10.1145/2897356.2897358

Ghoreishi, M., & Happonen, A. (2022). The case of fabric and textile industry: The emerging role of digitalization, internet-of-things and industry 4.0 for circularity. In X. S. Yang, S. Sherratt, N. Dey, A. Joshi, (Eds), Lecture Notes in Networks and Systems: Vol. 216. Proceedings of Sixth International Congress on Information and Communication Technology (pp. 189–200). Springer Nature. DOI: 10.1007/978-981-16-1781-2_18

Giessmann, A., & Stanoevska-Slabeva, K. (2012). Business models of platform as a service (PaaS) providers: Current state and future directions. *Journal of Information Technology Theory and Application*, 13(4), 31.

Gladkikh, A. A., Volkov, A. K., & Ulasyuk, T. G. (2021). Development of biometric systems for passenger identification based on noise-resistant coding means. *Civil Aviation High Technologies*, 24(2), 93–104. DOI: 10.26467/2079-0619-2021-24-2-93-104

Gladkikh, A. A., Volkov, A. K., Volkov, A. K., Andriyanov, N. A., & Shakhtanov, S. V. (2019, November). Development of network training complexes using fuzzy models and noise-resistant coding. In *International Conference on Aviamechanical Engineering and Transport (AviaENT 2019)* (pp. 373-379). Atlantis Press. https://doi.org/DOI: 10.2991/aviaent-19.2019.69

Glazner, C. G. (2006). Enterprise integration strategies across virtual extended enterprise networks: a case study of the F-35 Joint Strike Fighter Program enterprise (Doctoral dissertation, Massachusetts Institute of Technology).

Glebova, I., Berman, S., Khafizova, L., Biktimirova, A., & Alhasov, Z. (2023). Digital Divide of Regions: Possible Growth Points for Their Digital Maturity. *International Journal of Sustainable Development and Planning*, 18(5), 1457–1465. DOI: 10.18280/ijsdp.180516

Gong, Y., Mitchell, P., & Krishnamurthy, A. (2021). The impact of digitalization on the logistics industry: Opportunities and challenges. *Journal of Business Logistics*, 42(1), 121–135. DOI: 10.1111/jbl.12264

Goyal, R., Pokhriyal, D. S., Nechully, D. S., & Gupta, D. S. (2020). Management Barriers in Implementation of Integrated Operations Solutions in Indian Upstream Companies. *International Journal of Management*, 11(10).

Grigoriev, E. K., Nenashev, V. A., Sergeev, A. M., & Nenashev, S. A. (2020, September). Research and analysis of methods for generating and processing new code structures for the problems of detection, synchronization and noise-resistant coding. In *Image and Signal Processing for Remote Sensing XXVI* (Vol. 11533, pp. 319–328). SPIE., DOI: 10.1117/12.2574238

Gromova, E. A., & Pupentsova, S. V. (2020). Simulation modelling as a method of risk analysis in real estate valuation. *IOP Conference Series. Materials Science and Engineering*, 898(1), 012048. DOI: 10.1088/1757-899X/898/1/012048

Gungor, V. C., Sahin, D., Kocak, T., Ergut, S., Buccella, C., Cecati, C., & Hancke, G. P. (2011). Smart grid technologies: Communication technologies and standards. *IEEE Transactions on Industrial Informatics*, 7(4), 529–539. DOI: 10.1109/TII.2011.2166794

Gupta, C., Jindal, P., & Malhotra, R. K. (2022, November). A study of increasing adoption trends of digital technologies-An evidence from Indian banking. In *AIP Conference Proceedings* (Vol. 2481, No. 1). AIP Publishing. DOI: 10.1063/5.0104572

Habibi, M. R., Davidson, A., & Laroche, M. (2017). What managers should know about the sharing economy. *Business Horizons*, 60(1), 113–121. DOI: 10.1016/j.bushor.2016.09.007

Haleem, A., Javaid, M., Qadri, M. A., & Suman, R. (2022). Understanding the role of digital technologies in education: A review. *Sustainable operations and computers*, 3(4), 275-285. DOI: DOI: 10.1016/j.susoc.2022.05.004

Hamam, M. D., Layyinaturrobaniyah, L., & Herwany, A. (2020). Capital Structure and Firm's Growth in Relations to Firm Value at Oil and Gas Companies Listed in Indonesia Stock Exchange. *Journal of Accounting Auditing and Business*, 3(1), 14–28. DOI: 10.24198/jaab.v3i1.24760

Hamid, A. R. A., Yusof, S. M., Rahman, S. A., & Idris, M. A. (2019). The implementation of ISO 9001:2015 in the aerospace sector: Issues and challenges. *International Journal of Productivity and Performance Management*, 68(3), 504–523. DOI: 10.1108/IJPPM-09-2017-0218

Hampton, K., & Wellman, B. (2003). Neighboring in Netville: How the Internet supports community and social capital in a wired suburb. *City & Community*, 2(4), 277–311. DOI: 10.1046/j.1535-6841.2003.00057.x

Haouel, C., & Nemeslaki, A. (2024). Digital transformation in oil and gas industry: Opportunities and challenges. *Periodica Polytechnica Social and Management Sciences*, 32(1), 1–16. DOI: 10.3311/PPso.20830

Harahap, M. A. K., Kraugusteeliana, K., Pramono, S. A., Jian, O. Z., & Ausat, A. M. A. (2023). The role of information technology in improving urban governance. *Jurnal Minfo Polgan*, 12(1), 371–379. DOI: 10.33395/jmp.v12i1.12405

Hashmi, M. H. A., Khan, M., & Ajmal, M. M. (2020). The impact of internal and external factors on sustainable procurement: A case study of oil and gas companies. *International Journal of Procurement Management*, 13(1), 42–62. DOI: 10.1504/IJPM.2020.105189

He, H., & Harris, L. (2020). The Impact of COVID-19 Pandemic on Corporate Social Responsibility and Marketing Philosophy. *Journal of Business Research*, 116(August), 176–182. DOI: 10.1016/j.jbusres.2020.05.030 PMID: 32457556

Hein, A., Weking, J., Schreieck, M., Wiesche, M., Böhm, M., & Krcmar, H. (2019). Value co-creation practices in business-to-business platform ecosystems. *Electronic Markets*, 29(3), 503–518. DOI: 10.1007/s12525-019-00337-y

Heleniak, T. (2021). The future of the Arctic populations. *Polar Geography*, 44(2), 136–152. DOI: 10.1080/1088937X.2019.1707316

Hernandez, R. J. (2019). Sustainable product-service systems and circular economies. *Sustainability (Basel)*, 11(19), 5383. DOI: 10.3390/su11195383

Higher School of Economics. (2021). Market research of the market of transport-and-logistic services in Russia in 2016-2020, forecast till 2025 [In Russian]. Retrieved September 29, 2021, from https://marketing.rbc.ru/research/40161/

Holman, H. R., & Dutton, D. B. (1977). A case for public participation in science policy formation and practice. *Southern California Law Review*, 51(6), 1505–1534. 10822/715792 PMID: 11661665

Howlett, M. (2009). Government communication as a policy tool: A framework for analysis. *Canadian Political Science Review*, 3(2), 23–37. DOI: 10.24124/c677/2009134

Huang, J., Koroteev, D. D., Kharun, M., & Maksimenko, R. V. (2022, August). Impact analysis of digital technology on smart city and renewable energy management. In *AIP Conference Proceedings,* Vol. 2559(1). AIP Publishing. https://doi.org/DOI: 10.1063/5.0099010

Huang, Q., Wu, J., Zhao, C., & You, X. (2007, June). Waterfilling-like multiplicity assignment algorithm for algebraic soft-decision decoding of Reed-Solomon codes. In *2007 IEEE International Conference on Communications* (pp. 6210-6213). IEEE. https://doi.org/DOI: 10.1109/ICC.2007.1028

Huang, W. (2020). The study on the relationships among film fans' willingness to pay by film crowdfunding and their influencing factors. *Ekonomska Istrazivanja*, 33(1), 804–827. DOI: 10.1080/1331677X.2020.1734849

Hubbard, P. D. (2015). Fault management via dynamic reconfiguration for integrated modular avionics (Doctoral dissertation, Loughborough University).

Humpert, M., & Raspotnik, A. (2012). The future of Arctic shipping along the transpolar sea route. *Arctic Yearbook*, 2012(1), 281–307.

Hushko, S., Botelho, J. M., Maksymova, I., Slusarenko, K., & Kulishov, V. (2021). Sustainable development of global mineral resources market in Industry 4.0 context. []. IOP Publishing.]. *IOP Conference Series. Earth and Environmental Science*, 628(1), 012025. DOI: 10.1088/1755-1315/628/1/012025

Iacoviello, G., Bruno, E., & Cappiello, A. (2019). A theoretical framework for managing intellectual capital in higher education. *International Journal of Educational Management*, 33(5), 919–938.

Ibn-Mohammed, T., Mustapha, K. B., Godsell, J., Adamu, Z., Babatunde, K. A., Akintade, D. D., Acquaye, A., Fujii, H., Ndiaye, M. M., Yamoah, F. A., & Koh, S. C. L. (2021). A critical analysis of the impacts of COVID-19 on the global economy and ecosystems and opportunities for circular economy strategies. *Resources, Conservation and Recycling*, 164, 105169. DOI: 10.1016/j.resconrec.2020.105169 PMID: 32982059

Ikundi, F., Islam, R., & White, P. (2017, October). Platform as a Service (PaaS) in public cloud: Challenges and mitigating strategy. In R. Deng, J. Weng, K. Ren, & V. Yegneswaran, (Eds.), *Security and Privacy in Communication Networks:12th International Conference Proceedings* (pp. 296-304), [Lecture Notes of the Institute for Computer Sciences, Social Informatics and Telecommunications Engineering], Vol 198. Springer, Cham. https://doi.org/DOI: 10.1007/978-3-319-59608-2_17

Iliinskij, A., Afanasiev, M., Wei, T. X., Ishel, B., & Metkin, D. (2020, November). Organizational and management model of smart field technology on the arctic shelf. In Proceedings of the International Scientific Conference-Digital Transformation on Manufacturing, Infrastructure and Service (pp. 1-5). DOI: 10.1145/3446434.3446510

Imoize, A. L., Adedeji, O., Tandiya, N., & Shetty, S. (2021). 6G enabled smart infrastructure for sustainable society: Opportunities, challenges, and research roadmap. *Sensors (Basel)*, 21(5), 1709. DOI: 10.3390/s21051709 PMID: 33801302

Inés, A., & Moleskis, M. (2021). Beyond financial motivations in crowdfunding: A systematic literature review of donations and rewards. *Voluntas*, 32(2), 276–287. DOI: 10.1007/s11266-019-00173-w

Ingenhoff, D., Buhmann, A., White, C., Zhang, T., & Kiousis, S. (2018). Reputation Spillover: Corporate Crises' Effects on Country Reputation. *Journal of Communication Management (London)*, 22(1), 96–112. DOI: 10.1108/JCOM-08-2017-0081

Iqbal, K. M. J., Barykin, S. Y., Kharlamov, A. V., Kharlamova, T. L. V., & Khan, M. I. (2021). Innovative multivariate energy governance model for climate compatible development: The case of Pakistan. *Academy of Strategic Management Journal*, 20, 1–22.

Ishuk, T., Ulyanova, O., & Savchitz, V. (2015). Approaches of Russian oil companies to optimal capital structure. In *Proceeding of the IOP Conference Series: Earth and Environmental Science* (pp.1-5). IOP Publishing. DOI: DOI: 10.1088/1755-1315/27/1/012066

Iturralde, K., Pan, W., Linner, T., & Bock, T. (2022). Automation and robotic technologies in the construction context: research experiences in prefabricated façade modules. In *Rethinking building skins* (pp. 475–493). Woodhead Publishing. DOI: 10.1016/B978-0-12-822477-9.00009-7

Ivanova, I., Smorodinskaya, N., & Leydesdorff, L. (2020). On measuring complexity in a post-industrial economy: The ecosystem's approach. *Quality & Quantity*, 54(1), 197–212. DOI: 10.1007/s11135-019-00844-2

Ivanov, D. (2021). Digital supply chain and resilience in the context of COVID-19: A dynamic capability approach. *Supply Chain Management Review*, 27(1), 38–49. DOI: 10.1108/SCMR-06-2020-0235

Ivanov, D., & Dolgui, A. (2020). A digital supply chain twin for managing the disruption risks and resilience in the era of Industry 4.0. *Production Planning and Control*, 32(9), 775–788. DOI: 10.1080/09537287.2020.1768450

Izzo, M. F., Fasan, M., & Tiscini, R. (2022). The role of digital transformation in enabling continuous accounting and the effects on intellectual capital: The case of Oracle. *Meditari Accountancy Research*, 30(4), 1007–1026.

Jabbar, M. A., Shandilya, S. K., Kumar, A., & Shandilya, S. (2022). Applications of cognitive internet of medical things in modern healthcare. *Computers & Electrical Engineering*, 102, 108276. DOI: 10.1016/j.compeleceng.2022.108276 PMID: 35958351

Jacobides, M. G., Cennamo, C., & Gawer, A. (2018). Towards a theory of ecosystems. *Strategic Management Journal*, 39(8), 2255–2276. DOI: 10.1002/smj.2904

Jacobides, M. G., & Lianos, I. (2021). Regulating platforms and ecosystems: An introduction. *Industrial and Corporate Change*, 30(5), 1131–1142. DOI: 10.1093/icc/dtab060

Janssen, M., & Estevez, E. (2013). Lean government and platform-based governance—Doing more with less. *Government Information Quarterly*, 30, S1–S8. DOI: 10.1016/j.giq.2012.11.003

Jensen, J. P. (2015). Routes for extending the lifetime of wind turbines. In T. Cooper, N. Braithwaite, M. Moreno, & G. Salvia (Eds.), *Product Lifetimes and The Environment: Conference Proceedings* (pp. 152-157). Nottingham Trent University.

Jensen, F., & Whitfield, L. (2022). Leveraging participation in apparel global supply chains through green industrialization strategies: Implications for low-income countries. *Ecological Economics*, 194(1), 107331. DOI: 10.1016/j.ecolecon.2021.107331

Jia, Q., Jiao, L., Lian, X., & Wang, W. (2023). Linking supply-demand balance of ecosystem services to identify ecological security patterns in urban agglomerations. *Sustainable Cities and Society*, 92, 104497. DOI: 10.1016/j.scs.2023.104497

Jin, K. G., & Drozdenko, R. G. (2010). Relationships among Perceived Organizational Core Values, Corporate Social Responsibility, Ethics, and Organizational Performance Outcomes: An Empirical Study of Information Technology Professionals. *Journal of Business Ethics*, 92(3), 341–359. DOI: 10.1007/s10551-009-0158-1

Jones, L. A. (2020). *Reputation risk and potential profitability: Best practices to predict and mitigate risk through amalgamated factors*. Capitol Technology University.

Josefy, M., Dean, T. J., Albert, L. S., & Fitza, M. A. (2017). The role of community in crowdfunding success: Evidence on cultural attributes in funding campaigns to "save the local theater.". *Entrepreneurship Theory and Practice*, 41(2), 161–182. DOI: 10.1111/etap.12263

Kagan, E. S., Goosen, E. V., Pakhomova, E. O., & Goosen, O. K. (2021, July). Industry 4.0. and an upgrade of the business models of large mining companies. []). IOP Publishing.]. *IOP Conference Series. Earth and Environmental Science*, 823(1), 012057. DOI: 10.1088/1755-1315/823/1/012057

Kanishcheva, N. A. (2021, February). Current state of commercial banks in a digital economy. In International Scientific and Practical Conference "Russia 2020-a new reality: economy and society"(ISPCR 2020) (pp. 169-172). Atlantis Press. DOI: 10.2991/aebmr.k.210222.033

Kantabutra, S., & Ketprapakorn, N. (2020). Toward a theory of corporate sustainability: A theoretical integration and exploration. *Journal of Cleaner Production*, 270(4), 122292. DOI: 10.1016/j.jclepro.2020.122292

Kapoor, R., & Lee, J. M. (2013). Coordinating and competing in ecosystems: How organizational forms shape new technology investments. *Strategic Management Journal*, 34(3), 274–296. DOI: 10.1002/smj.2010

Kapranova, E. A., Nenashev, V. A., Sergeev, A. M., Burylev, D. A., & Nenashev, S. A. (2019, November). Distributed matrix methods of compression, masking and noise-resistant image encoding in a high-speed network of information exchange, information processing and aggregation. In *SPIE Future Sensing Technologies* (Vol. 11197, pp. 104–110). SPIE., DOI: 10.1117/12.2542677

Karanina, E. V., & Karaulov, V. M. (2023). Differentiated approach to the diagnostics of economic security and resilience of Russian regions (case of the Volga federal district). *R-Economy. 2023. Vol. 9. Iss. 1*, 9(1), 19-37.

Karapetrovic, S., & Casadesús, M. (2009). Implementing environmental management systems in Spanish universities. *The TQM Journal*, 21(5), 507–519. DOI: 10.1108/17542730910983396

Kassen, M. (2020). E-participation actors: Understanding roles, connections, partnerships. *Knowledge Management Research and Practice*, 18(1), 16–37. DOI: 10.1080/14778238.2018.1547252

Kathlene, L., & Martin, J. A. (1991). Enhancing citizen participation: Panel designs, perspectives, and policy formation. *Journal of Policy Analysis and Management*, 10(1), 46–63. DOI: 10.2307/3325512

Kazadaev, L. M., Sattarova, A. I., & Ryumshin, K. Y. (2024, July). Investigation of digital communication lines noise immunity in hydroacoustic channel using various types of product turbo code. In *2024 Systems of Signal Synchronization, Generating and Processing in Telecommunications (SYNCHROINFO)* (pp. 1-5). IEEE. https://doi.org/DOI: 10.1109/SYNCHROINFO61835.2024.10617661

Khalili, Y., Fakhari, H., Basti, E. M. K., & Aghajani, H. (2017). Intellectual capital indicators ranking in the universities of Iran using Delphi fuzzy technique. *RISK GOVERNANCE & CONTROL: Financial markets and institutions*, 147.

Khalimon, E. A., Guseva, M. N., Kogotkova, I. Z., & Brikoshina, I. S. (2019). Digitalization of the Russian economy: first results. European Proceedings of Social and Behavioural Sciences.

Khan, A., Rezaei, S., & Valaei, N. (2022). Social commerce advertising avoidance and shopping cart abandonment: A fs/QCA analysis of German consumers. *Journal of Retailing and Consumer Services*, 67, 102976. DOI: 10.1016/j.jretconser.2022.102976

Khan, K., Khurshid, A., & Cifuentes-Faura, J. (2023). Investigating the relationship between geopolitical risks and economic security: Empirical evidence from central and Eastern European countries. *Resources Policy*, 85, 103872. DOI: 10.1016/j.resourpol.2023.103872

Khan, S. N., & Ali, E. I. E. (2017). The moderating role of intellectual capital between enterprise risk management and firm performance: A conceptual review. *American Journal of Social Sciences and Humanities*, 2(1), 9–15.

Kharitonova, T., Grokhotova, M., & Bogdanova, O. (2021). Accelerated digitalization in logistics: A forced response to the pandemic. *Digital Logistics Journal*, 7(2), 45–58.

Kharlamova, T. L., Kharlamov, A. V., & Antohina, Y. A. (2020). Influence of information technologies on the innovative development of the economic system. In I. V. Kovalev, A. A. Voroshilova, G. Herwig, U. Umbetov, A. S. Budagov, & Y. Y. Bocharova (Eds.), *Economic and Social Trends for Sustainability of Modern Society (ICEST 2020), Vol 90. European Proceedings of Social and Behavioural Sciences* (pp. 391-401). European Publisher. DOI: 10.15405/epsbs.2020.10.03.44

Kharlamova, T., Kharlamov, A., & Gavrilova, R. (2020). The development of the Russian economy under the influence of the Fourth Industrial Revolution and the use of the potential of the Arctic. IOP Conference Series: Materials Science and Engineering, 940(1), 0–7. https://doi.org/DOI: 10.1088/1757-899X/940/1/012113

Khoso, A. K., Darazi, M. A., Mahesar, K. A., Memon, M. A., & Nawaz, F. (2022). The impact of ESL teachers' emotional intelligence on ESL students academic engagement, reading and writing proficiency: Mediating role of ESL students motivation. *International Journal of Early Childhood Special Education (INT-JECSE), 14*, 3267-3280. https://doi.org/DOI: 10.9756/INT-JECSE/V14I1.393

Khoso, A. K., Darazi, M. A., Mahesar, K. A., Memon, M. A., & Nawaz, F. (2022). The impact of ESL teachers' emotional intelligence on ESL Students academic engagement, reading and writing proficiency: Mediating role of ESL students motivation. *Int. J. Early Childhood Spec. Educ*, 14, 3267–3280.

Khoso, A. K., Khurram, S., & Chachar, Z. A. (2024). Exploring the effects of embeddedness-emanation feminist identity on language learning anxiety: A case study of female english as a foreign language (EFL) learners in higher education institutions of Karachi. *International Journal of Contemporary Issues in Social Sciences*, 3(1), 1277–1290.

Khoso, A. K., Khurram, S., & Chachar, Z. A. (2024). Exploring the Effects of Embeddedness-Emanation Feminist Identity on Language Learning Anxiety: A Case Study of Female English as A Foreign Language (EFL) Learners in Higher Education Institutions of Karachi. *International Journal of Contemporary Issues in Social Sciences*, 3(1), 1277–1290.

Khun, J. A. (2021). Relationship between macroambient factors, circular economy, and sustainability. In *No Poverty* [ENUNSDG Series]. (pp. 771–782). Springer., DOI: 10.1007/978-3-319-95714-2_31

Kikkas, K. N. (2020, September). Analysis of social development in the Arctic countries. In *IOP Conference Series: Materials Science and Engineering* (Vol. 940, No. 1, p. 012115). IOP Publishing. **DOI** DOI: 10.1088/1757-899X/940/1/012115

Kim, J. (2016). The platform business model and business ecosystem: Quality management and revenue structures. *European Planning Studies*, 24(12), 2113–2132. DOI: 10.1080/09654313.2016.1251882

Kim, K. Y., & Patel, P. C. (2017). Employee ownership and firm performance: A variance decomposition analysis of European firms. *Journal of Business Research*, 70, 248–254. DOI: 10.1016/j.jbusres.2016.08.014

Kim, S. T., & Choi, B. (2019). Price risk management and capital structure of oil and gas project companies: Difference between upstream and downstream industries. *Energy Economics*, 83, 361–374. DOI: 10.1016/j.eneco.2019.07.008

Kim, W. G., Li, J. J., & Brymer, R. A. (2016). The Impact of Social Media Reviews on Restaurant Performance: The Moderating Role of Excellence Certificate. *International Journal of Hospitality Management*, 55(May), 41–51. DOI: 10.1016/j.ijhm.2016.03.001

Kline, R. R. (2001). Technological determinism.

Kloppenburg, S., & Boekelo, M. (2019). Digital platforms and the future of energy provisioning: Promises and perils for the next phase of the energy transition. *Energy Research & Social Science*, 49, 68–73. DOI: 10.1016/j.erss.2018.10.016

Knill, C., Schulze, K., & Tosun, J. (2012). Regulatory policy outputs and impacts: Exploring a complex relationship. *Regulation & Governance*, 6(4), 427–444. DOI: 10.1111/j.1748-5991.2012.01150.x

Kobzev, A. S., Smirnova, E. N., & Fedorov, V. I. (2020). The implementation of integrated management systems in aerospace organizations using digital technologies. *Russian Engineering Research*, 40(1), 67–72. DOI: 10.3103/S1068798X20010136

Koch-Baumgarten, S., & Voltmer, K. (Eds.). (2010). *Public policy and the mass media: The interplay of mass communication and political decision making* (Vol. 66). Routledge. DOI: 10.4324/9780203858493

Koike-Akino, T., Wang, Y., Millar, D. S., Kojima, K., & Parsons, K. (2019, September). Neural turbo equalization to mitigate fiber nonlinearity. In *45th European Conference on Optical Communication (ECOC 2019)* (pp. 1-4). IET. https://doi.org/DOI: 10.1049/cp.2019.0803

Kolios, A., Mytilinou, V., Lozano-Minguez, E., & Salonitis, K. (2016). A comparative study of multiple-criteria decision-making methods under stochastic inputs. *Energies*, 9(7), 566. DOI: 10.3390/en9070566

Kolloch, M., & Dellermann, D. (2018). Digital innovation in the energy industry: The impact of controversies on the evolution of innovation ecosystems. *Technological Forecasting and Social Change*, 136, 254–264. DOI: 10.1016/j.techfore.2017.03.033

Kolmykova, T. S., Sirotkina, N. V., Serebryakova, N. A., Sitnikova, E. V., & Tretyakova, I. N. (2022). Modern Tendencies of Digitalization of Banking Activities in the Russian Economy. In *Business 4.0 as a Subject of the Digital Economy* (pp. 469–474). Springer International Publishing. DOI: 10.1007/978-3-030-90324-4_75

Komarova, A., Filimonova, I., Nemov, V. Y., & Provornaya, I. (2020). *Integrated indicators of companies' efficiency in the petroleum industry*. In *Proceedings of the 20th. International Multidisciplinary Scientific GeoConference SGEM*, 2020, 311–318. DOI: 10.5593/sgem2020/5.2/s21.038

Konakhina, N. A., Baranova, T. A., Mokhorov, D. A., & Prischepa, A. S. (2019, July). Innovative and technological development of the Russian Arctic. In *IOP Conference Series: Earth and Environmental Science* (Vol. 302, No. 1, p. 012101). IOP Publishing. **DOI** DOI: 10.1088/1755-1315/302/1/012101

Korobitsyn, B. A. (2016). Regional resilience of the Ural Federal District in economic shocks and crises: medico-demographic and environmental aspects. Ekonomika Regiona= Economy of Regions, (3), 790-801. .DOI: 10.17059/2016-3-15

Korovyakovsky, V. P., & Kupriyanovsky, O. N. (2018). Digitalization in logistics: Challenges and opportunities. *Journal of Logistics and Supply Chain Management*, 6(4), 233–245. DOI: 10.1016/j.jlscm.2018.06.004

Kottmeyer, B. (2021). Digitisation and sustainable development: The opportunities and risks of using digital technologies for the implementation of a circular economy. *Journal of Entrepreneurship and Innovation in Emerging Economies*, 7(1), 17–23. DOI: 10.1177/2393957520967799

Koval, Y. (2023). Directions management of economic security of the state in the context of globalization processes. *Public administration and law review*, (3), 39-48.

Kovrigin, E. A., & Vasiliev, V. A. (2020, September). Barriers in the integration of modern digital technologies in the system of quality management of enterprises of the aerospace industry. In 2020 International Conference Quality Management, Transport and Information Security, Information Technologies (IT&QM&IS) (pp. 331-335). IEEE. DOI: 10.1109/ITQMIS51053.2020.9322960

Kozlov, A., Kankovskaya, A., & Teslya, A. (2020, July). Digital infrastructure as the factor of economic and industrial development: case of Arctic regions of Russian North-West. In *IOP Conference Series: Earth and Environmental Science* (Vol. 539, No. 1, p. 012061). IOP Publishing. **DOI** DOI: 10.1088/1755-1315/539/1/012061

Kozlov, A. V., Teslya, A. B., & Zhang, X. (2017). Principles of assessment and management Approaches to innovation potential of coal industry enterprises. *Записки Горного института, 223* (англ.), 131-138. DOI: DOI: 10.18454/PMI.2017.1.131

Kozlov, A., Kankovskaya, A., Teslya, A., & Zharov, V. (2019). Comparative study of socio-economic barriers to development of digital competences during formation of human capital in Russian Arctic. *IOP Conference Series. Earth and Environmental Science*, 302(1), 012125. DOI: 10.1088/1755-1315/302/1/012125

Kozlov, A., Pavlova, E., & Królas, A. (2021). Development of integrated management systems in the aerospace industry: A Russian perspective. *International Journal of Engineering Research & Technology (Ahmedabad)*, 10(2), 110–117. DOI: 10.17577/IJERTV10IS020067

Kozlovtseva, V., Demianchuk, M., Koval, V., Hordopolov, V., & Atstaja, D. (2021). Ensuring sustainable development of enterprises in the conditions of digital transformations. In *E3S Web of Conferences* (No. 280, p. 02002).

Krasnov, A., Chargaziya, G., Griffith, R., & Draganov, M. (2019). Dynamic and Static Elements of a Consumer's Digital Portrait and Methods of Their Studying. *IOP Conference Series. Materials Science and Engineering*, 497(April), 012123. DOI: 10.1088/1757-899X/497/1/012123

Kraus, S., Jones, P., Kailer, N., Weinmann, A., Chaparro-Banegas, N., & Roig-Tierno, N. (2021). Digital transformation: An overview of the current state of the art of research. *SAGE Open*, 11(3), 21582440211047576. DOI: 10.1177/21582440211047576

Kraus, S., Richter, C., Brem, A., Cheng, C. F., & Chang, M. L. (2016). Strategies for reward-based crowdfunding campaigns. *Journal of Innovation and Knowledge*, 1(1), 13–23. DOI: 10.1016/j.jik.2016.01.010

Kricorian, K., Seu, M., Lopez, D., Ureta, E., & Equils, O. (2020). Factors influencing participation of underrepresented students in STEM fields: Matched mentors and mindsets. *International Journal of STEM Education*, 7(1), 1–9. DOI: 10.1186/s40594-020-00219-2

Krotov, A. V., Karatabanov, R. A., & Zan, V. M. (2019, November). Problems and prospects of competitiveness of the territories of Greater Altai in the context of sustainable developmen. In *International Conference on Sustainable Development of Cross-Border Regions: Economic, Social and Security Challenges (ICSDCBR 2019)* (pp. 257-260). Atlantis Press. DOI: 10.2991/icsdcbr-19.2019.54

Kruk, M., Semenov, A., Cherepovitsyn, A., & Nikulina, A. Y. (2018). Environmental and economic damage from the development of oil and gas fields in the Arctic shelf of the Russian Federation. https://www.um.edu.mt/library/oar//handle/123456789/37958

Kryshtanovych, M., Panfilova, T., Khomenko, A., Dziubenko, O., & Lukashuk, L. (2023). Optimization of state regulation in the field of safety and security of business: A local approach. *Business: Theory and Practice*, 24(2), 613–621. DOI: 10.3846/btp.2023.19563

Kryukov, V., & Moe, A. (2018). Does Russian unconventional oil have a future? *Energy Policy*, 119, 41–50. DOI: 10.1016/j.enpol.2018.04.021

Kryvovyazyuk, I., Britchenko, I., Smerichevskyi, S., Kovalska, L., Dorosh, V., & Kravchuk, P. (2023). Digital transformation and innovation in business: the impact of strategic alliances and their success factors.

Kulibanova, V. V., & Teor, T. R. (2017). Identifying Key Stakeholder Groups for Implementing a Place Branding Policy in Saint Petersburg. *Baltic Region*, 9(3), 99–115. DOI: 10.5922/2079-8555-2017-3-7

Kumar, Y., Kaul, S., & Hu, Y. C. (2022). Machine learning for energy-resource allocation, workflow scheduling and live migration in cloud computing: State-of-the-art survey. *Sustainable Computing : Informatics and Systems*, 36, 100780. DOI: 10.1016/j.suscom.2022.100780

Kunkel, S., & Matthess, M. (2020). Digital transformation and environmental sustainability in industry: Putting expectations in Asian and African policies into perspective. *Environmental Science & Policy*, 112, 318–329. DOI: 10.1016/j.envsci.2020.06.022

Kwilinski, A., Lyulyov, O., & Pimonenko, T. (2023). The impact of digital business on energy efficiency in EU countries. *Information (Basel)*, 14(9), 480. DOI: 10.3390/info14090480

LA, M., ZA, S., RS, A., IT, M., & Yu, A. N. (2024). Issues of improving the basis of tax administration. HOLDERS OF REASON, 2(6), 184-189.

Lamba, R., & Subramanian, A. (2020). Dynamism with incommensurate development: The distinctive Indian model. *The Journal of Economic Perspectives*, 34(1), 3–30. DOI: 10.1257/jep.34.1.3

Lampkin, N., Foster, C., & Padel, S. (1999). *The policy and regulatory environment for organic farming in Europe: Country reports*. Organic Farming in Europe: Economics and Policy, Vol. 2. Universität Hohenheim, Stuttgart-Hohenheim. https://doi.org/DOI: 10.52825/gjae.v50i7.1484

Langley, P. (2016). Crowdfunding in the United Kingdom: A cultural economy. *Economic Geography*, 92(3), 301–321. DOI: 10.1080/00130095.2015.1133233

Laužikas, M., & Miliūtė, A. (2020). Impacts of modern technologies on sustainable communication of civil service organizations. *Entrepreneurship and sustainability issues, 7*(3), 2494-2509. DOI: DOI: 10.9770/jesi.2020.7.3(69)

Lavrov, N., Druzhinin, A., & Alekseeva, N. (2020). Description and Analysis of Economic Efficiency of the Real Estate Model Transformed in the Framework of Digitalization. *IOP Conference Series. Materials Science and Engineering*, 940(1), 012042. DOI: 10.1088/1757-899X/940/1/012042

Lawrence, R., & Larsen, R. K. (2017). The politics of planning: Assessing the impacts of mining on Sami lands. *Third World Quarterly*, 38(5), 1164–1180. DOI: 10.1080/01436597.2016.1257909

Lee, C. S., & Wong, K. Y. (2019). Advances in intellectual capital performance measurement: A state-of-the-art review. *The Bottom Line (New York, N.Y.)*, 32(2), 118–134.

Lee, I., & Shin, Y. J. (2018). Fintech: Ecosystem, business models, investment decisions, and challenges. *Business Horizons*, 61(1), 35–46. DOI: 10.1016/j.bushor.2017.09.003

Leitner, K. H. (2002). Intellectual Capital Reporting for Universities: Conceptual background and application within the reorganisation of Austrian universities. *The Value of Intangibles. Autonomous University of Madrid Ministry of Economy*, (November), 25–26.

Leitner, K. H. (2004). Intellectual capital reporting for universities: Conceptual background and application for Austrian universities. *Research Evaluation*, 13(2), 129–140.

Liang, X., Hu, X., & Jiang, J. (2020). Research on the effects of information description on crowdfunding success within a sustainable economy—The perspective of information communication. *Sustainability (Basel)*, 12(2), 650. DOI: 10.3390/su12020650

Lieberson, S., & O'Connor, J. F. (1972). Leadership and organizational performance: A study of large corporations. *American Sociological Review*, 37(2), 117–130. DOI: 10.2307/2094020 PMID: 5020209

Li, H., & Shi, Y. (2013). Network-based predictive control for constrained nonlinear systems with two-channel packet dropouts. *IEEE Transactions on Industrial Electronics*, 61(3), 1574–1582. DOI: 10.1109/TIE.2013.2261039

Lihua, W. U., Tianshu, M. A., Yuanchao, B. I. A. N., Sijia, L. I., & Zhaoqiang, Y. I. (2020). Improvement of regional environmental quality: Government environmental governance and public participation. *The Science of the Total Environment*, 717(5), 137265. DOI: 10.1016/j.scitotenv.2020.137265 PMID: 32092810

Li, L., Huang, X., Wu, D., & Yang, H. (2023). Construction of ecological security pattern adapting to future land use change in Pearl River Delta, China. *Applied Geography (Sevenoaks, England)*, 154, 102946. DOI: 10.1016/j.apgeog.2023.102946

Li, N., Binti Mohd Ariffin, S. Z., & Gao, H. (2024). Optimizing Ecotourism in North Taihu Lake, Wuxi City, China: Integrating Back Propagation Neural Networks and Ant-Colony Algorithm for Sustainable Route Planning. *International Journal of Management Thinking*, 2(1), 1–15. DOI: 10.56868/ijmt.v2i1.53

Litvinenko, V. S. (2020). Digital economy as a factor in the technological development of the mineral sector. *Natural Resources Research*, 29(3), 1521–1541. DOI: 10.1007/s11053-019-09568-4

Liu, B., Li, S., Xie, Y., & Yuan, J. (2019 a, August). Deep learning assisted sum-product detection algorithm for faster-than-Nyquist signaling. In *2019 IEEE Information Theory Workshop (ITW)* (pp. 1-5). IEEE. https://doi.org/DOI: 10.1109/ITW44776.2019.8989271

Liu, R., Gailhofer, P., Gensch, C. O., Köhler, A., Wolff, F., Monteforte, M., ... & Williams, R. (2019). Impacts of the digital transformation on the environment and sustainability. *Issue Paper under Task, 3*.

Liu, C., Dou, X., Li, J., & Cai, L. A. (2020). Analyzing government role in rural tourism development: An empirical investigation from China. *Journal of Rural Studies*, 79, 177–188. DOI: 10.1016/j.jrurstud.2020.08.046

Liu, H., Zhu, Q., Khoso, W. M., & Khoso, A. K. (2023). Spatial pattern and the development of green finance trends in China. *Renewable Energy*, 211, 370–378. DOI: 10.1016/j.renene.2023.05.014

Liu, Y. H., & Poulin, D. (2019b). Neural belief-propagation decoders for quantum error-correcting codes. *Physical Review Letters*, 122(20), 200501. DOI: 10.1103/PhysRevLett.122.200501 PMID: 31172756

Liu, Z., Liu, J., & Osmani, M. (2021). Integration of digital economy and circular economy: Current status and future directions. *Sustainability (Basel)*, 13(13), 7217. DOI: 10.3390/su13137217

Loucks, J., Macaulay, J., Noronha, A., & Wade, M. (2016). *Digital Vortex: How Today's Market Leaders Can Beat Disruptive Competitors at Their Own Game.* DBT Center Press.

Luca, M. 2011. "Reviews, Reputation, and Revenue: The Case of Yelp.Com." SSRN *Electronic Journal.* https://doi.org/DOI: 10.2139/ssrn.1928601

Luhmann, N. (1996). *Social Systems.* Stanford University Press.

Lukashevich, N., Garanin, D., & Konnikov, E. (2019). Modeling of the parameters of the investment project based on the information-statistical approach. In *Proceedings of the 33rd International Business Information Management Association Conference, IBIMA 2019: Education Excellence and Innovation Management through. Vision (Basel)*, 2020, 8743–8752.

Lund, S., Madgavkar, A., Manyika, J., Smit, S., Ellingrud, K., Meaney, M., & Robinson, O. (2021). The future of work after COVID-19. *McKinsey global institute, 18.*

Lüthi, S., & Prässler, T. (2011). Analyzing policy support instruments and regulatory risk factors for wind energy deployment—A developers' perspective. *Energy Policy*, 39(9), 4876–4892. DOI: 10.1016/j.enpol.2011.06.029

Lu, Y., Zheng, H., Chand, S., Xia, W., Liu, Z., Xu, X., Wang, L., Qin, Z., & Bao, J. (2022). Outlook on human-centric manufacturing towards Industry 5.0. *Journal of Manufacturing Systems*, 62, 612–627. DOI: 10.1016/j.jmsy.2022.02.001

Lyshchikova, J. V., Stryabkova, E. A., Glotova, A. S., & Dobrodomova, T. N. (2019). The'Smart Region'concept: The implementation of digital technology. *Journal of Advanced Research in Law and Economics*, 10(4 (42)), 1338–1345. DOI: 10.14505//jarle.v10.4(42).34

Lyubimov, I. L., Lysyuk, M. V., & Gvozdeva, M. A. (2018). Atlas of economic complexity, Russian regional pages. *Voprosy Ekonomiki*, (6), 71–91. Advance online publication. DOI: 10.32609/0042-8736-2018-6-71-91

Magdich, I. A., Petrov, V. P., & Pyatibrat, A. O. (2019). Analiz sanitarnykh i bezvozvratnykh poter'v zavisimosti ot kharaktera i usloviy chrezvychaynykh situatsiy na zheleznoy doroge= Analysis of Sanitary and Irreparable Losses Depending on the Nature and Conditions of Railway Emergencies. *Medical-biological and Socio-psychological Problems of Safety in Emergency Situations*, 1, 72–80. DOI: 10.25016/2541-7487-2019-0-1-72-80

Makarov, I. A. (2016). Russia's participation in international environmental cooperation. *Strategic Analysis*, 40(6), 536–546. DOI: 10.1080/09700161.2016.1224062

Mamonov, S., Koufaris, M., & Benbunan-Fich, R. (2016). The role of the sense of community in the sustainability of social network sites. *International Journal of Electronic Commerce*, 20(4), 470–498. DOI: 10.1080/10864415.2016.1171974

Marcovecchio, I., Thinyane, M., Estevez, E., & Janowski, T. (2019). Digital government as implementation means for sustainable development goals. [IJPADA]. *International Journal of Public Administration in the Digital Age*, 6(3), 1–22. DOI: 10.4018/IJPADA.2019070101

Maroufkhani, P., Desouza, K. C., Perrons, R. K., & Iranmanesh, M. (2022). Digital transformation in the resource and energy sectors: A systematic review. *Resources Policy*, 76, 102622. DOI: 10.1016/j.resourpol.2022.102622

Martín, J. M. M., & Fernández, J. A. S. (2022). The effects of technological improvements in the train network on tourism sustainability. An approach focused on seasonality. *Sustainable Technology and Entrepreneurship*, 1(1), 100005. DOI: 10.1016/j.stae.2022.100005

Matsumoto, O., Kawai, K., & Takeda, T. (2017). FUJITSU cloud service K5 PaaS digitalizes enterprise systems. *Fujitsu Scientific and Technical Journal*, 53(1), 17–24.

Ma, X., Tong, T. W., & Fitza, M. (2013). How much does subnational region matter to foreign subsidiary performance? Evidence from Fortune Global 500 Corporations' investment in China. *Journal of International Business Studies*, 44(1), 66–87. DOI: 10.1057/jibs.2012.32

Mazzoleni, S., Turchetti, G., & Ambrosino, N. (2020). The COVID-19 Outbreak: From 'Black Swan' to Global Challenges and Opportunities. *Pulmonology*, 26(3), 117–118. DOI: 10.1016/j.pulmoe.2020.03.002 PMID: 32291202

McConnell, J. J., & Muscarella, C. J. (1985). Corporate capital expenditure decisions and the market value of the firm. *Journal of Financial Economics*, 14(3), 399–422. DOI: 10.1016/0304-405X(85)90006-6

McKinsey & Company. (2021). *The future of logistics: Digitalization and automation post-COVID-19*. Retrieved from https://www.mckinsey.com/industries/travel-transport-and-logistics/our-insights/the-future-of-logistics-post-covid-19

Melnyk, V. (2024). Transforming the nature of trust between banks and young clients: From traditional to digital banking. *Qualitative Research in Financial Markets*, 16(4), 618–635. DOI: 10.1108/QRFM-08-2022-0129

Ménard, C. (2004). The economics of hybrid organizations. [JITE]. *Journal of Institutional and Theoretical Economics*, 160(3), 345–376. DOI: 10.1628/0932456041960605

Mergel, I., Edelmann, N., & Haug, N. (2019). Defining digital transformation: Results from expert interviews. *Government Information Quarterly*, 36(4), 101385. DOI: 10.1016/j.giq.2019.06.002

Merkulov, V. I. (2020, September). Analysis of Russian Arctic LNG projects and their development prospects. In *IOP Conference Series: Materials Science and Engineering* (Vol. 940, No. 1, p. 012114). IOP Publishing. **DOI** DOI: 10.1088/1757-899X/940/1/012114

Metaxas, T. (2010). Place Marketing, Place Branding and Foreign Direct Investments: Defining Their Relationship in the Frame of Local Economic Development Process. *Place Branding and Public Diplomacy*, 6(3), 228–243. DOI: 10.1057/pb.2010.22

Meyskens, M., & Bird, L. (2015). Crowdfunding and value creation. *Entrepreneurship Research Journal*, 5(2), 155–166. DOI: 10.1515/erj-2015-0007

Minakir, P. A. (2020). Political value of expectations in economy. Spatial Economics= Prostranstvennaya Ekonomika, (3), 7-23. .DOI: 10.14530/se.2020.3.007-023

Mishchuk, I. (2023). Conceptual model of economic security formation and the place of the security process in this model. *Economics. Finance and Management Review*, (1), 40–49. DOI: 10.36690/2674-5208-2023-1-40

Mitchell, D. G., Lentmaier, M., & Costello, D. J. (2015). Spatially coupled LDPC codes constructed from protographs. *IEEE Transactions on Information Theory*, 61(9), 4866–4889. DOI: 10.1109/TIT.2015.2453267

Mitrova, T., & Melnikov, Y. (2019). Energy transition in Russia. *Energy Transitions*, 3(1-2), 73–80. DOI: 10.1007/s41825-019-00016-8

Mkumbuzi, W. P. (2016). Influence of intellectual capital investment, risk, industry membership and corporate governance mechanisms on the voluntary disclosure of intellectual capital by UK listed companies. *Asian Social Science*, 12(1), 42.

Mokyr, J., Vickers, C., & Ziebarth, N. L. (2015). The history of technological anxiety and the future of economic growth: Is this time different? *The Journal of Economic Perspectives*, 29(3), 31–50. DOI: 10.1257/jep.29.3.31

Molchan, A. S., Osadchuk, L. M., Anichkina, O. A., Ponomarev, S. V., & Kuzmenko, N. I. (2023). The'Digitalisation trap'of Russian regions. International Journal of Technology. *Policy and Management*, 23(1), 20–41.

Mollick, E. (2016, April 21). The unique value of crowdfunding is not money — It's community. *Harvard Business Review*. https://hbr.org/

Mollick, E. (2014). The dynamics of crowdfunding: An exploratory study. *Journal of Business Venturing*, 29(1), 1–16. DOI: 10.1016/j.jbusvent.2013.06.005

Mondejar, M. E., Avtar, R., Diaz, H. L. B., Dubey, R. K., Esteban, J., Gómez-Morales, A., Hallam, B., Mbungu, N. T., Okolo, C. C., Prasad, K. A., She, Q., & Garcia-Segura, S. (2021). Digitalization to achieve sustainable development goals: Steps towards a Smart Green Planet. *The Science of the Total Environment*, 794, 148539. DOI: 10.1016/j.scitotenv.2021.148539 PMID: 34323742

Moon, M. J. (2020). Fighting COVID-19 with agility, transparency, and participation: Wicked policy problems and new governance challenges. *Public Administration Review*, 80(4), 651–656. DOI: 10.1111/puar.13214 PMID: 32836434

Moore, J. F. (2006). Business ecosystems and the view from the firm. *Antitrust Bulletin*, 51(1), 31–75. DOI: 10.1177/0003603X0605100103

Morgan, N., & Pritchard, A. (2014). Destination Reputations and Brands: Communication Challenges. *Journal of Destination Marketing & Management*, 3(1), 1. DOI: 10.1016/j.jdmm.2014.02.001

Moskovchenko, D. V., Babushkin, A. G., & Yurtaev, A. A. (2020). The impact of the Russian oil industry on surface water quality (a case study of the Agan River catchment, West Siberia). *Environmental Earth Sciences*, 79(14), 355. DOI: 10.1007/s12665-020-09097-x

Mullin, A. E., Coe, I. R., Gooden, E. A., Tunde-Byass, M., & Wiley, R. E. (2021, November). Inclusion, diversity, equity, and accessibility: From organizational responsibility to leadership competency. []. Sage CA: Los Angeles, CA: SAGE Publications.]. *Healthcare Management Forum*, 34(6), 311–315. DOI: 10.1177/08404704211038232 PMID: 34535064

Mulska, O. P., Vasyltsiv, T. G., Kunytska-Iliash, M. V., & Baranyak, I. Y. (2023). Innovative Empirics of Migration-Economic Security Causal Nexus. *Science and innovation, 19*(3), 48-64.

Munsamy, M., & Telukdarie, A. (2019, December). Digital HRM model for process optimization by adoption of industry 4.0 technologies. In 2019 IEEE international conference on industrial engineering and engineering management (IEEM) (pp. 374-378). IEEE.

Nachmani, E., Bachar, Y., Marciano, E., Burshtein, D., & Be'ery, Y. (2018). Near maximum likelihood decoding with deep learning. *ArXiv, abs/1801.02726.* https://doi.org//arXiv.1801.02726 DOI: 10.48550

Naeem, M., De Pietro, G., & Coronato, A. (2021). Application of reinforcement learning and deep learning in multiple-input and multiple-output (MIMO) systems. *Sensors (Basel)*, 22(1), 309. DOI: 10.3390/s22010309 PMID: 35009848

Nasiopoulos, P., & Ward, R. K. (1994). A noise resistant synchronization scheme for HDTV images. *IEEE Transactions on Broadcasting*, 40(4), 228–237. DOI: 10.1109/11.362935

Nazari, Z., & Musilek, P. (2023). Impact of digital transformation on the energy sector: A review. *Algorithms*, 16(4), 211. DOI: 10.3390/a16040211

Nemov, V. Y., Filimonova, I. V., & Komarova, A. V. (2020, April). Assessment of the Mutual Influence of Energy Intensity of the Economy and Pollutant Emissions. In *IOP Conference Series: Earth and Environmental Science* (Vol. 459, No. 6, p. 062025). IOP Publishing. **DOI** DOI: 10.1088/1755-1315/459/6/062025

Nesadurai, H. E. (2004). Introduction: Economic security, globalization and governance. *The Pacific Review*, 17(4), 459–484. DOI: 10.1080/0951274042000326023

Nielsen, A. E., & Thomsen, C. (2011). Sustainable development: The role of network communication. *Corporate Social Responsibility and Environmental Management*, 18(1), 1–10. DOI: 10.1002/csr.221

Nikitina, M. G., Pobirchenko, V. V., Shutaieva, E. A., & Karlova, A. I. (2018). The investment component in a nation's economic security: the case of the Russian Federation. *Entrepreneurship and sustainability issues*, 6(2), 958.

Ning, J., & Xiong, L. (2024). Analysis of the dynamic evolution process of the digital transformation of renewable energy enterprises based on the cooperative and evolutionary game model. *Energy*, 288, 129758. DOI: 10.1016/j.energy.2023.129758

Nižetić, S., Šolić, P., Gonzalez-De, D. L. D. I., & Patrono, L. (2020). Internet of Things (IoT): Opportunities, issues and challenges towards a smart and sustainable future. *Journal of Cleaner Production*, 274(5), 122877. DOI: 10.1016/j.jclepro.2020.122877 PMID: 32834567

Notteboom, T., Pallis, A. A., & Rodrigue, J. P. (2021). Disruptions and resilience in global container shipping and ports: A critical review. *Maritime Economics & Logistics*, 23(2), 179–210. DOI: 10.1057/s41278-020-00180-5

Nurulin, Y. R., Skvortsova, I., & Vinogradova, E. (2020). On the issue of the green energy markets development. In Arseniev, D., Overmeyer, L., Kälviäinen, H., & Katalinić, B. (Eds.), *Cyber-Physical Systems and Control* [Lecture Notes in Networks and Systems]. Vol. 95). Springer., DOI: 10.1007/978-3-030-34983-7_34

Nurulin, Y., Skvortsova, I., Tukkel, I., & Torkkeli, M. (2019). Role of Knowledge in Management of Innovation. *Resources*, 8(2), 87. DOI: 10.3390/resources8020087

O'Dwyer, E., Pan, I., Charlesworth, R., Butler, S., & Shah, N. (2020). Integration of an energy management tool and digital twin for coordination and control of multi-vector smart energy systems. *Sustainable Cities and Society*, 62, 102412. DOI: 10.1016/j.scs.2020.102412

Oche, P. A., Ewa, G. A., & Ibekwe, N. (2021). Applications and challenges of artificial intelligence in space missions. *IEEE Access : Practical Innovations, Open Solutions*, 12, 44481–44509. DOI: 10.1109/ACCESS.2021.3132500

OECD. (2020). *The territorial impact of COVID-19: Managing the crisis across levels of government*. Organisation for Economic Co-operation and Development.

Øien, K., Hauge, S., & Grøtan, T. O. (2020). Barrier Management Digitalization in the Oil and Gas Industry-Status and Challenges. In *e-proceedings of the 30th European Safety and Reliability Conference and 15th Probabilistic Safety Assessment and Management Conference (ESREL2020 PSAM15)*. Research Publishing Services.

Okorokov, R., Timofeeva, A., & Kharlamova, T. (2019). Building intellectual capital of specialists in the context of digital transformation of the Russian economy. *IOP Conference Series. Materials Science and Engineering*, 497(1), 012015. DOI: 10.1088/1757-899X/497/1/012015

Olaru, M., Maier, D., Nicoara, D., & Maier, A. (2014). Establishing the basis for development of an organization by adopting the integrated management systems: Comparative study of various models and concepts. *Procedia: Social and Behavioral Sciences*, 109, 693–697. DOI: 10.1016/j.sbspro.2013.12.531

Olisaeva, A. V. (2019). Technological development of Russia: HR policy, digital transformation, Industry 4.0. *Planning and Teaching Engineering Staff for the Industrial and Economic Complex of the Region*, 86-89. Advance online publication. DOI: 10.17816/PTES26310

Orazalin, N., & Mahmood, M. (2018). Economic, environmental, and social performance indicators of sustainability reporting: Evidence from the Russian oil and gas industry. *Energy Policy*, 121, 70–79. DOI: 10.1016/j.enpol.2018.06.015

Osei, L. K., Cherkasova, Y., & Oware, K. M. (2023). Unlocking the full potential of digital transformation in banking: A bibliometric review and emerging trend. *Future Business Journal*, 9(1), 30. DOI: 10.1186/s43093-023-00207-2

Osembe, L., & Padayachee, I. (2016). Perceptions on benefits and challenges of cloud computing technology adoption by IT SMEs: A case of Gauteng province. *Journal of Contemporary Management*, 13(1), 1255–1297.

Özkan, P., & Yücel, E. K. (2020). Linear economy to circular economy: Planned obsolescence to cradle-to-cradle product perspective. In *Handbook of research on entrepreneurship development and opportunities in circular economy* (pp. 61–86). IGI Global., DOI: 10.4018/978-1-7998-5116-5.ch004

Pakhucha, E., Sievidova, I. O., Romaniuk, I., Bilousko, T., Tkachenko, S. O., Diadin, A. S., & Babko, N. (2023). Investigating the impact of structural changes: the socio-economic security framework.

Pałaka, D., Paczesny, B., Gurdziel, M., & Wieloch, W. (2020). Industry 4.0 in development of new technologies for underground mining. In *E3S Web of Conferences* (Vol. 174, p. 01002). EDP Sciences.

Paloma Sánchez, M., Elena, S., & Castrillo, R. (2009). Intellectual capital dynamics in universities: A reporting model. *Journal of Intellectual Capital*, 10(2), 307–324.

Pangsri, P. (2015). Application of the multi criteria decision making methods for project selection. *Universal Journal of Management*, 3(1), 15–20. DOI: 10.13189/ujm.2015.030103

Panychev, A. Y., Sigova, M. V., & Sokolov, G. V. (2020). The impact of digital transformation on logistics. *Rivista Internazionale di Economia dei Trasporti*, 47(3), 123–134. DOI: 10.1080/00207543.2020.1818273

Park, M. J., Kang, D., Rho, J. J., & Lee, D. H. (2016). Policy role of social media in developing public trust: Twitter communication with government leaders. *Public Management Review*, 18(9), 1265–1288. DOI: 10.1080/14719037.2015.1066418

Paschen, J. (2017). Choose wisely: Crowdfunding through the stages of the startup life cycle. *Business Horizons*, 60(2), 179–188. DOI: 10.1016/j.bushor.2016.11.003

Paschke, M., Lebedeva, O., Shabalov, M., & Ivanova, D. (2020). Economic and legal aspects of digital transformation in mining industry. In *Advances in raw material industries for sustainable development goals* (pp. 492–500). CRC Press. DOI: 10.1201/9781003164395-61

Pavlov, N., Kalyazina, S., Bagaeva, I., & Iliashenko, V. (2021). Key digital technologies for national business environment. In Murgul, V., & Pukhkal, V. (Eds.), *Energy management of municipal facilities and sustainable energy technologies EMMFT 2019. Advances in Intelligent Systems and Computing* (pp. 151–158). Springer., DOI: 10.1007/978-3-030-57453-6_12

Payne, J. E., Truong, H. H. D., Chu, L. K., Doğan, B., & Ghosh, S. (2023). The effect of economic complexity and energy security on measures of energy efficiency: Evidence from panel quantile analysis. *Energy Policy*, 177, 113547. DOI: 10.1016/j.enpol.2023.113547

Peng, Y., & Tao, C. (2022). Can digital transformation promote enterprise performance?—From the perspective of public policy and innovation. *Journal of Innovation & Knowledge*, 7(3), 100198. DOI: 10.1016/j.jik.2022.100198

Pereira, P., Bašić, F., Bogunovic, I., & Barcelo, D. (2022). Russian-Ukrainian war impacts the total environment. *The Science of the Total Environment*, 837, 155865. DOI: 10.1016/j.scitotenv.2022.155865 PMID: 35569661

Pereira, V., Jayawardena, N. S., Sindhwani, R., Behl, A., & Laker, B. (2024). Using firm-level intellectual capital to achieve strategic sustainability: Examination of phenomenon of business failure in terms of the critical events. *Journal of Intellectual Capital*.

Petrov, A. N. (2016). Exploring the Arctic's "other economies": Knowledge, creativity and the new frontier. *The Polar Journal*, 6(1), 51–68. DOI: 10.1080/2154896X.2016.1171007

Petruzzelli, A. M., Natalicchio, A., Panniello, U., & Roma, P. (2019). Understanding the crowdfunding phenomenon and its implications for sustainability. *Technological Forecasting and Social Change*, 141, 138–148. DOI: 10.1016/j.techfore.2018.10.002

Phaphoom, N., Wang, X., Samuel, S., Helmer, S., & Abrahamsson, P. (2015). A survey study on major technical barriers affecting the decision to adopt cloud services. *Journal of Systems and Software*, 103, 167–181. DOI: 10.1016/j.jss.2015.02.002

Pichtel, J. (2016). Oil and gas production wastewater: Soil contamination and pollution prevention. *Applied and Environmental Soil Science*, 2016(1), 2707989. DOI: 10.1155/2016/2707989

Pietro, F. D. (2019). Deciphering crowdfunding. In Lynn, T., Mooney, J., Rosati, P., & Cummins, M. (Eds.), *Disrupting Finance* [Series: Palgrave Studies in Digital Business and Enabling Technologies]. (pp. 1–14). Palgrave Pivot., DOI: 10.1007/978-3-030-02330-0_1

Pigola, A., da Costa, P. R., Carvalho, L. C., Silva, L. F. D., Kniess, C. T., & Maccari, E. A. (2021). Artificial intelligence-driven digital technologies to the implementation of the sustainable development goals: A perspective from Brazil and Portugal. *Sustainability (Basel)*, 13(24), 13669. DOI: 10.3390/su132413669

Plotnikov, V., & Pirogova, O. (2018).Key Competencies as an Enterprise Value Management Tool." In 31 IBIMA Conference (Milan: IBIMA), 1716–21.

Podvalny, S. L., Podvalny, E. S., & Provotorov, V. V. (2017). The Controllability of Parabolic Systems with Delay and Distributed Parameters on the Graph. Procedia Computer Science, 103(October 2016), 324–330. https://doi.org/DOI: 10.1016/j.procs.2017.01.115

Pokrovskaia, N. N., Korableva, O. N., Cappelli, L., & Fedorov, D. A. (2021). Digital regulation of intellectual capital for open innovation: Industries' expert assessments of tacit knowledge for controlling and networking outcome. *Future Internet*, 13(2), 44.

Poltorak, A., Volosyuk, Y., Tyshchenko, S., Khrystenko, O., & Rybachuk, V. (2023). DEVELOPMENT OF DIRECTIONS FOR IMPROVING THE MONITORING OF THE STATE ECONOMIC SECURITY UNDER CONDITIONS OF GLOBAL INSTABILITY. *Eastern-European Journal of Enterprise Technologies*, 122(13).

Popelo, O., Shaposhnykov, K., Popelo, O., Hrubliak, O., Malysh, V., & Lysenko, Z. (2023). The Influence of Digitalization on the Innovative Strategy of the Industrial Enterprises Development in the Context of Ensuring Economic Security. *International Journal of Safety and Security Engineering*, 13(1), 39–49. DOI: 10.18280/ijsse.130105

Popkova, E. G., De Bernardi, P., Tyurina, Y. G., & Sergi, B. S. (2022). A theory of digital technology advancement to address the grand challenges of sustainable development. *Technology in Society*, 68(4), 101831. DOI: 10.1016/j.techsoc.2021.101831

Popov, E., Dolghenko, R., Simonova, V., & Chelak, I. (2021). Analytical model of innovation ecosystem development. In 1st Conference on Traditional and Renewable Energy Sources: Perspectives and Paradigms for the 21st Century (TRESP 2021): Vol. 250, 01004. EDP Sciences. DOI: 10.1051/e3sconf/202125001004

Potapova, E. A., Iskoskov, M. O., & Mukhanova, N. V. (2022). The impact of digitalization on performance indicators of Russian commercial banks in 2021. *Journal of Risk and Financial Management*, 15(10), 452. DOI: 10.3390/jrfm15100452

Provotorov, V. V., Sergeev, S. M., & Part, A. A. (2019). Solvability of hyperbolic systems with distributed parameters on the graph in the weak formulation. *Vestnik of Saint Petersburg University Applied Mathematics Computer Science Control Processes*, 15(1). Advance online publication. DOI: 10.21638/11702/spbu10.2019.108

Pupentsova, S. V., Alekseeva, N. S., & Stroganova, O. A. (2020, February). Foreign and domestic experience in environmental planning and territory management. [IOP Publishing.]. *IOP Conference Series. Materials Science and Engineering*, 753(3), 032026. DOI: 10.1088/1757-899X/753/3/032026

Qiu, X. H., Wang, Z., & Xue, Q. (2015). Investment in deepwater oil and gas exploration projects: A multi-factor analysis with a real options model. *Petroleum Science*, 12(3), 525–533. DOI: 10.1007/s12182-015-0039-4

Radicic, D., Pugh, G., & Douglas, D. (2020). Promoting cooperation in innovation ecosystems: Evidence european traditional manufacturing SMEs. *Small Business Economics*, 54(1), 257–283. DOI: 10.1007/s11187-018-0088-3

Ramirez, Y., Lorduy, C., & Rojas, J. A. (2007). Intellectual capital management in Spanish universities. *Journal of Intellectual Capital*, 8(4), 732–748.

Ramskyi, A., Gontar, Z., Kazak, O., Podzihun, S., & Naumchuk, K. (2023). Formation of the security environment through manimization of the negative impact of threats in the socio-economic system.

Raza, M. H. (2024). The Effect of Remittances on Economic Expansion and Poverty Reduction: Evidence from Pakistan. *International Journal of Management Thinking*, 2(1), 75–92. DOI: 10.56868/ijmt.v2i1.47

RBC Market Research. (2021). Market research of the transport and logistics services market in Russia (2016-2020) with a forecast until 2025 [In Russian]. Retrieved September 29, 2021, from https://marketing.rbc.ru/research/40161/

Renwick, M. J., & Mossialos, E. (2017). Crowdfunding our health: Economic risks and benefits. *Social Science & Medicine*, 191, 48–56. DOI: 10.1016/j.socscimed.2017.08.035 PMID: 28889030

Rigger, E., Shea, K., & Stanković, T. (2022). Method for identification and integration of design automation tasks in industrial contexts. advanced engineering informatics, 52, 101558..

Roberts, R., Flin, R., Millar, D., & Corradi, L. (2021). Psychological factors influencing technology adoption: A case study from the oil and gas industry. *Technovation*, 102, 102219. DOI: 10.1016/j.technovation.2020.102219

Robinson, S. C. (2020). Trust, transparency, and openness: How inclusion of cultural values shapes Nordic national public policy strategies for artificial intelligence (AI). *Technology in Society*, 63, 101421. DOI: 10.1016/j.techsoc.2020.101421

Roblek, V., Meško, M., Bach, M. P., Thorpe, O., & Šprajc, P. (2020). The interaction between internet, sustainable development, and emergence of society 5.0. *Data*, 5(3), 80. DOI: 10.3390/data5030080

Rodrigues, L. F., Oliveira, A., & Rodrigues, H. (2023). Technology management has a significant impact on digital transformation in the banking sector. *International Review of Economics & Finance*, 88, 1375–1388. DOI: 10.1016/j.iref.2023.07.040

Rohani, A., Keshavarz, E., & Keshavarz, A. (2015). Prioritising (ranking) of indexes for measuring intellectual capital using FAHP and fuzzy TOPSIS techniques. *International Journal of Industrial and Systems Engineering*, 21(3), 356–376.

Romanyuk, M., Sukharnikova, M., & Chekmareva, N. (2021, March). Trends of the digital economy development in Russia. []. IOP Publishing.]. *IOP Conference Series. Earth and Environmental Science*, 650(1), 012017. DOI: 10.1088/1755-1315/650/1/012017

Romasheva, N., & Dmitrieva, D. (2021). Energy resources exploitation in the russian arctic: Challenges and prospects for the sustainable development of the ecosystem. *Energies*, 14(24), 8300. DOI: 10.3390/en14248300

Rong, K., Hu, G., Hou, J., Ma, R., & Shi, Y. (2013). Business ecosystem extension: Facilitating the technology substitution. *International Journal of Technology Management*, 63(3/4), 268–294. DOI: 10.1504/IJTM.2013.056901

Rong, K., Li, B., Peng, W., Zhou, D., & Shi, X. (2021). Sharing economy platforms: Creating shared value at a business ecosystem level. *Technological Forecasting and Social Change*, 169(2), 120804. DOI: 10.1016/j.techfore.2021.120804

Rosset, V., Paulo, M. A., Cespedes, J. G., & Nascimento, M. C. (2017). Enhancing the reliability on data delivery and energy efficiency by combining swarm intelligence and community detection in large-scale WSNs. *Expert Systems with Applications*, 78, 89–102. DOI: 10.1016/j.eswa.2017.02.008

Rouissi, C. (2020). The influence of the enterprise resource Planning (ERP) on management controllers: A study in the Tunisian context. *International Journal of Business and Management*, 15(4), 25–35. DOI: 10.5539/ijbm.v15n4p25

Royal, C., Evans, J., & Windsor, S. S. (2014). The missing strategic link–human capital knowledge, and risk in the finance industry–two mini case studies. *Venture Capital*, 16(3), 189–206.

Rudakova, E., Panshin, I., & Vlasov, A. (2021). Evaluating trends in modern logistics: Post-pandemic challenges and opportunities. *Transport and Logistics Journal*, 15(3), 22–35.

Ryu, S., & Kim, Y. G. (2016). A typology of crowdfunding sponsors: Birds of a feather flock together? *Electronic Commerce Research and Applications*, 16, 43–54. DOI: 10.1016/j.elerap.2016.01.006

S. Dhiman, H., Deb, D., S. Dhiman, H., & Deb, D. (2020). Multi-criteria decision-making: An overview. Decision and control in hybrid wind farms, 19-36.

Saitova, A. A., Ilyinsky, A. A., & Fadeev, A. M. (2022). Scenarios for the development of oil and gas companies in Russia in the context of international economic sanctions and the decarbonization of the energy sector. *Sever i Rynok: Formirovanie Ekonomicheskogo Poryadka*, 25(3), 134–143. DOI: 10.37614/2220-802X.3.2022.77.009

Samimi, A. (2020). Risk management in oil and gas refineries. *Progress in Chemical and Biochemical Research*, 3(2), 140–146. DOI: 10.33945/SAMI/PCBR.2020.2.8

Sampaio, P., Saraiva, P., & Domingues, P. (2012). Management systems: Integration or addition? *International Journal of Quality & Reliability Management*, 29(4), 402–424. DOI: 10.1108/02656711211224857

Sánchez, F., & Hartlieb, P. (2020). Innovation in the mining industry: Technological trends and a case study of the challenges of disruptive innovation. *Mining, Metallurgy & Exploration*, 37(5), 1385–1399. DOI: 10.1007/s42461-020-00262-1

Sanchez, J., Caire, R., & Hadjsaid, N. (2013, June). *ICT and power distribution modeling using complex networks. In 2013 IEEE Grenoble Conference.* IEEE., DOI: 10.1109/PTC.2013.6652388

Sandi, H., Afni Yunita, N., Heikal, M., Nur Ilham, R., & Sinta, I. (2021). Relationship Between Budget Participation, Job Characteristics, Emotional Intelligence and Work Motivation As Mediator Variables to Strengthening User Power Performance: An Emperical Evidence From Indonesia Government. *Morfai Journal*, 1(1), 36–48. DOI: 10.54443/morfai.v1i1.14

Savas, S. (2022). Digital Transformation from Data Mining to Big Data and Its Effects on Productivity. *Current Studies in Digital Transformation and Productivity*, 54.

Sayfuddin, A. T. M., & Chen, Y. (2021). The Signaling and Reputational Effects of Customer Ratings on Hotel Revenues: Evidence from TripAdvisor. *International Journal of Hospitality Management*, 99(October), 103065. DOI: 10.1016/j.ijhm.2021.103065

Saygıner, C. (2023). Software as a service (SaaS) adoption as a disruptive technology: Understanding the challenges and the obstacles of non-SaaS adopters. *Uluslararası Yönetim İktisat ve İşletme Dergisi*, 19(3), 501–515. DOI: 10.17130/ijmeb.1249540

Schipachev, A. M., & Dmitrieva, A. S. (2021). Application of the resonant energy separation effect at natural gas reduction points in order to improve the energy efficiency of the gas distribution system. Записки Горного института, 248, 253-259. .DOI: 10.31897/PMI.2021.2.9

Schmidt, C. W. (2011). Arctic oil drilling plans raise environmental health concerns. https://doi.org/DOI: 10.1289/ehp.119-a116

Schmidt, L., Falk, T., Siegmund-Schultze, M., & Spangenberg, J. H. (2020). The objectives of stakeholder involvement in transdisciplinary research. A conceptual framework for a reflective and reflexive practise. *Ecological Economics*, 176, 106751. DOI: 10.1016/j.ecolecon.2020.106751

Schulz, K. A., Gstrein, O. J., & Zwitter, A. J. (2020). Exploring the governance and implementation of sustainable development initiatives through blockchain technology. *Futures*, 122, 102611. DOI: 10.1016/j.futures.2020.102611

Secundo, G., Dumay, J., Schiuma, G., & Passiante, G. (2016). Managing intellectual capital through a collective intelligence approach: An integrated framework for universities. *Journal of Intellectual Capital*, 17(2), 298–319.

Secundo, G., Perez, S. E., Martinaitis, Ž., & Leitner, K. H. (2017). An Intellectual Capital framework to measure universities' third mission activities. *Technological Forecasting and Social Change*, 123, 229–239.

Semenyuta, O. G., & Shapiro, I. E. (2023). Risks of Banking Transformation in Digitalization. In *Technological Trends in the AI Economy: International Review and Ways of Adaptation* (pp. 357–362). Springer Nature Singapore. DOI: 10.1007/978-981-19-7411-3_38

Serenko, A., & Bontis, N. (2022). Global ranking of knowledge management and intellectual capital academic journals: A 2021 update. *Journal of Knowledge Management*, 26(1), 126–145.

Sergi, B. S., & Berezin, A. (2018). Oil and gas industry's technological and sustainable development: Where does Russia stand? In *Exploring the future of Russia's economy and markets: Towards sustainable economic development* (pp. 161–182). Emerald Publishing Limited. DOI: 10.1108/978-1-78769-397-520181009

Shahbaz, M., Wang, J., Dong, K., & Zhao, J. (2022). The impact of digital economy on energy transition across the globe: The mediating role of government governance. *Renewable & Sustainable Energy Reviews*, 166, 112620. DOI: 10.1016/j.rser.2022.112620

Shanti, R., Siregar, H., Zulbainarni, N., & Tony, . (2023). Role of digital transformation on digital business model banks. *Sustainability (Basel)*, 15(23), 16293. DOI: 10.3390/su152316293

Shapovalova, D., Galimullin, E., & Grushevenko, E. (2020). Russian Arctic offshore petroleum governance: The effects of western sanctions and outlook for northern development. *Energy Policy*, 146, 111753. DOI: 10.1016/j.enpol.2020.111753

Sharma, V. S., & Santharam, A. (2013, December). Implementing a resilient application architecture for state management on a PaaS cloud. In *2013 IEEE 5th International Conference on Cloud Computing Technology and Science,* Vol. 1, (pp. 142-147). IEEE. DOI: 10.1109/CloudCom.2013.26

Sharma, R., Kamble, S., Gunasekaran, A., & Kumar, V. (2020). A systematic review of blockchain technology adoption in supply chains. *Journal of Cleaner Production*, 262, 121–344. DOI: 10.1016/j.jclepro.2020.121344

Sheng, R. (2019). *Systems engineering for aerospace: A practical approach.* Academic Press.

Sheresheva, M. Y., Valitova, L. A., Sharko, E. R., & Buzulukova, E. V. (2022). Application of social network analysis to visualization and description of industrial clusters: A case of the textile industry. *Journal of Risk and Financial Management*, 15(3), 1–17. DOI: 10.3390/jrfm15030129

Shinkevich, A. I., Kudryavtseva, S. S., & Samarina, V. P. (2023). Ecosystems as an Innovative Tool for the Development of the Financial Sector in the Digital Economy. *Journal of Risk and Financial Management*, 16(2), 72. DOI: 10.3390/jrfm16020072

Shin, Y., Sung, S. Y., Choi, J. N., & Kim, M. S. (2015). Top Management Ethical Leadership and Firm Performance: Mediating Role of Ethical and Procedural Justice Climate. *Journal of Business Ethics*, 129(1), 43–57. DOI: 10.1007/s10551-014-2144-5

Shiobara, T. (2006). Oversights in Russia's corporate governance: The case of the oil and gas industry. In Tabata, S. (Ed.), *Dependent on oil and gas: Russia's integration into the world economy* (pp. 85–114). Slavic Research Center.

Shirvani Dastgerdi, A., & De Luca, G. (2019). Strengthening the City's Reputation in the Age of Cities: An Insight in the City Branding Theory. *City, Territory and Architecture*, 6(1), 2. DOI: 10.1186/s40410-019-0101-4

Shmatok, A. N., Zolotenkova, M. K., & Egorov, V. V. (2024, May). Usage of noise-resistant coding for stable data transmission over household electrical networks. In *2024 IEEE Ural-Siberian Conference on Biomedical Engineering, Radioelectronics and Information Technology (USBEREIT)* (pp. 212-216). IEEE. https://doi.org/DOI: 10.1109/USBEREIT61901.2024.10583992

Shneor, R. (2020). Crowdfunding models, strategies, and choices between them. In Shneor, R., Zhao, L., & Flåten, B. T. (Eds.), *Advances in Crowdfunding* (pp. 21–42). Palgrave Macmillan., DOI: 10.1007/978-3-030-46309-0_2

Shneor, R., & Munim, Z. H. (2019). Reward crowdfunding contribution as planned behaviour: An extended framework. *Journal of Business Research*, 103(2), 56–70. DOI: 10.1016/j.jbusres.2019.06.013

Short, J. C., Ketchen, D. J.Jr, McKenny, A. F., Allison, T. H., & Ireland, R. D. (2017). Research on crowdfunding: Reviewing the (very recent) past and celebrating the present. *Entrepreneurship Theory and Practice*, 41(2), 149–160. DOI: 10.1111/etap.12270

Shvedina, S. A. (2020). Digital transformation of mining enterprises contributes to the rational use of resources. []. IOP Publishing.]. *IOP Conference Series. Earth and Environmental Science*, 408(1), 012064. DOI: 10.1088/1755-1315/408/1/012064

Silkina, G. (2019, April). From analogue to digital tools of business control: Succession and transformation. In *IOP Conference Series: Materials Science and Engineering,* Vol. 497, 012018. IOP Publishing. https://doi.org/DOI: 10.1088/1757-899X/497/1/012018

Singh, S., Sharma, P. K., Yoon, B., Shojafar, M., Cho, G. H., & Ra, I. H. (2020). Convergence of blockchain and artificial intelligence in IoT network for the sustainable smart city. *Sustainable Cities and Society*, 63(10), 102364. DOI: 10.1016/j.scs.2020.102364

Sinkevičius Virginijus. (2020, June). *Commissioner speech at post-COVID green deal technology powered recovery in Europe*. European Commission, Brussels.

Sishi, M., & Telukdarie, A. (2020). Implementation of Industry 4.0 technologies in the mining industry-a case study. *International Journal of Mining and Mineral Engineering*, 11(1), 1–22. DOI: 10.1504/IJMME.2020.105852

Smirnova, O., & Ponomaryova, A. (2021). Assessment of the Industrial Development Rates of Russian Regions in the Context of the Digitalization of the Economy. In Digital Transformation in Industry: Trends, Management, Strategies (pp. 283-290). Springer International Publishing. DOI: 10.1007/978-3-030-73261-5_26

Soare, A., & Militaru, C. (2018). Quality management systems in the aerospace industry: An approach based on AS/EN/JISQ 9100 standards. *Management of Sustainable Development*, 10(1), 39–44. DOI: 10.1515/msd-2018-0010

Sokolitsyn, A., Ivanov, M., Sokolitsyna, N., & Babkin, I. (2019). Analytical model of the Interrelation between Enterprise's Activity and its Financial Stability. In *Proceedings of the IOP Conference Series: Materials Science and Engineering* (pp.1-11). IOP Publishing Ltd. DOI: DOI: 10.1088/1757-899X/618/1/012086

Solodovnikov, A. Yu., Chistobaev, A. I., & Semenova, Z. A. (2016). Influence of Oil and Gas Facilities of Western Siberia on the Fauna. *Geography and Natural Resources*, (4). Advance online publication. DOI: 10.21782/GiPR0206-1619-2016-4(48-54)

Song, M., Xie, Q., Shahbaz, M., & Yao, X. (2023). Economic growth and security from the perspective of natural resource assets. *Resources Policy*, 80, 103153. DOI: 10.1016/j.resourpol.2022.103153

Spulber, D. F., & Yoo, C. S. (2005). Network regulation: The many faces of access. *Journal of Competition Law & Economics*, 1(4), 635–678. DOI: 10.1093/joclec/nhi024

Sridhar, S., Hahn, A., & Govindarasu, M. (2011). Cyber–physical system security for the electric power grid. *Proceedings of the IEEE*, 100(1), 210–224. DOI: 10.1109/JPROC.2011.2165269

Stanko, M. A., & Henard, D. H. (2017). Toward a better understanding of crowdfunding, openness and the consequences for innovation. *Research Policy*, 46(4), 784–798. DOI: 10.1016/j.respol.2017.02.003

SteadieSeifi, M., Dellaert, N. P., Nuijten, W., Van Woensel, T., & Raoufi, R.SteadieSeifi. (2014). Multimodal freight transportation planning: A literature review. *European Journal of Operational Research*, 233(1), 1–15. DOI: 10.1016/j.ejor.2013.06.055

Stepputat, M., Beuss, F., Pfletscher, U., Sender, J., & Fluegge, W. (2021). Automated one-off production in woodworking by Part-to-Tool. *Procedia CIRP*, 104, 307–312. DOI: 10.1016/j.procir.2021.11.052

Straßheim, H. (2020). The rise and spread of behavioral public policy: An opportunity for critical research and self-reflection. *International Review of Public Policy*, 2(2: 1), 115-128. DOI: DOI: 10.4000/irpp.897

Su, F., Liu, Y., Chen, S. J., & Fahad, S. (2023). Towards the impact of economic policy uncertainty on food security: Introducing a comprehensive heterogeneous framework for assessment. *Journal of Cleaner Production*, 386, 135792. DOI: 10.1016/j.jclepro.2022.135792

Sufi, A., & Taylor, A. M. (2022). Financial crises: A survey. Handbook of international economics, 6, 291-340.

Su, J., Yao, S., & Liu, H. (2022). Data governance facilitate digital transformation of oil and gas industry. *Frontiers in Earth Science (Lausanne)*, 10, 861091. DOI: 10.3389/feart.2022.861091

Sulyandziga, L. (2019). Indigenous peoples and extractive industry encounters: Benefit-sharing agreements in Russian Arctic. *Polar Science*, 21, 68–74. DOI: 10.1016/j.polar.2018.12.002

Sun, Y., Dong, Y., Chen, X., & Song, M. (2023). Dynamic evaluation of ecological and economic security: Analysis of China. *Journal of Cleaner Production*, 387, 135922. DOI: 10.1016/j.jclepro.2023.135922

Švarc, J., Lažnjak, J., & Dabić, M. (2021). The role of national intellectual capital in the digital transformation of EU countries. Another digital divide? *Journal of Intellectual Capital*, 22(4), 768–791.

Tabolina, A. V., Olennikova, M. V., Tikhonov, D. V., Kozlovskii, P., Baranova, T. A., & Gulk, E. B. (2020). Project Activities in Technical Institutes as a Mean of Preparing Students for Life and Professional Self-determination. In Advances in Intelligent Systems and Computing: Vol. 1134 AISC (pp. 800–807). Springer. https://doi.org/DOI: 10.1007/978-3-030-40274-7_77

Talapatra, S., Uddin, M. K., & Rahman, M. H. (2019). Development of an implementation framework for integrated management systems. *The TQM Journal*, 31(3), 472–489. DOI: 10.1108/TQM-08-2018-0107

Tang, S., Chen, L., & Ma, X. (2012). Progressive list-enlarged algebraic soft decoding of Reed-Solomon codes. *IEEE Communications Letters*, 16(6), 901–904. DOI: 10.1109/LCOMM.2012.042512.112511

TARANTOLA, S., ROSSOTTI, A., CONTINI, P., & CONTINI, S. (2018). A guide to the equipment, methods and procedures for the prevention of risks, emergency response and mitigation of the consequences of accidents: Part I.

Taylor, C. M., Gallagher, E. A., Pollard, S. J., Rocks, S. A., Smith, H. M., Leinster, P., & Angus, A. J. (2019). Environmental regulation in transition: Policy officials' views of regulatory instruments and their mapping to environmental risks. *The Science of the Total Environment*, 646, 811–820. DOI: 10.1016/j.scitotenv.2018.07.217 PMID: 30064107

Teece, D. J. (2016). Business ecosystem. In Augier, M., & Teece, D. (Eds.), *The Palgrave Encyclopedia of Strategic Management* (pp. 1–4). Palgrave Macmillan., DOI: 10.1057/978-1-349-94848-2_724-1

Thunberg, S., & Arnell, L. (2022). Pioneering the use of technologies in qualitative research–A research review of the use of digital interviews. *International Journal of Social Research Methodology*, 25(6), 757–768. DOI: 10.1080/13645579.2021.1935565

Thurner, T., & Proskuryakova, L. N. (2014). Out of the cold–the rising importance of environmental management in the corporate governance of Russian oil and gas producers. *Business Strategy and the Environment*, 23(5), 318–332. DOI: 10.1002/bse.1787

Timchenko, O., Nebrat, V. V., Lir, V., Bykonia, O., & Dubas, Y. (2019). Organizational and economic determinants of digital energy development in Ukraine. *Economy and forecasting*, Vol. 2019(3), 59-75. https://doi.org/DOI: 10.15407/econforecast2019.03.059

Timkina, T. A., Savelyeva, N. K., & Kryukova, A. D. (2023). Innovative Activity of a Commercial Bank During the Period of Economic Transition. In *Anti-Crisis Approach to the Provision of the Environmental Sustainability of Economy* (pp. 367–373). Springer Nature Singapore. DOI: 10.1007/978-981-99-2198-0_40

Tischner, U., Schmincke, E., Rubik, F., & Prösler, M. (2000). *How to do EcoDesign?: A guide for environmentally and economically sound design*. Verlag Form Publisher.

Tishkov, P. I., Khakimova, G. R., Diasamidze, M. A., & Minchenko, L. V. (2017). Organization and methodological basis for company's risk monitoring and control. *2017 XX IEEE International Conference on Soft Computing and Measurements (SCM)*, 799–800. https://doi.org/DOI: 10.1109/SCM.2017.7970728

Todeschini, B. V., Cortimiglia, M. N., de-Menezes, D. C., & Ghezzi, A. (2017). Innovative and sustainable business models in the fashion industry: Entrepreneurial drivers, opportunities, and challenges. *Business Horizons*, 60(6), 759–770. DOI: 10.1016/j.bushor.2017.07.003

Tohanean, D., & Toma, S. G. (2024). The impact of cloud systems on enhancing organizational performance through innovative business models in the digitalization era. In *Proceedings of the International Conference on Business Excellence,* Vol. 18(1), (pp. 3568-3577). Sciendo. https://doi.org/DOI: 10.2478/picbe-2024-0289

Tomczak, A., & Brem, A. (2013). A conceptualized investment model of crowdfunding. *Venture Capital, 15*(4), 335–59. da Cruz, J. V. (2018). Beyond financing: Crowdfunding as an informational mechanism. *Journal of Business Venturing, 33*(3), 371–93. https://doi.org/DOI: 10.1080/13691066.2013.847614

Tomic, S., Spasojevic-Brkic, V., & Klarin, M. (2012). The importance of AS/EN/JISQ 9100 quality standards in the aerospace industry. *Quality and Reliability Engineering International*, 28(4), 459–467. DOI: 10.1002/qre.1234

Triantaphyllou, E., & Triantaphyllou, E. (2000). *Multi-criteria decision making methods*. Springer Us., DOI: 10.1007/978-1-4757-3157-6_2

Trinh, T. H., Liem, N. T., & Kachitvichyanukul, V. (2014). A game theory approach for value co-creation systems. *Production & Manufacturing Research*, 2(1), 253–265. DOI: 10.1080/21693277.2014.913124

Troise, C., & Camilleri, M. A. (2021). The Use of Digital Media for Marketing, CSR Communication and Stakeholder Engagement. In Camilleri, M. A. (Ed.), *Strategic Corporate Communication in the Digital Age* (pp. 161–174). Emerald Publishing Limited., DOI: 10.1108/978-1-80071-264-520211010

Tsoy, D., Tirasawasdichai, T., & Kurpayanidi, K. I. (2021). Role of social media in shaping public risk perception during COVID-19 pandemic: A theoretical review. *International Journal of Management Science and Business Administration*, 7(2), 35–41. DOI: 10.18775/ijmsba.1849-5664-5419.2014.72.1005

Tyaglov, S. G., Sheveleva, A. V., Rodionova, N. D., & Guseva, T. B. (2021, March). Contribution of Russian oil and gas companies to the implementation of the sustainable development goal of combating climate change. []. IOP Publishing.]. *IOP Conference Series. Earth and Environmental Science*, 666(2), 022007. DOI: 10.1088/1755-1315/666/2/022007

Tyncherov, K. T., Mukhametshin, V. S., & Rakhimov, N. (2021, February). Theoretical basis for constructing special codes for a noise-resistant downhole telemetry system. In *Journal of Physics: Conference Series,* Vol. 1753(1), 012081. IOP Publishing. DOI: 10.1088/1742-6596/1753/1/012081

U.S. Census Bureau. (2021). Quarterly retail e-commerce sales. U.S. Department of Commerce. Retrieved from https://www.census.gov/

Ulewicz, R., KRSTIĆ, B., & Ingaldi, M. (2022). Mining Industry 4.0—Opportunities and Barriers. *Acta Montanistica Slovaca*, 27(2).

Uzquiano, A., & Arlotta, P. (2022). Brain organoids: The quest to decipher human-specific features of brain development. *Current Opinion in Genetics & Development*, 75, 101955. DOI: 10.1016/j.gde.2022.101955 PMID: 35816938

Vachha, B. A., & Middlebrooks, E. H. (2022). Brain Functional Imaging Anatomy. *Neuroimaging Clinics of North America*, 32(3), 491–505. DOI: 10.1016/j.nic.2022.04.001 PMID: 35843658

Van Hau, N., Khanh Ly, C. T., Quynh Nga, N., Hong Duyen, N. T., & Huong Hue, T. T. (2022). Digital Transformation in Mining Sector in Vietnam. *Inżynieria Mineralna*, (2), 21–30.

Vasiliev, D., Kuznetsov, I., Ivanov, V., & Smirnov, A. (2020). Digital transformation in aerospace: The role of integrated management systems. *Journal of Aerospace Engineering*, 233(1), 105–114. DOI: 10.1177/0954410020922057

Vatansever, A. (2020). Put over a barrel?"Smart" sanctions, petroleum and statecraft in Russia. *Energy Research & Social Science*, 69, 101607. DOI: 10.1016/j.erss.2020.101607

Veltri, S., Mastroleo, G., & Schaffhauser-Linzatti, M. (2014). Measuring intellectual capital in the university sector using a fuzzy logic expert system. *Knowledge Management Research and Practice*, 12(2), 175–192.

Verbivska, L., Abramova, M., Gudz, M., Lyfar, V., & Khilukha, O. (2023). Digitalization of the Ukrainian economy during a state of war is a necessity of the time. *Amazonia Investiga*, 12(68), 184–194. DOI: 10.34069/AI/2023.68.08.17

Vertakova, Y., Klevtsova, M., & Zadimidchenko, A. (2022). Multiplicative methodology for assessing investment attractiveness and risk for industries. *Journal of Risk and Financial Management*, 15(10), 419. DOI: 10.3390/jrfm15100419

Vetrova, E. N., Khakimova, G. R., Gladysheva, I. V., & Lapochkina, L. V. (2018). Development Strategy of University Interaction with Employers. *2018 XVII Russian Scientific and Practical Conference on Planning and Teaching Engineering Staff for the Industrial and Economic Complex of the Region (PTES)*, 245–248. https://doi.org/DOI: 10.1109/PTES.2018.8604249

Vetrova, E. N., Khakimova, G. R., Tihomirov, N. N., & Diasamidze, M. A. (2017). Structurization of intellectual capital risks in the conditions of labor market integration and globalization. *2017 XX IEEE International Conference on Soft Computing and Measurements (SCM)*, 801–803. https://doi.org/DOI: 10.1109/SCM.2017.7970729

Vetrova, M., & Ivanova, D. (2021). Closed Product Life Cycle as a Basis of the Circular Economy. [JBER]. *Journal of Business & Economics Research*, 5(4).

Vezzoli, C., & Manzini, E. (2008). *Design for environmental sustainability* (1st ed.). Springer Science and Business Media., DOI: 10.1007/978-1-4471-7364-9

Villarón-Peramato, O., García-Sánchez, I. M., & Martínez-Ferrero, J. (2018). Capital structure as a control mechanism of a CSR entrenchment strategy. *European Business Review*, 30(3), 340–371. DOI: 10.1108/EBR-03-2017-0056

Walthoff-Borm, X., Schwienbacher, A., & Vanacker, T. (2018). Equity crowdfunding: First resort or last resort? *Journal of Business Venturing*, 33(4), 513–533. DOI: 10.1016/j.jbusvent.2018.04.001

Wang, C. N., Tsai, H. T., Ho, T. P., Nguyen, V. T., & Huang, Y. F. (2020). Multicriteria decision making (MCDM) model for supplier evaluation and selection for oil production projects in Vietnam. *Processes (Basel, Switzerland)*, 8(2), 134. DOI: 10.3390/pr8020134

Wang, K. J., Dagne, T. B., Lin, C. J., Woldegiorgis, B. H., & Nguyen, H. P. (2021). Intelligent control for energy conservation of air conditioning system in manufacturing systems. *Energy Reports*, 7, 2125–2137. DOI: 10.1016/j.egyr.2021.04.010

Wang, Z., & Wang, L. (2012, August). Occupancy pattern based intelligent control for improving energy efficiency in buildings. In *2012 IEEE International Conference on Automation Science and Engineering (CASE)* (pp. 804-809). IEEE. DOI: 10.1109/CoASE.2012.6386336

Wanke, P., Azad, M. A. K., Emrouznejad, A., & Antunes, J. (2019). A dynamic network DEA model for accounting and financial indicators: A case of efficiency in MENA banking. *International Review of Economics & Finance*, 61, 52–68. DOI: 10.1016/j.iref.2019.01.004

Warszawski, A. (1996). Strategic planning in construction companies. *Journal of Construction Engineering and Management*, 122(2), 133–140. DOI: 10.1061/(ASCE)0733-9364(1996)122:2(133)

Weible, C. M., Nohrstedt, D., Cairney, P., Carter, D. P., Crow, D. A., Durnová, A. P., Heikkila, T., Ingold, K., McConnell, A., & Stone, D. (2020). COVID-19 and the policy sciences: Initial reactions and perspectives. *Policy Sciences*, 53(4), 225–241. DOI: 10.1007/s11077-020-09381-4 PMID: 32313308

Wei, Z., Zhu, S., Dai, X., Wang, X., Yapanto, L. M., & Raupov, I. R. (2021). Multi-criteria decision making approaches to select appropriate enhanced oil recovery techniques in petroleum industries. *Energy Reports*, 7, 2751–2758. DOI: 10.1016/j.egyr.2021.05.002

Whitt, R. S. (2004). A Horizontal Leap Forward: Formulating a New Communications Public Policy Framework Based on the Network Layers Model. *Federal Communications Law Journal*, 56(3), 587–673.

Wieninger, S., Gotzen, R., Gudergan, G., & Wenning, K. M. (2019). The strategic analysis of business ecosystems : New conception and practical application of a research approach. In 2019 IEEE International Conference on Engineering, Technology and Innovation (ICE/ITMC) [Conference Session]. Institute of Electrical and Electronics Engineers Inc. DOI: 10.1109/ICE.2019.8792657

Williamson, P. J., & De Meyer, A. (2012). Ecosystem advantage: How to successfully harness the power of partners. *California Management Review*, 55(1), 24–46. DOI: 10.1525/cmr.2012.55.1.24

Wilson, J. H., & Joye, S. W. 2020. "Research Designs and Variables." In *Research Methods and Statistics: An Integrated Approach*, 40–72. 2455 Teller Road, Thousand Oaks California 91320: SAGE Publications, Inc. https://doi.org/DOI: 10.4135/9781071802717.n3

Wilson, E., & Stammler, F. (2016). Beyond extractivism and alternative cosmologies: Arctic communities and extractive industries in uncertain times. *The Extractive Industries and Society*, 3(1), 1–8. DOI: 10.1016/j.exis.2015.12.001

Wood, D. A. (2016). Supplier selection for development of petroleum industry facilities, applying multi-criteria decision making techniques including fuzzy and intuitionistic fuzzy TOPSIS with flexible entropy weighting. *Journal of Natural Gas Science and Engineering*, 28, 594–612. DOI: 10.1016/j.jngse.2015.12.021

Wortmann, F., Gebauer, H., Lamprecht, C., & Fleisch, E. (2024). Drive the PaaS transformation. In *Understanding Products as Services: How the Internet and AI are Transforming Product Companies* (pp. 57–67). Emerald Publishing Limited., DOI: 10.1108/978-1-83797-823-620241009

Wudhikarn, R. (2017, March). Determining key performance indicators of intellectual capital in logistics business using Delphi method. In *2017 International Conference on Digital Arts, Media and Technology (ICDAMT)* (pp. 164-169). IEEE.

Wulf, F., Lindner, T., Strahringer, S., & Westner, M. (2021). IaaS, PaaS, or SaaS? The why of cloud computing delivery model selection: Vignettes on the post-adoption of cloud computing. In *Proceedings of the 54th Hawaii International Conference on System Sciences* (pp. 6285-6294). Ostbayerische Technische Hochschule Regensburg. https://doi.org/DOI: 10.24251/HICSS.2021.758

Xie, K. L., Zhang, Z., & Zhang, Z. (2014). The Business Value of Online Consumer Reviews and Management Response to Hotel Performance. *International Journal of Hospitality Management*, 43(October), 1–12. DOI: 10.1016/j.ijhm.2014.07.007

Xin-gang, Z., & Jin, Z. (2022). Industrial restructuring, energy consumption and economic growth: Evidence from China. *Journal of Cleaner Production*, 335, 130242. DOI: 10.1016/j.jclepro.2021.130242

Xu, C., Chen, X., & Dai, W. (2022). Effects of digital transformation on environmental governance of mining enterprises: Evidence from China. *International Journal of Environmental Research and Public Health*, 19(24), 16474. DOI: 10.3390/ijerph192416474 PMID: 36554353

Xu, W., Tan, X., Be'ery, Y., Ueng, Y. L., Huang, Y., You, X., & Zhang, C. (2020). Deep learning-aided belief propagation decoder for polar codes. *IEEE Journal on Emerging and Selected Topics in Circuits and Systems*, 10(2), 189–203. DOI: 10.1109/JETCAS.2020.2995962

Xu, Y., Wang, L., Xiong, Y., Wang, M., & Xie, X. (2023). Does digital transformation foster corporate social responsibility? Evidence from Chinese mining industry. *Journal of Environmental Management*, 344, 118646. DOI: 10.1016/j.jenvman.2023.118646 PMID: 37481916

Yakovlev, A. A. (2008). *State-business relations and improvement of corporate governance in Russia* (BOFIT Discussion Paper No. 26/2008). Bank of Finland. https://ssrn.com/abstract=1324548

Yakutseni, S. P., & Solov'ev, I. A. (2020). Calculation of environmental damage as a result of an accident at a fuel depot in Norilsk. Geograficheskaya sreda i zhivye sistemy, (4), 48-56. https://doi.org/DOI: 10.18384/2712-7621-2020-4-48-56

Yamashita, A. S., & Fujii, H. (2022). Trend and priority change of climate change mitigation technology in the global mining sector. *Resources Policy*, 78, 102870. DOI: 10.1016/j.resourpol.2022.102870

Yangibaevich, A. A. (2023). Transformation of the regulation of commercial banks in the conditions of the development of the digital economy of Uzbekistan. *SAARJ Journal on Banking & Insurance Research*, 12(2), 11–18. DOI: 10.5958/2319-1422.2023.00005.X

Yan, M. R., Hong, L. Y., & Warren, K. (2022). Integrated knowledge visualization and the enterprise digital twin system for supporting strategic management decision. *Management Decision*, 60(4), 1095–1115. DOI: 10.1108/MD-02-2021-0182

Young, A., & Rogers, P. (2019). A review of digital transformation in mining. *Mining, Metallurgy & Exploration*, 36(4), 683–699. DOI: 10.1007/s42461-019-00103-w

Young, O. R. (2016). Governing the arctic ocean. *Marine Policy*, 72, 271–277. DOI: 10.1016/j.marpol.2016.04.038

Younkin, P., & Kashkooli, K. (2016). What problems does crowdfunding solve? *California Management Review*, 58(2), 20–43. DOI: 10.1525/cmr.2016.58.2.20

Yurak, V. V., Polyanskaya, I. G., & Malyshev, A. N. (2023). The assessment of the level of digitalization and digital transformation of oil and gas industry of the Russian Federation. Gornye nauki i tekhnologii= Mining Science and Technology (Russia), 8(1), 87-110.

Yu, Z., Li, Y., & Dai, L. (2023). Digital finance and regional economic resilience: Theoretical framework and empirical test. *Finance Research Letters*, 55, 103920. DOI: 10.1016/j.frl.2023.103920

Zaboyev, A. I., & Zhuravleva, E. K. (2014). Challenges of digital technology integration in logistics. *Russian Transport Review*, 12(2), 45–56.

Zamira, J. (2024). ENSURING ECONOMIC SECURITY IN THE BANKING SECTOR. *Gospodarka i Innowacje.*, 47, 343–348.

Zhang, H., & Chen, W. (2019). Backer motivation in crowdfunding new product ideas: Is it about you or is it about me? *Journal of Product Innovation Management*, 36(2), 241–262. DOI: 10.1111/jpim.12477

Zhang, H., Zhu, L., Dai, T., Zhang, L., Feng, X., Zhang, L., & Zhang, K. (2023). Smart object recommendation based on topic learning and joint features in the social internet of things. *Digital Communications and Networks*, 9(1), 22–32. DOI: 10.1016/j.dcan.2022.04.025

Zhang, J., & Meng, J. (2009). Noise resistant OFDM for power-line communication systems. *IEEE Transactions on Power Delivery*, 25(2), 693–701. DOI: 10.1109/TPWRD.2009.2036626

Zhang, L., Mu, R., Zhan, Y., Yu, J., Liu, L., Yu, Y., & Zhang, J. (2022). Digital economy, energy efficiency, and carbon emissions: Evidence from provincial panel data in China. *The Science of the Total Environment*, 852, 158403. DOI: 10.1016/j.scitotenv.2022.158403 PMID: 36057314

Zhao, L., & Sun, Z. (2020). Pure donation or hybrid donation crowdfunding: Which model is more conducive to prosocial campaign success? *Baltic Journal of Management*, 15(2), 237–260. DOI: 10.1108/BJM-02-2019-0076

Zhironkina, O., & Zhironkin, S. (2023). Technological and intellectual transition to mining 4.0: A review. *Energies*, 16(3), 1427. DOI: 10.3390/en16031427

Zhironkin, S., & Ezdina, N. (2023). Review of transition from mining 4.0 to mining 5.0 innovative technologies. *Applied Sciences (Basel, Switzerland)*, 13(8), 4917. DOI: 10.3390/app13084917

Zhironkin, S., Gasanov, M., & Suslova, Y. (2022). Orderliness in Mining 4.0. *Energies*, 15(21), 8153. DOI: 10.3390/en15218153

Zhou, Y. (2024). Natural resources and green economic growth: A pathway to innovation and digital transformation in the mining industry. *Resources Policy*, 90, 104667. DOI: 10.1016/j.resourpol.2024.104667

Zhukova, O. (2019). *Education in modern Russia: policy and discourse* (Master's thesis).

Zolotuhin, A. B., & Stepin, Y. P. (2019, November). Problem and models of multicriteria decision making and risk assessment of the arctic offshore oil and gas field development. In *IOP Conference Series: Materials Science and Engineering* (Vol. 700, No. 1, p. 012050). IOP Publishing. **DOI**DOI: 10.1088/1757-899X/700/1/012050

Zuboff, S. (2019). *The Age of Surveillance Capitalism: The Fight for a Human Future at the New Frontier of Power, edn.* PublicAffairs.

Zutshi, A., & Creed, A. (2018). Declaring Talloires: Profile of sustainability communications in Australian signatory universities. *Journal of Cleaner Production*, 187, 687–698. DOI: 10.1016/j.jclepro.2018.03.225

About the Contributors

Ahdi Hassan has been Associate or Consulting Editor of numerous journals and also served the editorial review board from 2013- to till now. He has a number of publications and research papers published in various domains. He has given contribution with the major roles such as using modern and scientific techniques to work with sounds and meanings of words, studying the relationship between the written and spoken formats of various Asian/European languages, developing the artificial languages in coherence with modern English language, and scientifically approaching the various ancient written material to trace its origin. He teaches topics connected but not limited to communication such as English for Young Learners, English for Academic Purposes, English for Science, Technology and Engineering, English for Business and Entrepreneurship, Business Intensive Course, Applied Linguistics, interpersonal communication, verbal and nonverbal communication, cross cultural competence, language and humor, intercultural communication, culture and humor, language acquisition and language in use.

Shakir Ullah is a trained ethnographer. He received his Ph.D in cultural anthropology and M.Phil in Sociology. His research goal is to link anthropological theory and ethnographic data on development with issues related to mega-development projects. His broader areas of interest include anthropology of development, applied anthropology, social change, and development. Before joining the Society of Fellows at SUSTech, he worked as a lecturer at the University of Peshawar, as an assistant professor at the University of Swabi Pakistan, and as a research and reporting officer at Alkhidmat Foundation Pakistan. He has written extensively on state-led mega-development projects in Pakistan, China One Belt One Road, and China Pakistan Economic Corridor (CPEC).

Sergey Barykin is a doctor of economic sciences (specialities finances 08.00.10 and logistics 08.00.05, year of the dissertation defence - 2009), and a professor at Graduate School of Service and Trade, Institute of Industrial Management,

Economics and Trade, Peter the Great St. Petersburg Polytechnic University, Russia (since 2019). Since 2022 to present - a head of the Dissertation Council U.5.2.3.24 at Peter the Great St. Petersburg Polytechnic University, (SPbGEU) in the speciality 5.2.3. Regional and Industrial Economics: 4. Economics of services, 5. Transport and logistics, Marketing. Since 2018 to 2022 - an active member of the Dissertation Council D 212.354.02, created on the basis of the St. Petersburg State University of Economics (SPbGEU) in the specialty 08.00.05 - Economics and National Economy Management: Logistics. 2002 – 2017 - Associate Professor, St. Petersburg State University of Economics (SPbGEU) (Russia) ACHIEVEMENTS 2020 Grant funded by RFBR, project number № 20-014-00029 «Development of trade and economic cooperation within the framework of the Eurasian Economic Union on the basis of digital logistics platforms» (Number CITIS AAAA-A 20-120031390080-9, https://kias.rfbr.ru/index.php) FIELDS OF INTEREST Digital platforms, Smart and sustainable cities, digital and international networks, Industry 4.0, integrated marketing and logistics, digital echelons and consumer value chains, logistics and international trade network configuration, digital economy, BRICS trade and economic cooperation.

Elena de la Poza obtained her PhD from Polytechnic University of Valencia (UPV) in 2008. She is currently an Associate Professor of the Department of Economics and Social Sciences of the UPV. Her research is specialized in Sustainability, the empirical analysis of social phenomena and modeling the economic-financial impact of organizations. She has co-edited two books and has published more than 20 international scientific papers and 12 book chapters. She has presented papers at more than 50 national and international congresses, and has been guest speaker at 5 international conferences. She has done research at several prestigious centers such as University of North Carolina at Chapel Hill, Università degli studi de Padova, University of Knoxville, Michigan State University, University of Panama, Warsaw University of Life Sciences, Technical University of Ostrava. Currently she participates in three international European projects on Sustainability and Higher Education.

* * *

Vladimir Plotnikov is Professor of Department of General Economic Theory & History of Economic Thought in Saint-Petersburg State University of Economics (Russia). He has published over 50 scientific papers in the area of innovations, economic security, state regulation of economy, sustainability, industrial and regional development. His main field of interest include regulation in the social-economic systems.

Index

A

Aerospace 301, 302, 303, 304, 305, 306, 307, 308, 309, 310, 311, 312, 313, 314, 316, 317, 318, 319, 320
Aerospace industry 301, 302, 303, 305, 306, 307, 308, 310, 314, 316, 317, 319, 320
Automation 26, 33, 34, 37, 89, 90, 92, 93, 97, 98, 100, 105, 107, 117, 181, 263, 265, 266, 268, 269, 270, 275, 276, 278, 295, 322, 323, 324, 331, 335, 337, 339

B

Business ecosystem 25, 26, 27, 28, 30, 31, 32, 34, 36, 37, 41
Business ecosystem approach 25, 26, 30, 31, 32, 34, 36, 37
Business sector 35, 113, 114, 115, 117, 118, 119, 122, 130

C

Circular business models 1, 14
Circular economy 1, 2, 3, 4, 5, 6, 8, 9, 10, 11, 12, 13, 14, 15, 16, 17, 19, 20, 21, 22, 23, 82, 95, 233, 268, 276
Cluster 35, 74, 142, 143, 145, 146, 147, 158, 166, 168, 169, 170, 171, 216
Commercial banks 113, 114, 115, 117, 118, 119, 122, 128, 130, 131, 133
Contract choice 222
Credit institutions 113
Crowdfunding campaign 184, 192, 196

D

Decomposition 11, 202, 203, 212, 217
Desk research methods 43, 186
Digital communication 178, 222, 224, 229, 230, 233, 235
Digital development 33
Digital energy 157, 158, 159, 160, 162, 163, 164, 165, 166, 174, 175, 176, 180
Digitalization 5, 6, 12, 15, 18, 23, 26, 27, 29, 31, 33, 36, 37, 40, 45, 72, 73, 75, 78, 81, 84, 85, 88, 93, 94, 95, 100, 113, 114, 115, 116, 117, 118, 119, 120, 121, 122, 123, 124, 125, 126, 128, 129, 130, 131, 132, 133, 134, 155, 158, 159, 160, 175, 194, 202, 213, 216, 221, 238, 263, 272, 278, 297, 321, 322, 323, 324, 325, 326, 327, 328, 329, 330, 331, 333, 334, 335, 336, 337, 338, 339
Digitalization trends 36
Digital technologies 1, 5, 8, 10, 17, 19, 25, 31, 46, 65, 66, 67, 68, 69, 70, 71, 73, 74, 75, 76, 77, 78, 79, 80, 81, 82, 84, 87, 88, 89, 92, 94, 96, 97, 98, 100, 101, 102, 108, 109, 115, 117, 118, 119, 120, 130, 157, 161, 163, 174, 176, 184, 195, 218, 226, 235, 237, 263, 265, 267, 279, 280, 285, 293, 302, 304, 307, 308, 309, 318, 319, 323, 324, 325, 326, 327, 328, 329, 330, 331, 333, 336, 337
Digital transformation 2, 30, 32, 66, 79, 80, 81, 82, 84, 85, 87, 88, 90, 91, 92, 93, 94, 95, 96, 97, 98, 99, 102, 105, 108, 109, 110, 111, 112, 114, 115, 116, 117, 119, 120, 121, 132, 133, 134, 158, 179, 180, 201, 239, 257, 267, 277, 296, 297, 298, 304, 305, 307, 308, 309, 310, 319, 320, 324, 325, 327, 328, 329, 330, 335, 338, 339

E

Economic growth 2, 11, 15, 70, 112, 115, 136, 137, 138, 139, 140, 141, 143, 145, 146, 147, 150, 151, 155, 202, 203, 210, 212, 213, 242, 265, 266, 271, 272, 276, 277, 278
Economic security 135, 136, 137, 138, 139, 140, 141, 142, 143, 144, 145, 146, 148, 149, 150, 151, 152, 153, 154, 155
Electronic document flow 328, 329, 330, 331, 335, 336, 337
Engagement 62, 82, 131, 139, 153, 158,

165, 178, 197, 216, 221, 224, 225, 228, 230, 231, 235, 236, 240, 244, 247, 248, 267, 297, 319
Entrepreneur-customer relations 183, 186
Entrepreneurship 22, 35, 80, 82, 154, 184, 186, 194, 196, 197, 198, 199, 238
Environmental management system 302
Equity 165, 184, 187, 188, 196, 197, 199, 201, 202, 204, 205, 206, 207, 208, 209, 210, 211, 212, 213, 214, 223, 234, 263, 264, 276, 277

F

Factor analysis 135, 137, 210, 218

H

Human-centric technologies 174, 262, 264, 266, 268, 271, 272, 273, 275
Human-oriented technologies 157, 158, 159, 160, 161, 175

I

Innovative economy 15
Integrated management systems 301, 302, 303, 305, 308, 316, 318, 319, 320
Intellectual capital 280, 281, 282, 284, 285, 286, 287, 288, 289, 293, 294, 295, 296, 297, 298, 299
Interaction 7, 28, 29, 30, 33, 36, 38, 113, 114, 115, 117, 119, 120, 121, 122, 123, 130, 136, 184, 196, 234, 239, 262, 264, 267, 269, 275, 299
Involvement 10, 30, 35, 76, 96, 164, 185, 208, 221, 222, 230, 231, 239, 288

L

Light industry 25, 26, 27, 31, 32, 34, 35, 37
Logistics 19, 89, 241, 242, 299, 321, 322, 323, 324, 325, 326, 327, 328, 329, 330, 331, 333, 334, 335, 336, 337, 338, 339

M

Mining 4.0 87, 90, 91, 95, 96, 97, 98, 102, 108, 111, 112
Mining industry 87, 88, 89, 90, 91, 92, 93, 95, 96, 97, 98, 100, 102, 103, 105, 106, 108, 109, 110, 111, 112
Monitoring of place reputation 48, 49, 50

N

Neural network 108, 159, 162, 169, 170, 171

O

Occupational health and safety management systems 305, 309
Oil and gas business 65, 74
Oil and gas company 213
Online environment 45, 46
Optimization 5, 28, 69, 75, 90, 91, 92, 98, 154, 215, 263, 278, 289, 324, 334, 336

P

Place reputation 43, 46, 47, 48, 49, 50, 51, 58, 59, 60, 61
Product as a service 1, 2, 3, 5, 6, 12, 15, 17
Protection protocols 159
Public policy 44, 221, 222, 223, 224, 227, 229, 230, 234, 236, 237, 239, 240, 267, 333

Q

Quality management system 302
Questioning 113

R

Receiver operating characteristic 135
Redundant codes 159
Reward crowdfunding 183, 185, 186, 187, 188, 189, 190, 194, 195, 198
Risk assessment 117, 177, 259, 279, 280, 281, 292, 293

Risk management 90, 94, 108, 207, 208, 217, 244, 248, 258, 279, 280, 287, 289, 293, 296, 306, 307
Robotization 26, 37, 76, 102, 321

S

Split testing 135
Sustainability 1, 2, 5, 8, 11, 14, 15, 17, 18, 20, 21, 22, 23, 40, 68, 70, 78, 84, 87, 92, 94, 95, 96, 97, 98, 102, 106, 107, 109, 110, 133, 136, 137, 138, 142, 148, 154, 158, 160, 162, 163, 164, 165, 174, 175, 176, 198, 202, 203, 204, 209, 210, 212, 213, 215, 216, 217, 221, 222, 223, 226, 227, 228, 229, 230, 231, 232, 233, 234, 235, 236, 237, 238, 240, 244, 246, 248, 256, 267, 284, 286, 292, 294, 296, 298, 305, 317, 322
Sustainable development 1, 2, 9, 16, 19, 20, 65, 66, 67, 68, 70, 71, 72, 75, 78, 79, 80, 81, 82, 83, 84, 85, 88, 109, 110, 131, 138, 148, 160, 203, 204, 206, 208, 209, 213, 216, 226, 228, 229, 232, 236, 238, 239, 265, 320, 336

T

Traditional industries 25, 26, 29, 31, 32, 34, 35, 36, 37
Transformation 2, 19, 23, 25, 26, 29, 30, 31, 32, 33, 34, 35, 36, 37, 39, 66, 79, 80, 81, 82, 84, 85, 87, 88, 90, 91, 92, 93, 94, 95, 96, 97, 98, 99, 102, 105, 108, 109, 110, 111, 112, 114, 115, 116, 117, 119, 120, 121, 131, 132, 133, 134, 158, 160, 164, 179, 180, 185, 191, 195, 201, 216, 239, 257, 267, 277, 296, 297, 298, 304, 305, 307, 308, 309, 310, 319, 320, 324, 325, 327, 328, 329, 330, 334, 335, 338, 339
Transport-and-logistic activity 321

U

Unmanned vehicles 321, 328

V

Venture on demand 183

W

World economy 37, 218